Advances In San Juan Basin Paleontology

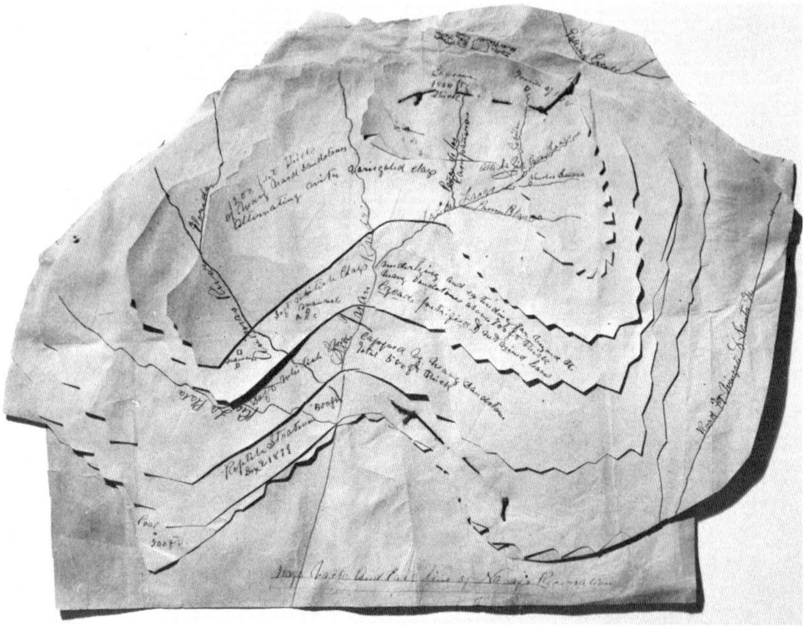

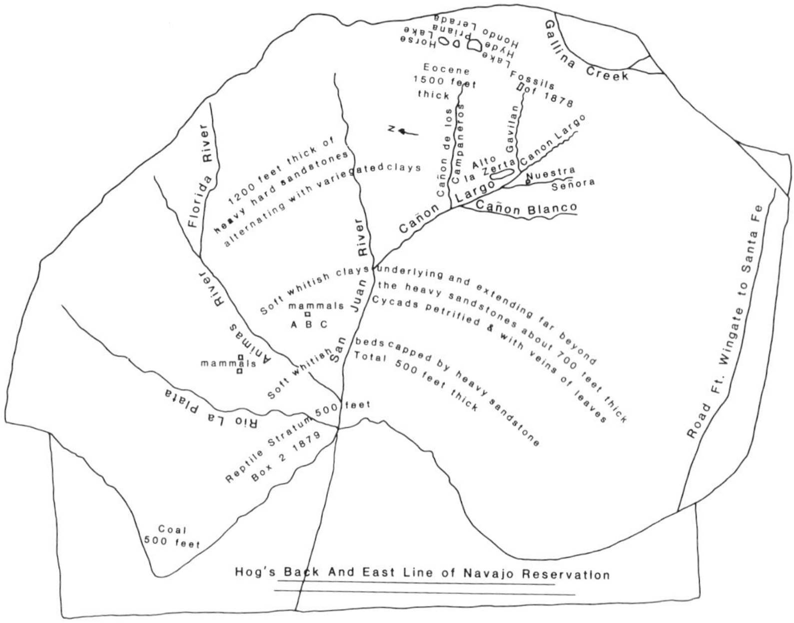

The first geologic map of the San Juan Basin, prepared by David Baldwin in 1879 and sent to O. C. Marsh. The map was created from pieces of brown wrapping paper laid one on top of the other and tied together with string. It indicates the major areas where Baldwin collected for Marsh, though he later collected from these and other areas in the Basin for Cope. The transcription was made from the original map, which is in the Manuscripts and Archives Collection of the Sterling Memorial Library, Yale University, New Haven, Connecticut (Courtesy of the Division of Vertebrate Paleontology, Yale Peabody Museum).

Advances in San Juan Basin Paleontology

Edited by
Spencer G. Lucas,
J. Keith Rigby, Jr.
and Barry S. Kues

UNIVERSITY OF NEW MEXICO PRESS
Albuquerque

Library of Congress Cataloging in Publication Data

Main entry under title:

Advances in San Juan Basin paleontology.

Includes bibliographies.
1. Vertebrates, Fossil—Addresses, essays, lectures. 2. Paleontology—San Juan Basin, N. M. and Colo.—Addresses, essays, lectures. I. Lucas, Spencer G. II. Rigby, J. Keith. III. Kues, Barry S.
QE841.A38 560'.9789 80-52276
ISBN 0-8263-0554-7

Manufactured in the United States of America.
Library of Congress Catalog Card Number 80-52276.
International Standard Book Number 0-8263-0554-7.
First edition

Contents

Acknowledgments

This book could not have been completed without the assistance of many people and institutions. The substantial time and effort spent by the contributors is evident in their papers. In addition to reviewers acknowledged in individual papers, we thank those who reviewed papers at the request of the editors; namely John Ostrom, Yale University; Coleman Robison and F. M. O'Neill, Albuquerque office, Bureau of Land Management; J. W. Froehlich, University of New Mexico; and Tom Lehman, University of Texas. The American Museum of Natural History, Yale–Peabody Museum of Natural History, and National Museum of Natural History generously provided photographs from their archives. The Bureau of Land Management and University of New Mexico helped to subsidize some of the typing, drafting, and other pre–publication costs associated with the completion of the manuscript. And finally, we acknowledge our intellectual debt to David Baldwin, and Edward D. Cope, whose discoveries and studies first stimulated interest in San Juan Basin paleontology with a force that has endured for more than a century with no signs of abating.

Introduction

The San Juan Basin of northwestern New Mexico and adjacent states is a large depositional basin in the Colorado Plateau, in which great thicknesses of Late Cretaceous and early Tertiary sediments were deposited. In Late Cretaceous time part of the western shoreline of the epeiric sea that covered the Midcontinent migrated back and forth across this area, producing a classic sequence of transgressive-regressive facies that records a large variety of depositional environments and marine and terrestrial faunas. With the final retreat of this sea near the close of the Cretaceous, and renewed subsidence of the basin, large volumes of lowland continental sediments were deposited, preserving some of the most abundant and diverse Late Cretaceous and early Tertiary vertebrate assemblages in the world.

It is difficult to overstate the role that the rocks and fossils of the San Juan Basin have played in our understanding of the evolution of life during the transition from Late Cretaceous to early Tertiary time. Any textbook, article, or lecture concerned with this period relies, directly or indirectly, on the fossil record of the San Juan Basin. The basin is one of a small number of places on the earth's surface where continental sediments containing Late Cretaceous dinosaurs are immediately overlain by rocks bearing some of the earliest Tertiary mammals. Therefore, it has figured prominently in discussions about the placement of the Cretaceous–Tertiary boundary. It has achieved additional biostratigraphic importance by producing three mammalian faunas—Puerco, Torrejon, and Tiffany—that are the standards for the North American land mammal "ages" that represent early, middle, and late Paleocene time. Indeed, the long succession of superposed vertebrate faunas ranging from the Late Cretaceous to early Eocene present in the San Juan Basin is absolutely unique; no sedimentary sequence as long, complete, and richly fossiliferous spanning the Mesozoic–Cenozoic boundary exists in a single basin anywhere else on earth. The importance of this sequence is augmented by the profound changes that occurred in vertebrate communities during this time. Mature terrestrial communities dominated by dinosaurs disappeared as the dinosaurs became extinct, giving way to the mammal-dominated communities that heralded the establishment and elaboration of mammalian hegemony.

The reconstruction of this complex transition has been aided by the work of some of America's greatest paleontologists and most perceptive stratigraphers—Bauer, Brown, Cope, Dane, Gilmore, Granger, Matthew, Osborn, Reeside, Simpson, Stanton, and Sternberg, among others. Since the first description of vertebrate fossils from the Basin in 1874, and David Baldwin's discovery in July, 1879 of Paleocene mammals, opening a hitherto unknown chapter in the history of life, a wealth of knowledge has emerged from the San Juan Basin. But in spite of this early surge of interest, paleontological work on the terrestrial faunas and floras has been sporadic over the greater part of the last century. Consequently, new information still awaits discovery, interpretation, and integration into existing knowledge, and some of the older work needs revision.

This book presents a survey of recent work on the paleontology of the terrestrial deposits adjoining the Cretaceous–Tertiary boundary, through papers covering a broad range of topics in vertebrate paleontology, paleobotany, stratigraphy, biostratigraphy, and paleoecology. It reflects an increase in interest in San Juan Basin paleontology during the past few years, and our observation that a significant portion of recent work, and the new ideas and interpretations that flow from this work, was in a form or quantity that prohibited rapid dissemination through the paleontological community. Thus, the papers presented here range from reviews and revisions based on recent studies, to analyses of new data, to progress reports and summaries of large-scale studies now in progress. Our intent has been to provide an entry into the "state of the art" of knowledge of the Late Cretaceous–early Tertiary continental paleontology of the San Juan Basin. Obviously, though this book presents a large amount of information on San Juan Basin paleontology, it is not the last word on the subject. Rather, by emphasizing significant recent discoveries and ideas, the articles included here underscore the tremendous amount of new data, new approaches to long-standing problems, and great scope for new discoveries that remain for future workers in the San Juan Basin.

It is also of interest to note that paleontologists presently studying the San Juan Basin faunas and floras have been encountering situations outside the usual and traditional aspects of paleontology. The time is past, even in the vast reaches of the western United States, when a paleontologist could leave an area knowing that it would remain intact until he returned the next collecting season. Increasingly diverse and widespread types of land use, from strip-mining to the building of roads and other facilities for a rapidly increasing population to off-road vehicle use, just to mention a few, have a tremendous potential for disturbing or destroying paleontological information. The San Juan Basin, which is rich in coal, gas, oil and uranium, and undergoing a population explosion, is experiencing many of these impacts now; other areas in the West also have or will begin to encounter them in the near future.

In response to this, various agencies of the Federal Government, such as the Bureau of Land Management, Bureau of Indian Affairs, and Forest Service, have assumed the responsibility for managing scientific resources, including fossils, within their areas of jurisdiction. The role of these agencies in preserving paleontological information and stimulating research in areas where many other land use activities are occurring or imminent is still being defined. Over the past four years the Bureau of Land Management has commissioned or required of private industry several surveys of the paleontology of various parts of the San Juan Basin. Paleontologists from the University of New Mexico, Museum of Northern Arizona, University of Texas at El Paso, Louisiana State University, Brigham Young University, Michigan State University, Yale University, University of Texas at Austin, and the Albuquerque Office of the Bureau of Land Management have participated. These surveys, while providing the Bureau of Land Management with the information it needs, have also discovered much new information about the fossil biotas of the Basin.

Both of these phenomena—the effects of large-scale land use, and federal management of paleontological materials—have profound implications for the discipline of paleontology, and they have been manifested most clearly in recent years in the San Juan Basin. It has become evident that the responsibility of paleontologists extends beyond the scientific study of their fossils to an active role in providing advice on how best to protect the rocks and fossils vital to their research and to the research of workers who will follow. If this is done, further study of this incomparable fossil field will produce new information about Late Cretaceous and early Tertiary continental biotas that will be a fitting continuation of the important contributions made by past workers.

Part 1
History

1 History of vertebrate paleontology in the San Juan Basin

George Gaylord Simpson
Tucson, Arizona

The San Juan Basin contains a tremendous mass of sedimentary rocks ranging in age from Carboniferous to Eocene and in kind from gypsum to conglomerate. For more than 120 years these have attracted the interest of geologists. A whole library of studies about them has been published. Much of the interest has been economic, and in recent years it has been increasingly concerned with exploring and exploiting the fossil fuels in which the basin is extraordinarily rich. For more than a century, however, there has been another overriding attraction for everyone interested in the history of the earth and of life on it. Although that attraction involves the entire mass of rocks, many investigators find a relatively late part of the sequence particularly interesting and instructive.

This sequence comprises, in the main, the Late Cretaceous Fruitland and Kirtland Formations and above them the Paleocene Nacimiento and early Eocene San José Formations, with various more or less distinguishable members and some intercalations. They are all composed of continental sediments. The many fossil vertebrates they contain are mainly those of the land, but some are also of freshwater streams. These animals are involved in one of the most fascinating parts of the whole history of life: the last years of the Age of Reptiles and the early years of the Age of

Mammals. The record in these formations is almost continuous; it covers some tens of millions of years during which there was an almost complete turnover in continental faunas.

The present account is devoted to the human history of the discovery and study of the record of those events in the course of earth history. I shall summarize this part of human history more or less in its own chronological order, even though that requires some dodging about in the geological chronology among Cretaceous, Paleocene, and Eocene and in zoological content among the various groups of reptiles and mammals. Although some vertebrates of other kinds have been found in these rocks, since they have not added significantly to the fossil record, for present purposes they can be ignored. There have certainly been dozens (and perhaps hundreds) of people not named here who have picked up some fossils from these beds; however, this account is limited to the collections and studies that have contributed most significantly to present knowledge. San Juan Basin geographic names mentioned here are indicated in Text-figure 1.1.

The history starts in 1859 with the first examination of some of these rocks by a trained geologist, J. S. Newberry (1822–1892), who accompanied a party of army engineers under J. N. Macomb in a traverse across the San Juan Basin. Newberry (1876, an account written in 1860 and published without later change) crossed the entire Fruitland to San José geological section, from well down in the Cretaceous to well up in the Tertiary. He, however, considered it all Late Cretaceous in age. He found some poorly preserved traces of fossil plants in these beds, but no animal fossils.

Some otherwise unknown Indians and at least one Mexican- (or Spanish-) American are known to have picked up bits of fossil mammals earlier, but the scientific discovery of fossil vertebrates was made by Edward Drinker Cope (1840–1897; Text-fig. 1.2) on 7 September 1874 in what is now called Arroyo Blanco, on the east side of the continental divide between the places now known as Regina, to the south, and Llaves, to the north. There on that date Cope found remains of eight different mammals, several turtles and crocodiles, and some shark teeth (the latter redeposited from Cretaceous marine beds, which Cope could hardly know). He immediately recognized that the mammals were Eocene in age, but he correlated the beds with middle Eocene. (They are now known to be early Eocene in age.) Cope remained in that vicinity until 17 October and made a large collection mainly of fossil mammals. He later reported on the fossil vertebrates (Cope, 1877) and the geology (1875) of the Eocene of what we now call the San Juan Basin. Both were published in reports of the Geographic Survey West of the 100th Meridian, generally called the Wheeler Survey. Cope was nominally attached

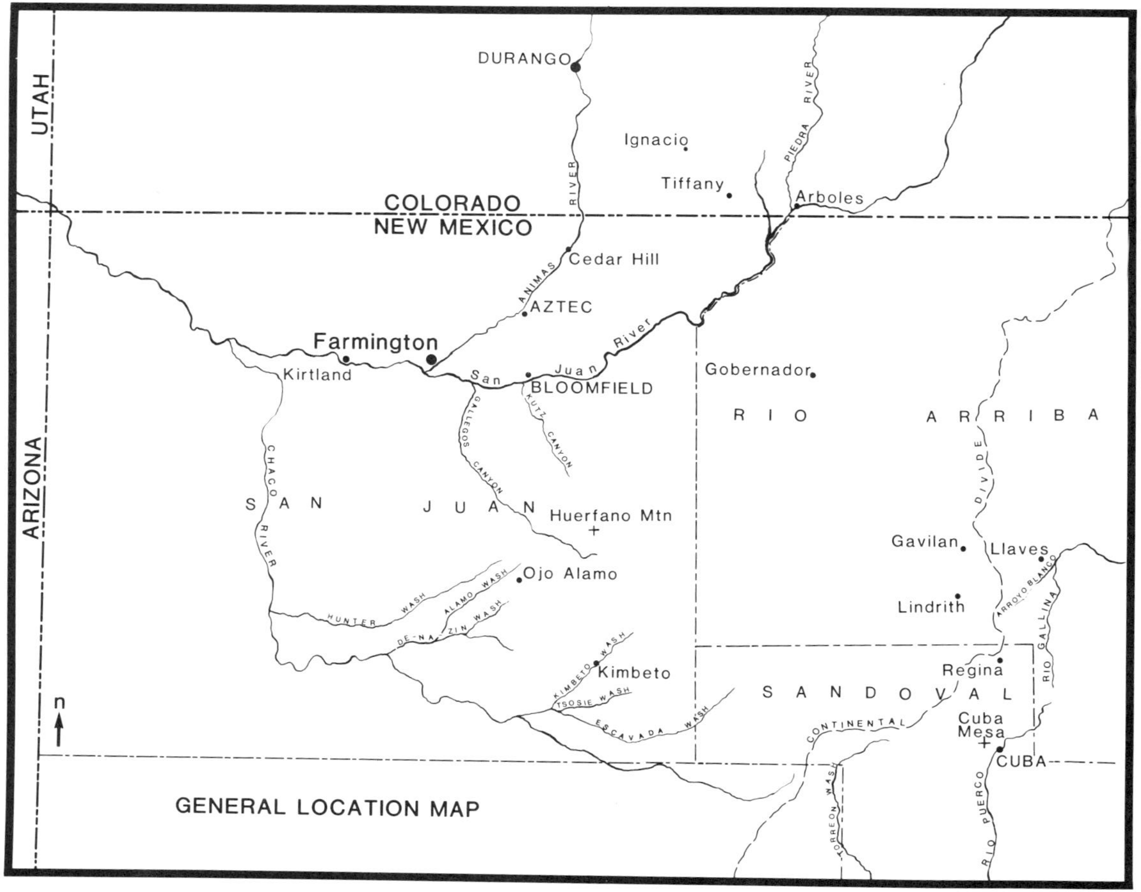

TEXT-FIG. *1.1*—Map of San Juan Basin geographic names. Names used are those on U.S. Geological Survey topographic maps. Slight variations of some names (e.g. Kimbetoh for Kimbeto) are found in some papers in this volume, reflecting usage or spelling preferred by individual authors. 10 miles = .4 inch

TEXT-FIG. *1.2*—Edward Drinker Cope (courtesy of the American Museum of Natural History).

to this survey when he made these discoveries, although he had left the main party and gone off on his own. (For further details see Simpson, 1948 and 1951.)

In 1875 a New Mexican resident named David Baldwin worked in the Eocene with Lieutenant W. L. Carpenter for the Wheeler Survey. It is not clear whether he collected fossil vertebrates at that time, but in May 1876, he started doing so for Professor Othniel Charles Marsh (1831–1899), Cope's bitter rival, who was following up Cope's discovery of this fossil field. Baldwin continued to collect for Marsh intermittently until 1880. In the San Juan Basin he collected fossil mammals mostly from the red beds (Largo facies of later workers) on the west scarp of the continental divide in the general region of the localities now known as Gavilân and Lindrith. Like all collections from those beds in that area, these consisted mostly of a single genus and species. Eventually, in 1892, on the basis of the specimens sent him by Baldwin, Marsh called this species *Hyracops socialis*, but both the name of the genus and the name of the species are invalid because synonymous with those of *Meniscotherium chamense*, already found and named by Cope in 1874. Marsh also proposed a new order of mammals, Mesodactyla, for this animal, but that supposed order has never been accepted by any other authority, all others placing *Meniscotherium* in the order Condylarthra, which includes the earliest and most primitive hoofed mammals. In the meantime Marsh had become annoyed by the lack of novelty (or other supposed novelties) in the mammals collected for him by Baldwin. Baldwin had become more than annoyed by the difficulty of getting his pay from Marsh, a common problem for Marsh's hired collectors. Therefore, in 1880, with mutual dissatisfaction and dislike, Baldwin and Marsh broke off their connection, and Baldwin transferred his allegiance to Marsh's bête noire Cope. In retrospect there is an aspect of irony (or perhaps justice) in the fact that Marsh thus lost what would have been one of his greatest coups, and Cope gained what was definitely his greatest. (These incidents are briefly treated in Schuchert & LeVene, 1940; there is further unpublished information in the files of Marsh at the Yale Peabody Museum and some in those of Osborn and Cope at the American Museum of Natural History.)

It was Baldwin who made the most important paleontological discovery in the San Juan Basin, the discovery that eventually made the basin famous among paleontologists everywhere. He found the first known early Paleocene mammalian fauna in the world and the beginnings of the faunal sequence that led to the now accepted definition and standard for a worldwide Paleocene epoch, the crucial epoch when mammals finally became the dominant land vertebrates. The discovery not yet of the oldest such fauna (early Paleocene in present terms) but of the oldest at the time (middle Paleocene in present terms) might have dated from 1879, when Baldwin sent to Marsh some teeth that were later found to be older than the early Eocene. Neither Baldwin nor Marsh recognized the nature or importance of these finds, and Marsh probably did not even look at them until 1894 or so (see Simons, 1963). The discovery in the sense of understanding it and carrying it down to the earliest Paleocene may more clearly be said to have developed through the year 1880 to 1886 while Baldwin was collecting for Cope. It was Cope and a whole sequence of later students, Matthew foremost among them, who made the erudite and crucial studies involved, but it was Baldwin, an untutored but brilliant collector, who found the first materials and who soon understood the essentials of their occurrence and significance. Baldwin was a great collector. He made a superb early collection of Permian reptiles, which he shipped to Marsh, who never bothered to open the boxes. Little is publicly known about Baldwin's life, although some (and perhaps much) unpublished documentation exists. I have no information on his birth or death. He lived much of his life in Abiquiu, on the Chama River of New Mexico not far from the San Juan Basin, and spent his late years in Farmington, the metropolis of the San Juan Basin. Carpenter said that when they met in 1875, Baldwin "was equipped like a Mexican with a burro, some corn meal, and a pickaxe" (Schuchert & LeVene, 1940, p. 179). Much later memories affirmed that such were his usual transport, supplies, and equipment on all his lone collecting trips, often made in winter because then snow on the ground could be melted for water in the vast waterless stretches of the San Juan Basin.

Baldwin knew that he had found at least one Tertiary mammalian fauna older than previously known, underlying the known Eocene but above recognizable Cretaceous. Cope also realized that he was receiving the oldest known Tertiary mammals. From 1881 through 1888, the years while Baldwin was collecting them, he described and named them as fast as they came in. Most of Cope's 41 papers bearing on these finds in those years were brief (under four pages long). However, material on the early finds up to the summer of 1883 was incorporated in the enormous volume long known as "Cope's Bible." In its more than a thousand pages this volume also included numerous other Tertiary faunas (Cope, 1885). A synopsis of almost the whole Baldwin collection was published in 1888 (Cope, 1888), and finally an isolated last note was published in the year of Cope's death (Cope, 1897; all these works are listed in Matthew, 1937).

Until 1882 Cope called this fauna (or these faunas, their number not established) and the beds from which it (or they) came "Lower" or "Lowest Eocene." In 1874 he had observed that near the then village of Nacimiento and the Rio Puerco (one of several of that name in New Mexico) at what is now known as the point of Cuba Mesa, there were sediments below what he was then calling Eocene or Bridger. He thought these beds possibly older Eocene or possibly equivalent to the Fort Union beds of the Upper Missouri. He found no fossils in them but applied the name Puerco to them. In 1882 he decided that the fossils being collected by Baldwin were from beds equivalent to his Puerco (although none came from his type locality for the Puerco), and he started to call the fauna "Puerco Eocene" and the rocks "Puerco formation." In 1884 and later he used the odd expression "Puerco Epoch" and the less odd but still unorthodox "Puerco Series." He evidently anticipated the conclusion, which became accepted and general only much later, that he was dealing with a distinct major subdivision of the Tertiary Period.

When Henry Fairfield Osborn (1857–1935; Text-fig. 1.3) went to the American Museum of Natural History in 1891, he organized a Department of Mammalian Paleontology, soon changed to Department of Vertebrate Paleontology. He set up an ambitious program for collecting a complete sequence of North American mammalian faunas starting with "the Older Tertiaries" under the direction of J. L. Wortman (Text-fig. 1.4) as Osborn's assistant curator. Wortman went to the San Juan Basin in 1892 with O. A. Peterson (later a prominent vertebrate paleontologist at the Carnegie Museum) and T. Rafferty (not otherwise known to me) as his field help. They made a considerable collection, probably mostly from Baldwin's localities, although that was not made clear. This collec-

TEXT-FIG. *1.3*—Henry Fairfield Osborn (courtesy of the American Museum of Natural History).

TEXT-FIG. *1.4*—Jacob L. Wortman (courtesy of the American Museum of Natural History).

tion was described by Osborn and Charles Earle (1895), the publication stating that "the larger part" was by the nominal junior author. Wortman had noted that the specimens were concentrated in the extreme upper and lower strata of the "Puerco," and a list ascribed many of the known species (mostly Cope's) to lower or upper levels or faunas.

In 1896 Wortman returned to the San Juan Basin and again collected in the "Puerco" beds, now with young Walter Granger as his field assistant. Granger (1872–1941; Text-fig. 1.5) was destined to have a great career. He eventually became Curator of Fossil Mammals and Department Chairman at the American Museum and collected innumerable fossils in the San Juan Basin and many other parts of the world. The 1896 collection was placed in the capable hands of W. D. Matthew, Granger's elder by one year.

In 1895 the American Museum purchased the Cope Collection of fossil vertebrates, including most of the San Juan Basin fossils collected by Cope himself and by Baldwin for Cope. In that same year young William Diller Matthew (1871–1930; Text-fig. 1.6) joined the staff as "second assistant." (As is well known, Matthew became one of the greatest paleomammalogists, advanced in the Museum to the rank of Curator-in-Chief of Mineralogy, Geology, Geography, and Astronomy, and in 1926 went on to a professorship at the University of California in Berkeley, where he died only four years later.) With all of the then known American mammals from what he later called the Paleocene epoch in his hands, Matthew revised them carefully, dividing them into "two absolutely distinct faunas" with no species in common (Matthew 1897). He therefore restricted the name "Puerco" to the earlier fauna and lower beds and, at Wortman's suggestion, gave the name "Torrejon" to the later fauna and upper beds. It has been generally assumed that the type locality for the Torrejon is Arroyo Torreon (not Torrejon), although Matthew did not say so, and Wortman (in Osborn & Earle, 1895, p. 2) did not give it as one of the "principal localities" for his "upper strata." Apparently it did not occur to anyone until much later that the Puerco of Cope has a type locality where no fossils had yet been found and which might prove to be equivalent to either "Puerco" or "Torrejon" of Matthew, or both, or perhaps neither.

As early as 1885 Cope had mentioned the presence of dinosaurs in Baldwin's collections from the San Juan Basin, but as the specimens were scraps of no particular importance, this lead was not followed up. However, in 1902, George H. Pepper, who was with the Hyde Exploring Expedition, noticed numerous dinosaur bones near Ojo Alamo, a former trading post at a spring that is still a landmark in the Basin. Pepper picked up some scraps and submitted them to the American Museum. They were not exactly identifiable, but it was decided to look further into

TEXT-FIG. *1.5*—Collecting trip, 3 miles south of Ojo San Jose, 1912. Walter Granger at far right; Stein and Towy to his left (courtesy of the American Museum of Natural History).

TEXT-FIG. *1.6*—William Diller Matthew (courtesy of the American Museum of Natural History).

this occurrence of dinosaurs. That task was assigned to Barnum Brown, then an assistant curator in Osborn's department. Brown (1873–1963; Text-fig. 1.7) visited the San Juan Basin in 1904 and made a small but interesting collection of dinosaurs there, notably the skull of a then new genus of duckbill dinosaurs to which the name *Kritosaurus* was given (Brown, 1910). (Reeside, 1924, p. 31, erroneously wrote that this genus was based only on fragments later collected by Sinclair and Granger, and he questioned whether it was identifiable.) As is well known, Brown became a great (perhaps the greatest) collector of dinosaurs, but he did no further work in the San Juan Basin.

Brown specified that the known dinosaurs were from shales below the heavy conglomeratic sandstone near the then existing Ojo Alamo trading post. These shales were named the Ojo Alamo beds by Brown. Thus their dinosaurs and other vertebrates, augmented by later collecting, have long been known as the Ojo Alamo fauna. However, the same shales were later named Kirtland Shale by the U.S. Geological Survey (Bauer, 1916), the name that has become fixed in stratigraphic use. As a stratigraphic term, the name "Ojo Alamo" was transferred to the con-

TEXT-FIG. *1.7*—Barnum Brown (courtesy of the American Museum of Natural History).

glomeratic sandstone that had been explicitly excluded from the Ojo Alamo in the original publication of that name. Thus arose the anomaly that the Ojo Alamo fauna of almost all the paleontological literature is not from the Ojo Alamo Sandstone (or Formation) named in almost all the stratigraphic literature but from the Kirtland Shale (or Formation). It has been later found that the Ojo Alamo of the stratigraphic literature comprises more than one tongue of sandstone and is (at least in part) of early Paleocene (Puercan) age. Knotty stratigraphic problems still remain here, but they need not be further discussed in this short treatment of the history of more strictly paleontological collections and studies.

In 1908 and 1909 J. H. Gardner, then with the U.S. Geological Survey, visited the San Juan Basin and made small collections from the Paleocene beds. His work was significant primarily because it was he who held that the Puerco and Torrejon formations, as they were then being called, formed a stratigraphic unit for which he proposed the name "Nacimiento group" (Gardner, 1910). It has since been generally recognized, by paleontologists at least, that the beds in which the Puerco and Torrejon faunas occur are not separately mappable and, therefore, are not acceptable separate lithostratigraphic units. The Puerco and Torrejon are two distinct faunas, both of which occur in the single Nacimiento formation. Nacimiento, the name of a former village near the present town of Cuba, was not on the Nacimiento Formation, but Gardner's application of that name is now universally accepted, and it would be folly to change it merely because its nominal type locality is inappropriate.

In 1909 Gardner was accompanied by J. W. Gidley (1866–1931; Text-fig. 1.8), the distinguished paleomammalogist of the U.S. National Museum. Gidley then found a few bits of fossil mammals near Ignacio in the northern part of the San Juan Basin in southwestern Colorado. In 1917 Gidley published a brief note on these in a paper by Wegemann (see Wegemann, 1917). He gave approximate identifications of four mammals and concluded correctly that their age was between that of the Torrejon and the "Wasatch" (then a vague general term for early Eocene in the western United States). Gidley's discovery was part of what is now called the Tiffany fauna. A considerable collection of it was later made by Granger and finally described by me. As it turned out, Gidley's discovery essentially completed the Paleocene sequence in the San Juan Basin and indeed for North America as a whole.

From 1912 to 1914 Granger carried out an intensive collecting program in the Paleocene and Eocene of the San Juan Basin, making large collections from the Puerco and Torrejon faunas and from what was then sometimes called "New Mexican Wasatch," later named the San José Formation. Part of the time he was accompanied by W. J. Sinclair, a vertebrate paleontologist at Princeton University, by George Olsen, a preparator and collector at the American Museum, and by William Stein, a western cowhand who spent some summers as a camp man and fossil collector. The immediate results of that extensive field work were primarily stratigraphic papers on the early and middle Paleocene (Sinclair & Granger, 1914) and the early Eocene (Granger, 1914) of the San Juan Basin. Since the paper on the Paleocene was for a long time the basic source of stratigraphic and fossil locality information for what Sinclair and Granger were still calling the Puerco and Torrejon formations, it is still useful in that respect. Granger's paper on the Eocene divided what he called "New Mexico 'Wasatch' " into Almagre (below) and Largo (above) "beds," which were correlated, without claim to complete precision, with parts of the Gray Bull, Lysite, and Lost Cabin sequence in Wyoming. They are now considered partly successive rock and faunal facies.

In 1915 J. B. Reeside, Jr. worked with C. M. Bauer, and in 1916 he worked with F. R. Clark on stratigraphy in the San Juan Basin for the U.S. Geological Survey. They made considerable collections of fossil vertebrates, especially turtles, from the Fruitland, Kirtland, "Ojo Alamo" (not realizing the ambiguity of the name), Puerco, and Torrejon. The fossil reptiles, including eighteen new species of turtles, were described in three papers (Gilmore, 1916, 1919, 1922) by C. W. Gilmore (1874–1945; Text-fig. 1.8), the distinguished paleoherpetologist of the U.S. National Museum. For 26 years (1905–1931) he was contemporaneous with his somewhat older paleomammalogical colleague, J. W. Gidley. Gilmore's later (1928) monumental work on the fossil lizards of North America includes only two specimens from the San Juan Basin, both found by Sinclair and Granger in 1913 in the middle Paleocene (Torrejon fauna).

In 1915 Matthew and Granger published the first four parts of a "Revision of the lower Eocene Wasatch and Wind River [mammalian] faunas"; the fifth and last part appeared in 1918. Although the revision as a whole had that running title and joint authorship, parts I, II, IV, and V had as sole author Matthew (1915 a, b, c, and 1918), and III had as sole author Granger (1915). Although this series was based mainly on faunas from Wyoming, it also included what was called "the Wasatch of New Mexico," that is, our present San José Formation in the San Juan Basin. Even now, more than sixty years later, that is the nearest thing we have to a revision of the early Eocene faunas of the Basin. The taxa known from New Mexico were not brought together into a faunal list.

In 1916 Granger collected in the San Juan Basin for the last time. The principal results of his work that summer (again with George Olsen) were collections of Torrejon fossils at localities in "Coots" (properly

TEXT-FIG. *1.8*—From left to right: Oliver Perry Hay, Charles W. Gilmore and James Williams Gidley (courtesy of the Smithsonian Institution).

Kutz) Canyon and along the Animas River between Aztec and Cedar Hill. They also included collections of post-Torrejon Paleocene mammals in southern Colorado southeast of Ignacio and somewhat north of Tiffany and Arboles (Text-fig. 1.1). The latter area had been called to Granger's attention by Gidley, who had found a few fossil mammals there in 1908. In 1916 Granger found an extraordinarily rich little pocket of small mammals at a point then near the Mason schoolhouse (and hence called the Mason Pocket). Granger (1917) soon summarized that season's finds and named the rocks in which the Mason Pocket was found the Tiffany beds, after what was then a nearby stop on the Denver and Rio Grande Railway. Matthew (as reported by Granger) was inclined to consider the Tiffany beds as the top of the Paleocene; however, in 1917 Granger himself thought it more consistent to place them as the base of the lower Eocene.

Matthew's view on that subject has prevailed. There had been an increasing tendency to recognize the Paleocene as a distinct epoch of the Tertiary period and to redefine this originally European name in American and, indeed, in San Juan Basin terms. Recognition of a pre-Eocene Tertiary epoch and its designation as Paleocene was first proposed in 1874 by Wilhelm Philipp Schimper, a German paleobotanist, to include the times of various European floras, some now considered Eocene and some Paleocene. Schimper's proposal of a new epoch put some deposits now known to be of the same geological age in different epochs. This proposal was not well received; the concept and term "Paleocene" were not generally accepted in Schimper's sense. However, in the early 20th century, vertebrate paleontologists were increasingly using the term in an overlapping but different sense without being very explicit as to its current precise meaning. Matthew briefly specified the sense in which the term "Paleocene" had "been revived by several vertebrate paleontologists in recent years to cover the faunal zones previously known as Basal Eocene . . . Its upper limit is marked by the first appearance of the principal modern orders of mammals and of certain modern groups of reptiles simultaneously in Western America and in Western Europe" (Matthew, 1920). In his final masterpiece, Matthew (1937) made it even more explicit that the Paleocene in this sense comprises the span of time in which the Puerco, Torrejon, and Tiffany faunas lived and that of faunas elsewhere contemporaneous with those specified in the San Juan Basin. In non-paleontologist circles there was for a time some resistance to recognizing a Paleocene epoch, and a few paleontologists even considered returning to Schimper's original ill-defined concept of it. In 1968 this matter was considered in a *Colloque sur l'Eocène* in Paris. The conclusion, although expressed in different terms, was to recommend adoption of what was essentially Matthew's position and the San Juan Basin standard. (See Pomerol, 1969, and a summary in Gingerich, 1975.) Today even the U.S. Geological Survey (rarely "the first by whom the new are tried" and sometimes "the last to lay the old aside") recognizes the Paleocene in the Matthew–San Juan Basin sense.

From 1916 to 1928 there continued to be much stratigraphic work in the San Juan Basin; however, the most significant paleontological collecting was by Charles H. Sternberg. Sternberg (1850–1943) was a collector of fossil vertebrates who had worked off and on for Cope, for some institutions, and as a free-lance commercial collector, sometimes with his three sons. From 1921–1924, although in his 70's, he worked alone as a free lance in the San Juan Basin, collecting mainly dinosaurs in the Fruitland and Kirtland formations. He sold specimens to the University of Uppsala in Sweden and to the American Museum in New York. Among those sold to both were specimens which belonged to a then new genus of horned dinosaurs named *Pentaceratops* by Osborn in 1923. Sternberg also sold the Chicago Natural History Museum an almost complete skeleton of a new species of *Parasaurolophus*, eventually described by Ostrom (1963). On the life of this sturdy pioneer see Gilmore (1944) and Sternberg (1909, 1932) although the 1909 work antedates his work in the San Juan Basin.

As the specimens from the Mason pocket were being prepared, Matthew published brief papers on some of them. In 1921 Matthew and Granger published names and diagnoses, without figures, for eight new genera and species from the Mason pocket (Tiffany beds), two new genera and species from the Torrejon fauna, and three new genera and one new species from the Puerco fauna. These were published "in the course of the revision of the Paleocene faunas" as it seemed "advisable to place [them] on record in advance of the full description." The full descriptions were all eventually published, but for reasons outside the authors' will and control, this was not until 1935 for the Tiffany fauna and 1937 for the Puerco and Torrejon faunas.

In 1928 and again in 1930 C. L. Camp and V. L. VanderHoof of the University of California at Berkeley collected Paleocene mammals in the San Juan Basin. Their collection included two new genera and species, *Kimbetohia campi* and *Tiznatzinia vanderhoofi*, eventually described by me (Simpson, 1936).

In 1929 I spent my first summer in the San Juan Basin collecting in the Puerco fauna with A. C. Sawtelle, a student (who did not go on in paleontology), Jeff Roberson, camp assistant from Farmington, and (much of the time) T. E. Reynolds, then still a Jesuit priest. We made a large collection of mammals and also collected a nearly complete alligator skeleton from classic localities (especially Barrel Spring Arroyo and the richly fossiliferous arroyo that Sinclair and Granger found to be nameless, now usually called Tsosie or some reasonable approximation of that).

W. D. Matthew died on 24 September 1930, age 59, at the height of his powers and productivity. He was the most eminent paleomammalogist of his time and had made himself an incomparable authority on early Tertiary mammals, particularly those of the Puerco and Torrejon faunas. As opportunity presented itself amidst many other occupations, he had been studying those faunas since the 1890's. In 1916 and 1917 he had written most of a long technical monograph in the San Juan Basin Paleocene. From 1928 until near the time of his death in 1930 he had been revising, extending, and completing that massive work. His former associates and his successors at the American Museum spent several years putting the manuscript and illustrations in well-edited form for publication. It was finally published by the American Philosophical

Society in February 1937 (Matthew, 1937). Much intensive work of various kinds has been devoted to the San Juan Basin Paleocene since then, some of which is still going on. However, Matthew's monograph will always be the indispensable classic in its subject.

Matthew had intended to include the full treatment of the Mason Pocket faunule and the Tiffany fauna as a whole in his big San Juan Basin Paleocene monograph, but when death overtook him, he had not had time to do more than write a few, short preliminary papers. When that task fell to me, I completed it as well as I could (Simpson, 1935 a, b, c). (The 77 printed pages were written as a single monograph, but because of editorial restrictions on lengths of publications then in effect at the American Museum it had to appear as three almost simultaneous separate numbers of *Novitates.*) This fuller knowledge of the Tiffany fauna confirmed its position in time between the Torrejon fauna and those now recognized as definitely of early Eocene age. The Tiffany fauna was found to be intermediate between the Torrejon and the Clark Fork of Wyoming, in some respects nearer the latter.

In 1937 the Vertebrate Section of the Paleontological Society (which in 1940 became the separate Society of Vertebrate Paleontology) voted to set up a provincial time scale for the Tertiary of North America, based mainly on mammalian faunas. For that purpose, Granger (then chairman of the Section) appointed a committee of seven: four paleomammalogists, a paleoherpetologist, an invertebrate paleontologist–stratigrapher, and a paleobotanist. Almost all American paleontologists concerned with the continental Tertiary also worked on the project, and a final report was accepted at the 1939 meeting of the vertebrate paleontologists and published by the Geological Society of America in 1941 (Wood et al., 1941; all members of the committee but not other coworkers are named in the citation of this publication). A major feature of this report was the now well-known designation of provincial ages, subdivisions of the usual geological epochs, in principle covering the whole of Tertiary time without overlap or hiatus. Each age was given a geographic name in use for a vertebrate fauna, with *-ian*, *-an*, or *-n* appended. The ages were designated as "provincial" because, unlike the names of epochs, applied worldwide, they were considered applicable only to the North American Continental Tertiary, which is construed as designating a province. Because they were defined primarily in terms of faunas of land mammals, they are now usually called "land mammal" rather than "provincial" ages, a change that I consider unnecessary (and even to some extent confusing) but acceptable on the basis of usage. Following this lead, similar provincial (or land mammal) ages have since come into use for provinces on other continents.

Wood et al. (1941) is included in this history of vertebrate paleontology in the San Juan Basin not only because it has been important in the development of modern geochronology but also (and especially) because it has involved an authoritative definition and subdivision of the Paleocene epoch, based mainly on the Paleocene of the San Juan Basin. It divided the North American continental Paleocene into five provincial ages, from oldest to youngest: Puercan, Dragonian, Torrejonian, Tiffanian, and Clarkforkian. The sequence Puercan–Torrejonian–Tiffanian had already been well established. The inclusion in the Paleocene of the Clarkforkian age (based mainly on a mammalian fauna in the Bighorn Basin of Wyoming) and the interpolation of a Dragonian age (based entirely on a poorly known local fauna from Dragon Canyon in Utah) were then and still are debatable. The question was raised whether the so-called Clark Fork fauna was indeed an associated and distinctive fauna and whether, if so, it was not better placed in the Eocene. At present (1979) there seems to be a consensus that there is a distinguishable Clarkforkian age but that it is best considered as earliest Eocene, not latest Paleocene, in age. As for the Dragonian, that local fauna is probably intermediate between the classic Puerco and the classic Torrejon of Granger and Matthew (see Tomida, this volume), but the question is whether it is practical or desirable to recognize it as a separate age or to consider it as either late Puercan or early Torrejonian. (It should be remembered that in geochronology the technical term *age* applies not to a point in time but to a considerable span of time within an epoch.) The problems of Clarkforkian and Dragonian are now being worked on; however, since their bearing on the San Juan Basin is indirect, they need not be discussed further in this history.

In 1946 the American Museum resumed a collecting program (under my direction) in the Tertiary of the San Juan Basin. This continued every summer through 1951 and again in 1953 and 1958. (In 1959 I left the American Museum and have not been active in extensive field work in the San Juan Basin since.) During most of this campaign, George O. Whitaker, expert field and laboratory technician, worked with me. In 1958 he was in charge of the field work, which in that year I only planned and visited. In one season, 1947, William Fish was my field assistant; at various times a number of others (students, friends, local employees, family) became involved.

In 1946–1950 we worked mainly on the early Eocene, making stratigraphic studies and a large collection of mammals, strongly supplementing the Cope and Granger collections and with more precise data than had been possible for those forerunners (all finds now being located geographically on maps and air photographs and stratigraphically in measured sections). After the first two years I wrote a fairly detailed discussion of the history of study of this formation and of its stratigraphy (Simpson, 1948). Up until then this important and well-defined geological formation had never had a valid name. In 1948 I gave it the name "San José," by which it is now universally known.

At first we worked mainly on the splendid exposures of the San José in the great arroyo heads and rincones east of the continental divide. We also extended this work to the high exposures west of the divide, especially in 1947 and 1948. In 1947 Fish and I found concentrations of *Meniscotherium* at several places, including one especially rich pocket that was later extensively quarried. Although it included little articulated material, this pocket was unusual in including all parts of the disarticulated skeletons in nearly the same proportions as in life. We notified other paleontologists of this find, and several of them also collected at that locality. Among other remarkable finds were a herd of early ungulates, *Coryphodon*, buried and preserved whole, and a unique skeleton of an ancestral carnivore found by my wife Anne Roe.

Up until this time no identifiable fossils had been found in the type area of Cope's Puerco beds. In 1949 Whitaker and I spent some weary weeks looking for fossils there. We finally succeeded in finding identifiable mammals at several localities southwest and north of the present town of Cuba, at several levels in Cope's Puerco (Simpson, 1959). All of those mammals are definitely Torrejonian in age. It is possible that beds below our finds do include some of pre-Torrejonian Tertiary age, but at present there is no evidence that any of the type Puerco is Puercan in age. I emphasize that this is not a reason for abandoning the completely established (and among paleontologists worldwide) usage of the term "Puercan."

During this campaign Whitaker and I also made a reconnaissance of the Tiffany beds and found a second, much smaller and less rich pocket of small mammal remains near the (now completely cleaned out) Mason Pocket. In 1951 Whitaker, with David Kitts and John Ostrom, then student assistants and now both university professors (Oklahoma and Yale), collected from the Tiffany beds a skeleton of a large pantodont (early ungulate), found by Harley Barnes, Jr., then employed by the U.S. Geological Survey. They turned the skeleton over to us because we had been working in that area.

Throughout these years I planned to use the San José collections assembled in the American Museum for a full revision and discussion of the mammalian faunas in that formation. That plan was dropped when I left the Museum; only a few individual specimens of the many that we collected have been placed on published record.

In 1959 a considerable group with Whitaker worked in Tsosie Arroyo digging out matrix at the relatively rich Puercan fossil locality from which I had collected thirty years earlier. A large amount of this matrix was screen washed at the San Juan River some miles away, and a good collection of small bones and teeth was accumulated. That collection is still at the American Museum and has been mentioned but not enumerated or described in the twenty years since it was made.

This account must now backtrack to consider some of the important works by others in the San Juan Basin since the start of the American Museum's campaign of 1946–1959. In 1948 Robert W. Wilson, then and now again at the University of Kansas, collected Puercan and Torrejonian mammals in the San Juan Basin with Russell Camp and E. C. Galbreath. Their principal discovery was an extraordinarily rich pocket of Torrejonian mammals in Kutz Canyon, where Granger had prospected with lesser results. Wilson returned in 1950, 1956, and 1962, amassing some 300 identifiable mammals, many from his "Angels Peak Pocket" but some also from other localities in Kutz Canyon (Wilson, 1951; Wilson and Szalay, 1972). Included were the first primates from the middle Paleocene of the San Juan Basin. (Others of Torrejonian age were already known from more northern localities.)

In 1949 C. Lewis Gazin of the U.S. National Museum with his son Chester and Franklin Pierce, laboratory and field technician, made a noteworthy collection from the type Torrejonian. Only one specimen in this collection has been the subject of separate publication (Gazin, 1968), but the collection has served importantly in comparative studies.

The 1950 field conference of the Society of Vertebrate Paleontology deserves at least brief mention in a history of vertebrate paleontology in the San Juan Basin. That traveling conference visited the San José Eocene east of the continental divide, the Kirtland and Puercan in Barrel Spring Arroyo, and the Torrejonian at the rim of Kutz Canyon. Thus about 80 people (paleontologists and some of their family members) became at least superficially acquainted with much of the geological section involved in this history. (See Colbert et al., 1950, and Colbert, 1950.)

Some other paleontological developments in and about the San Juan Basin in the 1950's by the American Museum and others have been mentioned previously as a sequel to work started in the 1940's.

Surprisingly little in the field of this history seems to have gone on in the 1960's. The only really important event for which I have a note bearing on this subject during this decade is that in 1962 William A. Clemens, then of the University of Kansas and later of the University of California at Berkeley, started his successful search for Cretaceous mammals in the Fruitland and Kirtland formations. That continuing effort is said to have turned up more than a dozen mammal localities in those two formations, but few details have been published. Clemens (1973) did publish a brief and preliminary faunal list. There is another in a fuel resource paper by Fassett and Hinds (1971).

In 1970 what was reported in a friendly way as a horde of vertebrate paleontologists and students from the University of California at Berkeley descended on and spent about a week in the San Juan Basin, where they were joined by Malcolm McKenna and others from the American Museum and by Everett Lindsay from the University of Arizona. That signaled (and to an extent stimulated) continued interest in the Late Cretaceous and early Tertiary paleontology of the Basin, as seen in increasingly intensive and extensive work there through the 1970's by staff and students of the University of Arizona and by others. But the happenings of this decade are not so much history as current events. They will doubtless be chronicled by those taking part in them, and the present historical retrospect can stop at the round-figure date 1970.

As I am myself one of the characters in the history that I have been asked to summarize, perhaps in closing I may be excused a personal remark unusual in scientific publication. The years that I spent in the San Juan Basin with good companions, with heavy, productive physical labor, with mind-stretching problems, with exciting discoveries, with marvelous nature all around me, were among the happiest of the many happy years of my life.

REFERENCES

Bauer, C. M. 1916. Stratigraphy of a part of the Chaco River valley. U.S. Geological Survey Prof. Paper 98:271–278.

Brown, B. 1910. The Cretaceous Ojo Alamo beds of New Mexico with description of the new dinosaur genus *Kritosaurus*. Bull. Am. Mus. Nat. Hist. 28:267–274.

Clemens, W. A. 1973. The roles of fossil vertebrates in interpretation of Late Cretaceous stratigraphy of the San Juan Basin, New Mexico. Four Corners Geol. Soc. Memoir, p. 154–167.

Colbert, E. H. 1950. The fourth field conference of the Society of Vertebrate Paleontology. Soc. Vert. Paleontol. News Bull. 30:3–6.

Colbert, E. H., S. A. Northrop, A. S. Romer and G. G. Simpson. 1950. Guidebook for the fourth field conference of the Society of Vertebrate Paleontology in northwestern New Mexico. Distributed by the American Museum of Natural History, New York, 91 p.

Cope, E. D. 1875. Report on the geology of that part of northwestern New Mexico examined during the field season of 1874. *In* Ann. Rept. Geog. Explor. West of the 100th Mer., and Appendix LL, Ann. Rept. Chief of Engineers for 1875, p. 61–97 of separate issue, 981–1017 of full report.

———. 1877. Report upon the extinct Vertebrata obtained in New Mexico by parties of the expedition of 1874. Geog. Surveys West of the 100th Mer., pt. 2:i–iv, 1–370.

———. 1885. The Vertebrata of the Tertiary formations of the West. Book I. Rept. U.S. Geol. Survey Territories, F. V. Hayden in charge, 3:xi–xxxiv, 1–1009.

———. 1888. Synopsis of the vertebrate fauna of the Puerco series. Trans. Am. Phil. Soc., n.s., 16:298–361.

———. 1897. The position of the Periptychidae. Am. Nat. 31:335–336.

Fassett, J. E. and J. S. Hinds. 1971. Geology and fuel resources of Fruitland Formation and Kirtland Shale of the San Juan Basin, New Mexico and Colorado. U.S. Geol. Survey Prof. Paper 676:1–76.

Gardner, J. H. 1910. The Puerco and Torrejon Formations of the Nacimiento group. Jour. Geol. 18:702–741.

Gazin, C. L. 1968. A new primate from the Torrejon middle Paleocene of the San Juan Basin, New Mexico. Proc. Biol. Soc. Washington 81:629–634.

Gilmore, C. W. 1916. Contributions to the geology and paleontology of San Juan County, New Mexico. 2. Vertebrate faunas of the Ojo Alamo, Kirtland and Fruitland Formations. U.S. Geol. Survey Prof. Paper 98-Q:279–308.

———. 1919. Reptilian faunas of the Torrejon, Puerco, and underlying Upper Cretaceous formations of San Juan County, New Mexico. U.S. Geol. Survey Prof. Paper 119: 1–71.

———. 1922. A new sauropod dinosaur from the Ojo Alamo Formation of New Mexico. Smithsonian Misc. Coll. 72, No. 14:1–9.

———. 1928. Fossil lizards of North America. Nat. Acad. Sci. Mem. 22, No. 3:i–ix, 1–201.

———. 1944. Charles H. Sternberg. Soc. Vert. Paleontol. News Bull. 11:12–15.

Gingerich, P. D. 1975. What is type Paleocene? Am. Jour. Sci. 275:284–285.

Granger, W. 1914. On the names of lower Eocene faunal horizons of Wyoming and New Mexico. Bull. Am. Mus. Nat. Hist. 33:201–207.

———. 1915. A revision of the lower Eocene Wasatch and Wind River faunas. Part III. Bull. Am. Mus. Nat. Hist. 34:329–361.

———. 1917. Notes on Paleocene and lower Eocene mammal horizons of northern New Mexico and southern Colorado. Bull. Am. Mus. Nat. Hist. 37:821–830.

Matthew, W. D. 1897. A revision of the Puerco fauna. Bull. Am. Mus. Nat. Hist. 9: 259–323.

———. 1915a. A revision of the lower Eocene Wasatch and Wind River faunas. Introduction and Part I. Bull. Am. Mus. Nat. Hist. 34:1–103.

———. 1915b. A revision of the lower Eocene Wasatch and Wind River faunas. Part II. Bull. Am. Mus. Nat. Hist. 34:311–328.

———. 1915c. A revision of the lower Eocene Wasatch and Wind River faunas. Part IV. Bull. Am. Mus. Nat. His. 34:429–483.

———. 1918. A revision of the lower Eocene Wasatch and Wind River faunas. Part V. Bull. Am. Mus. Nat. Hist. 38:565–657.

———. 1920. Status and limits of the Paleocene. Bull. Geol. Soc. Am. 31:221.

———. 1937. Paleocene faunas of the San Juan Basin, New Mexico. Trans. Am. Phil. Soc., n.s. 30:i–viii, 1–510.

Matthew, W. D. and W. Granger. 1921. New genera of Paleocene mammals. Am. Mus. Novitates 13:1–7.

Newberry, J. S. 1876,. Geological report. *In* J. N. Macomb, Report of the exploring expedition from Santa Fe, New Mexico, to the junction of the Grand and Green rivers of the great Colorado of the West in 1859, Eng. Dept., U.S. Army: 9–118.

Osborn, H .F. and C. Earle. 1895. Fossil mammals of the Puerco Beds. Collection of 1892. Bull. Am. Mus. Nat. Hist. 7:1–70.

Ostrom, J. H. 1963. *Parasaurolophus cyrtocristatus*, a crested hadrosaurian dinosaur from New Mexico. Fieldiana, Geology 14:142–168.

Pomerol, C. 1969. Rapport sur la limite Paléocène–Eocène. Mem. Bur. Recherches Géol. Minières 69:447–449.

Reeside, J. B., Jr. 1924. Upper Cretaceous and Tertiary formations of the western part of the San Juan Basin of Colorado and New Mexico. U.S. Geol. Survey Prof. Paper 134:i–iv, 1–70.

Schuchert, C. and C. M. LeVene. 1940. O. C. Marsh. Pioneer in Paleontology. Yale Univ. Press, New Haven, 541 p.

Simons, E. L. 1963. David Baldwin, O. C. Marsh and the discovery of the first continental Paleocene faunas of the New World. Postilla, No. 75:1–11.

Simpson, G. G. 1935a,b,c. The Tiffany fauna, upper Paleocene. Am. Mus. Novitates 795: 1–19; 816:1–30; 817:1–28.

————. 1936. Additions to the Puerco fauna, lower Paleocene. Am. Mus. Novitates 849:1–11.

————. 1948. The Eocene of the San Juan Basin, New Mexico. Am. Jour. Sci. 246: 257–282, 363–385.

————. 1951. Hayden, Cope, and the Eocene of New Mexico. Proc. Acad. Nat. Sci. Philadelphia 103:1–21.

————. 1959. Fossil mammals from the type area of the Puerco and Nacimiento strata, Paleocene of New Mexico. Am. Mus. Novitates 1957:1–22.

Sinclair, W. J. and W. Granger. 1914. Paleocene deposits of the San Juan Basin, New Mexico. Bull. Am. Mus. Nat. Hist. 33:297–316.

Sternberg, C. H. 1909. The Life of a Fossil Hunter. Holt, New York, 286 p.

————. 1932. Hunting dinosaurs in the badlands of the Red Deer River, Alberta Canada. Published by the author, San Diego, California, 261 p.

Wegemann, C. H. 1917. Wasatch fossils in so-called Fort Union beds of the Powder River Basin, Wyoming. U.S. Geol. Survey Prof. Paper 108D:57–60.

Wilson, R. W. 1951. Preliminary survey of a Paleocene faunule from the Angels Peak area, New Mexico. Univ. Kansas Pub. Mus. Nat. Hist. 5:1–11.

Wilson, R. W. and F. S. Szalay. 1972. New paromomyid primate from middle Paleocene beds, Kutz Canyon area, San Juan Basin, New Mexico. Am. Mus. Novitates 2499:1–18.

Wood, H. E., 2nd, R. W. Chaney, J. Clark, E. H. Colbert, G. L. Jepsen, J. B. Reeside, Jr. and C. Stock. 1941. Nomenclature and correlation of the North American continental Tertiary. Bull. Geol. Soc. Am. 52:1–48.

Part 2
Morphology and Systematics

2

The occurrence of *Pentaceratops* (Ornithschia: Ceratopsia) with a description of its frill

Timothy Rowe
Department of Geology,
Museum of Northern Arizona
Flagstaff, Arizona

Edwin H. Colbert
Department of Geology,
Museum of Northern Arizona
Flagstaff, Arizona

J. Dale Nations
Department of Geology,
Northern Arizona University
Flagstaff, Arizona

INTRODUCTION

The genus *Pentaceratops* was established by Henry Fairfield Osborn in 1923, the type species being *Pentaceratops sternbergii.* The name '*Pentaceratops*' was suggested to Osborn by W. D. Matthew, being based on the rather pointed jugal bones which, in addition to the usual three horns of the skull, distinguished this ceratopsian dinosaur from others by the possession of five "horns." The name is perhaps more fanciful than real, since the jugals in some specimens are not particularly hornlike.

At the present time *Pentaceratops* is known from five specimens that include substantial portions of the skull and frill, in addition to a number of more fragmentary specimens referred (with varying degrees of confidence) to this genus. This sparse representation of *Pentaceratops* illustrates a problem commonly encountered by paleontologists, especially when working with dinosaurs, namely that of small samples. As a result, certain anatomical features may remain completely unknown, ranges and significance of morphological variation may be difficult to assess, and classification methods may often be specially tailored to suit the material available. In the case of *Pentaceratops*, the paucity of fossils has left our knowledge, particularly of the frill, far from complete. Although seemingly complete frills are figured in the literature (e.g. Wiman, 1930; Ostrom, 1966; Farlow & Dodson, 1975), their posterior parts were drawn largely from plaster restorations based on other ceratopsian genera.

This situation was recently improved when the Museum of Northern Arizona discovered near Burnham, New Mexico an excellent skull with a complete frill, jaws, and a partial postcranial skeleton. The specimen was discovered during the course of a predevelopment inventory survey of the paleontological resources of a proposed surface coal mine, on a lease held jointly by Consolidation Coal Company and El Paso Natural Gas Company. This specimen revealed that the top of the frill has been improperly restored in all published accounts of *Pentaceratops*, and it forms the basis for the description of the frill in this paper.

MATERIAL

Material referred to *Pentaceratops* is in the following institutions:

American Museum of Natural History (AMNH)

AMNH 6325 *Pentaceratops sternbergii*, Type, skull with anterior two-thirds of frill.

AMNH 1624 *Pentaceratops sternbergii*, skull with anterior two-thirds of frill.

AMNH 1625 cf. *Pentaceratops*, rear part of frill.

AMNH 1622 cf. *Pentaceratops*, skull without frill.

Museum of Northern Arizona (MNA)

MNA Pl. 1747 *Pentaceratops sternbergii*, skull with complete frill, jaws, partial vertebral column and ribs.

MNA Pl. 1668 *Pentaceratops* cf. *sternbergii*, right lower jaw, partial squamosal, tibia, fibula, proximal femur, 2 vertebrae, 5 ribs.

Museum of Natural History, University of Kansas (UKVP)

UKVP 16100 *Pentaceratops* sp., skull with frill.

Laboratory of Paleontology, University of Arizona (UALP)

UALP 13342 cf. *Pentaceratops*, skull without frill.

University of New Mexico, Albuquerque, (UNM)

UNM-FKK-035 cf. *Pentaceratops*, fragmentary squamosal and parietal of juvenile. (see Lehman, this volume)

University of Uppsala, Sweden (PMU)

PMU.R200 *Pentaceratops fenestratus*, Cotype, laterally crushed skull and frill.

PMU.R268 *Pentaceratops fenestratus*, Cotype, nearly complete post-cranial skeleton questionably associated with skull (PMU. R200).(see Mateer, this volume)

PMU.————*Pentaceratops* cf. *fenestratus*, partial squamosal and parietal; distal quadrate; proximal part of supraorbital horn core.

United States National Museum (USNM)

USNM 12002 *Pentaceratops sternbergii*, left squamosal.

USNM 12741 cf. *Pentaceratops*, supraorbital horn core, braincase. (see Lehman, this volume)

USNM 12743 cf. *Pentaceratops*, supraorbital horn core, partial squamosal.

Current ceratopsian classifications are based on cranial characters, and most published studies of *Pentaceratops* deal exclusively with the skull and frill. Only a single postcranial skeleton, designated as the cotype of *Pentaceratops fenestratus* Wiman 1930, has been described (Wiman, 1930; Lull, 1933), but Mateer (this volume) points out that there is no direct evidence that this postcranial skeleton in fact belongs to *Pentaceratops*. At present, *Pentaceratops* postcranial material can not be distinguished from that of most other large ceratopsians. This is owing largely to the scarcity of postcranial material found in association with identifiable cranial remains. However, a lack of detailed comparative study of the ceratopsian postcranium also contributes to the problem. It is likely, therefore, that future work will lead to the recognition of more *Pentaceratops* material in existing museum collections.

OCCURRENCE OF *PENTACERATOPS*

Gilmore (1933) reported a ceratopsian epoccipital (AMNH 6591) from Baiying Bologai, Mongolia, which he tentatively referred to *Pentaceratops*. As the base of Gilmore's specimen is missing, he states that "owing to the lack of this surface a positive identification of the bone is not possible" (p. 19). While this specimen probably is a ceratopsian epoccipital, it does not resemble epoccipitals from other specimens referred to *Pentaceratops* more closely than it resembles these bones from other ceratopsians, especially *Chasmosaurus*. Furthermore, no other *Pentaceratops* material has been reported from outside of New Mexico. There seems to be no clear reason to refer Gilmore's specimen positively to *Pentaceratops*, and it will not be considered further in this paper. All known *Pentaceratops* material has come from the western edge of the San Juan Basin, in San Juan County, New Mexico.

The Museum of Northern Arizona specimen (MNA Pl. 1747) was collected in 1977 from the upper part of the Fruitland Formation in the NW¼SW¼ Sec. 28, T24N, R13W, approximately 5 km east of the Burnham Trading Post (the trading post was destroyed by fire in 1978). This specimen was found in a 2 m thick lens of grayish-white, fine to medium grained, heterogeneous, cross-bedded sandstone. Carbonized remains of palms, conifers, and deciduous angiosperms were found throughout the sandstone, with accumulations on cross-bedding planes. The sandstone grades laterally into and is interbedded with yellowish-gray mudstones, all of which make up a 6 m thick relatively weather resistant unit located approximately 145 m above the base of the Fruitland Formation. The mudstone below the sandstone yielded the remains of turtle (*Trionyx*?) bone and shell fragments, a single isolated large carnosaur (*Albertosaurus*?) tooth, and pelecypods ("*Unio*"). The skull was found lying upside-down with a number of disarticulated vertebrae and ribs lying beneath it and was apparently buried on the slope of a stream point bar or channel bar.

A second nearly complete skull and frill (UKVP 16100) was collected by the Museum of Natural History, University of Kansas, in 1966 from the upper part of the Fruitland Formation in the SW¼ Sec. 34, T24N, R13W, along Hunter Wash. The University of Arizona collected a skull lacking the frill (UALP 13342) in 1975 from the Kirtland Shale in the SE¼NE¼ Sec. 14, T24N, R12W, along Alamo Wash.

The type skull and partial frill of *Pentaceratops sternbergii* (AMNH 6325) was collected from the Fruitland Formation in 1922 by Charles H. Sternberg from a locality 14.4 km east of Tsaya, probably from the NW¼ T22N, R11W. Sternberg also collected the postcranial skeleton, but it was later discarded (Lull, 1933). That same year Sternberg collected another ceratopsian skull without the frill (AMNH 1622), probably belonging to *Pentaceratops*, from near Tsaya, and in 1923 he obtained a third skull and partial frill (AMNH 1624) from the same area. Precise locality data for the latter specimen have been lost.

The skull of the type of the second species of *Pentaceratops*, *P. fenestratus* Wiman 1930, was collected from the Kirtland Shale in 1921 by C. H. Sternberg. The locality of this specimen was described by Wiman (1930) and Lull (1933) as one mile south of Kimbeto Wash, on the south branch of Meyers Creek (Myers Creek in Lull, 1933). Gilmore (1935) reported the locality of this specimen as one mile south of Kimbeto Wash, but on the north branch of Meyers Creek. Meyers or Myers Creek is not labeled on USGS topographic maps of this area. It is, however, labeled on Plate 64 of Bauer (1916), and has apparently been renamed Ah-shi-sle-pah Wash on USGS maps (Kues, et al., 1977, p. 120). Kimbeto Wash, however, lies south of Ah-shi-sle-pah Wash, not to the north as implied by the locality data presented by Wiman, Lull, and Gilmore. If Ah-shi-sle-pah Wash is in fact "Meyers Creek," the locality data presented by Wiman, Lull, and Gilmore are in error and should read one mile (1.6 km) *north* of

Kimbeto Wash, on the south branch of Ah-shi-sle-pah (Meyers) Wash. The south branch of Ah-shi-sle-pah Wash cuts through the Kirtland Shale about 1.6 km north of Kimbeto Wash in the NW¼ T22N, R10W. This is probably where the skull of *Pentaceratops fenestratus* was collected. A postcranial skeleton (PMU.R268), designated by Wiman (1930) as the cotype of *Pentaceratops fenestratus* despite the fact that it was not directly associated with the skull, was collected from Ah-shi-sle-pah Wash in T22N, R11W (Mateer, this volume). These specimens were sold to the University of Uppsala, Sweden, together with other material referred to this species. This material includes a partial squamosal and parietal, also from the Kirtland Shale along the south branch of "Meyers Creek," 6.4 km above "Cattle Ranch"; and a distal quadrate, partial squamosal and proximal part of a supraorbital horn core collected from the Fruitland Formation on the south side of Alamo Wash, 4.8 km above Hunter's Store (Lull, 1933).

In 1929 Gilmore collected a supraorbital horn core and braincase (USNM 12741) which he tentatively referred to *Pentaceratops* (Lehman, this volume). This material was collected from Hunter Wash in T25N, R11W, from exposures of purple shale which Gilmore believed to be part of the McDermott Formation. Baltz et al. (1966) stated that the equivalence of these purple shales in the Hunter Wash area with the type McDermott could not be demonstrated with any certainty and assigned them to the upper shale member of the Kirtland Shale.

Lehman (this volume) has reported a squamosal and parietal fragment of a juvenile ceratopsian (UNM-FKK-035), which he referred to *Pentaceratops*. This material was collected from De-na-zin Wash in T24N, R11W, from exposures of the Naashoibito Member, which has been designated as the uppermost member of the Kirtland Shale (Baltz et al., 1966; Fassett & Hinds, 1971; O'Sullivan et al., 1972). More recently it has been considered to be the lowermost member of the Ojo Alamo Sandstone (Powell, 1973; Lindsay, et al., 1981). It is doubtful that such fragmentary material, particularly of a juvenile individual, can be confidently assigned to *Pentaceratops*, and Lehman's identification is admittedly tentative. In lieu of more complete material, the occurrence of *Pentaceratops* in the Naashoibito Member must remain questionable.

Other material referred to *Pentaceratops* includes a left squamosal (USNM 12002; plate 13(2) in Gilmore, 1935) collected from the upper part of the Fruitland Formation or lower part of the Kirtland Shale in the SW¼ T24N, R13W; and a supraorbital horn core with a partial squamosal (USNM 12743) from the lower part of the Kirtland Shale in the SW¼ T24N, R12W. All of this material was collected in 1929 by G. F. Sternberg. In 1924 C. H. Sternberg collected a specimen (AMNH 1625) identified as the rear part of the crest of *Pentaceratops* from an unrecorded locality in the Fruitland Formation.

Although locality data for much of the *Pentaceratops* material are vague, sufficient information is available to permit at least a general stratigraphic location of all fossils. *Pentaceratops* occurs in the Fruitland Formation and extends vertically into the Kirtland Shale where it reaches into the upper shale member, and has been questionably reported from the overlying Naashoibito Member (Lehman, this volume).

The Fruitland Formation is composed predominantly of interbedded sandstone, shale, carbonaceous sandstone and shale, coal, and local thin beds of limestone near the bottom of the section. Coal and organic material become progressively less abundant towards the top of the section as the Fruitland grades into the overlying Kirtland Shale. The Kirtland Shale is composed predominantly of gray shales with interbedded lenses of sandstone and mudstone, and is distinguished from the Fruitland largely by its lack of coal. The highest laterally persistent coal bed is commonly considered to mark the top of the Fruitland (e.g. Fassett & Hinds, 1971), but in the field it is often difficult to distinguish between the formations in their thick transition zone. The Fruitland intertongues with the underlying Pictured Cliffs Sandstone, which is a littoral deposit. Overlying the Kirtland Shale is the Ojo Alamo Sandstone. The Fruitland–Kirtland Formations represent coastal swamp, river, floodplain, and lake deposits laid down shoreward of the regressing Pictured Cliffs Sea (Fassett & Hinds, 1971). Fruitland–Kirtland deposition appears to have been continuous, as no physical evidence has been found to indicate anything other than local significant periods of erosion or non-deposition, and no differences in faunal or floral composition have yet been recognized laterally or vertically between the base of the Fruitland Formation and the base of the Naashoibito Member (however, see Lucas, this volume).

The lowermost occurrence of *Pentaceratops* in the Fruitland section is uncertain. The lower third of the Fruitland, to the west of Burnham where the Fruitland is at its thickest, was carefully prospected by the Museum of Northern Arizona in 1974, 1977, and 1978 but proved to be poorly fossiliferous, and it is unlikely that any of the *Pentaceratops* material from unrecorded localities in the Fruitland came from this region. The lower half of the Fruitland along the southwestern edge of the San Juan Basin has yielded a number of vertebrates including a partial, articulated skeleton of *Kritosaurus*. Fruitland exposures in this area were prospected in detail by the Sternbergs and it is possible that *Penta-*

ceratops material has come from the lower part of the Fruitland in this region. However, it is not clear whether the base of the Fruitland in this region correlates with the base of the Fruitland near Burnham or some point higher in the section. The *Pentaceratops* localities in the Fruitland Formation for which precise data are available occur in the upper two-thirds of the section.

Pentaceratops extends vertically into the upper shale member of the Kirtland Shale, and Lehman (this volume) has tentatively identified *Pentaceratops* from the overlying Naashoibito Member. Both the immaturity and the fragmentary nature of Lehman's specimen (UNM FKK-035) leave the occurrence of *Pentaceratops* in the Naashoibito Member questionable. However, if the discovery of more complete material confirms the occurrence of *Pentaceratops* in the Naashoibito, its range would then extend through the Kirtland Shale into what Powell (1973) and Lindsay et al. (1981) considered to be the lowermost member of the Ojo Alamo Sandstone.

There is general agreement that the Fruitland and Kirtland Formations are of Late Cretaceous age, although their precise correlation and age have been the subject of considerable debate. Early attempts at correlating these formations with other Late Cretaceous non-marine sediments from the Western Interior of North America relied upon a very few vertebrate taxa for comparison, and such attempts yielded a variety of interpretations. Clemens (1973) reviewed the early literature pertaining to the age of the Fruitland and Kirtland Formations and stated that "In large part the differences in interpretation appear traceable to inadequate data or faulty application of biostratigraphic technique or both" (p. 164). Clemens' (1973) preliminary study of the mammals of the Hunter Wash local fauna indicated that:

> Although . . . some elements of the Hunter Wash local fauna appear identical or closely related to members of local faunas recovered from the Judith River, Oldman, and Milk River Formations, there are differences in overall composition. The mammals of the Hunter Wash local fauna do not simply record a southern extension of one of these older, northern mammalian faunal units. Analysis of the microvertebrates of the Hunter Wash local fauna emphasizes its unique composition, and holds open the question: do these dissimilarities in composition reflect temporal or ecologically based, biogeographic differences, or some combination of these factors? (p. 164)

Until this question is more fully answered, doubt will exist regarding the validity of attempts to determine the age of the Fruitland and Kirtland Formations using vertebrates alone; the same objection applies to the use of paleobotanical data without independent lines of evidence. Despite this confusion in older accounts, recent detailed studies of the Fruitland–Kirtland fauna and flora together with paleomagnetic and radiometric data are resolving the existing ambiguities. Thus a clearer understanding of their age is beginning to emerge.

Powell (1973) noted that the age significance of the dinosaurs of the Hunter Wash local fauna is equivocal, and that they show similarities to the Judith River, Belly River, Two Medicine, and Edmonton dinosaur faunas. Clemens (1973) concluded that the Hunter Wash local fauna is probably younger than the Judith River and Oldman faunas. Russell (1975) noted that almost all mammalian genera present in the Fruitland and Kirtland Formations occur in both the Judithian and Lancian faunas. Lindsay et al. (1981) noted that the Hunter Wash local fauna lacks some but includes other mammals which are characteristic of the Lancian Land Mammal "Age" of Russell (1975), and concluded that the Hunter Wash local fauna must be close to the Edmontonian–Lancian transition.

Russell (1975) mentioned that:

> The evidence of the fossil invertebrates is relevant to the problem of the Fruitland–Kirtland age. The marine Pictured Cliffs Sandstone and Lewis Shale, which underlie the Fruitland Formation, contain marine molluscs similar to those of the Bearpaw Formation of the northern plains. This suggests that the Fruitland–Kirtland fauna cannot be older than Edmontonian. The Fruitland Formation contains the fresh-water mussel *Proparreysia holmsiana* (White), which occurs in the Lancian fauna, as does *Proparreysia pyramidatoides* (Whitfield), doubtfully identified from the Kirtland Shale by Stanton (1916). (p. 152)

Anderson (1960) and Baltz et al. (1966) used palynological data to confirm the Late Cretaceous age of the Kirtland Shale, but were unable to offer a more precise age assignment. Tschudy (1973) concluded that the Fruitland Formation is of Campanian–Maastrichtian age on the basis of an unspecified number of samples from a single core through the Fruitland. Zavada (1976) examined sixteen samples of coals and shales from the Fruitland which yielded 152 species belonging to 97 genera of spores and pollen, and concluded that the Fruitland is of Maastrichtian age.

Recent paleomagnetic studies in the San Juan Basin (Butler et al., 1977; Lindsay et al., 1978; Lindsay et al., 1981) have determined that the Fruitland–Kirtland Formations correlate with anomalies 31–30 of the magnetic polarity column (LaBrecque et al., 1977) of middle to late Maastrichtian age. Alvarez and Vann (1979), Fassett (1979), and Lucas and Rigby (1979) have argued that an unconformity at the base of the Kimbeto Member of the overlying Ojo Alamo Sandstone, which was not recognized by Butler et al. (1977) or Lindsay et al. (1978), casts doubt on the validity of this correlation. However, Lindsay et al. (1979a,b) presented arguments that we feel rule out any significant unconformity in their section. Furthermore, Lindsay et al. (1981) have identified ash in their section through the Kirtland which yields a radiometric date of 64.6 ± 3.0 million years, clearly placing the Kirtland in the Maastrichtian. This date is very close to the radiometric age of the Kneehills Tuff (65 million years according to Folinsbee et al., 1965; Lambert, 1971, cited in Lindsay et al., 1981). The Kneehills Tuff separates an Edmontonian Land Mammal "Age" fauna from an overlying Lancian age fauna. Significantly, the University of Arizona *Pentaceratops* came from a horizon immediately above Lindsay's ash (L. Jacobs, oral comm., 1979).

Thus, a considerable body of evidence including vertebrate, invertebrate, paleobotanic, paleomagnetic, and radiometric data argues that the Fruitland–Kirtland Formations are of Maastrichtian, probably Edmontonian to earliest Lancian, age. The temporal range of *Pentaceratops* would therefore have overlapped with that of *Anchiceràtops*, *Arrhinoceratops* and *Leptoceratops* from the Edmonton of Alberta, and possibly with *Torosaurus* and *Triceratops* from the Lance Formation. Fragmentary ceratopsian material from the Fruitland–Kirtland Formations has been referred to *Ceratops* and *Monoclonius* (Lull, 1933) and *Chasmosaurus* (Gilmore, 1935), but these identifications are admittedly quite tentative (Lull, 1933; Gilmore, 1935). In lieu of more complete material, their co-occurrence with *Pentaceratops* remains questionable, although it is not unreasonable to expect other ceratopsian genera in addition to *Pentaceratops* in the Fruitland and Kirtland Formations.

THE FRILL IN *PENTACERATOPS*

In his diagnosis of *Pentaceratops* Osborn (1923) noted that the frill is characterized by an "Elongated narrow fontanelle in the dermosupraoccipital crest" and that the "borders of the crest are moderately sculptured." He also pointed out that because of the elongated frill and the large size of its contained fontanelles, this genus shows resemblances to *Torosaurus* from the Lance Formation of Wyoming.

It should be emphasized that the type specimen of *Pentaceratops sternbergii* (AMNH 6325), as figured by Osborn (1923), has an incomplete frill; its posterior medial and lateral portions evidently were not preserved. Subsequent to its description the skull was placed on exhibition with the missing parts of the frill restored in plaster. A second skull

TEXT-FIG. *2.1*—Posterior view of skull and frill of *Pentaceratops* cf. *sternbergii* (MNA Pl. 1747). One-twelfth natural size.

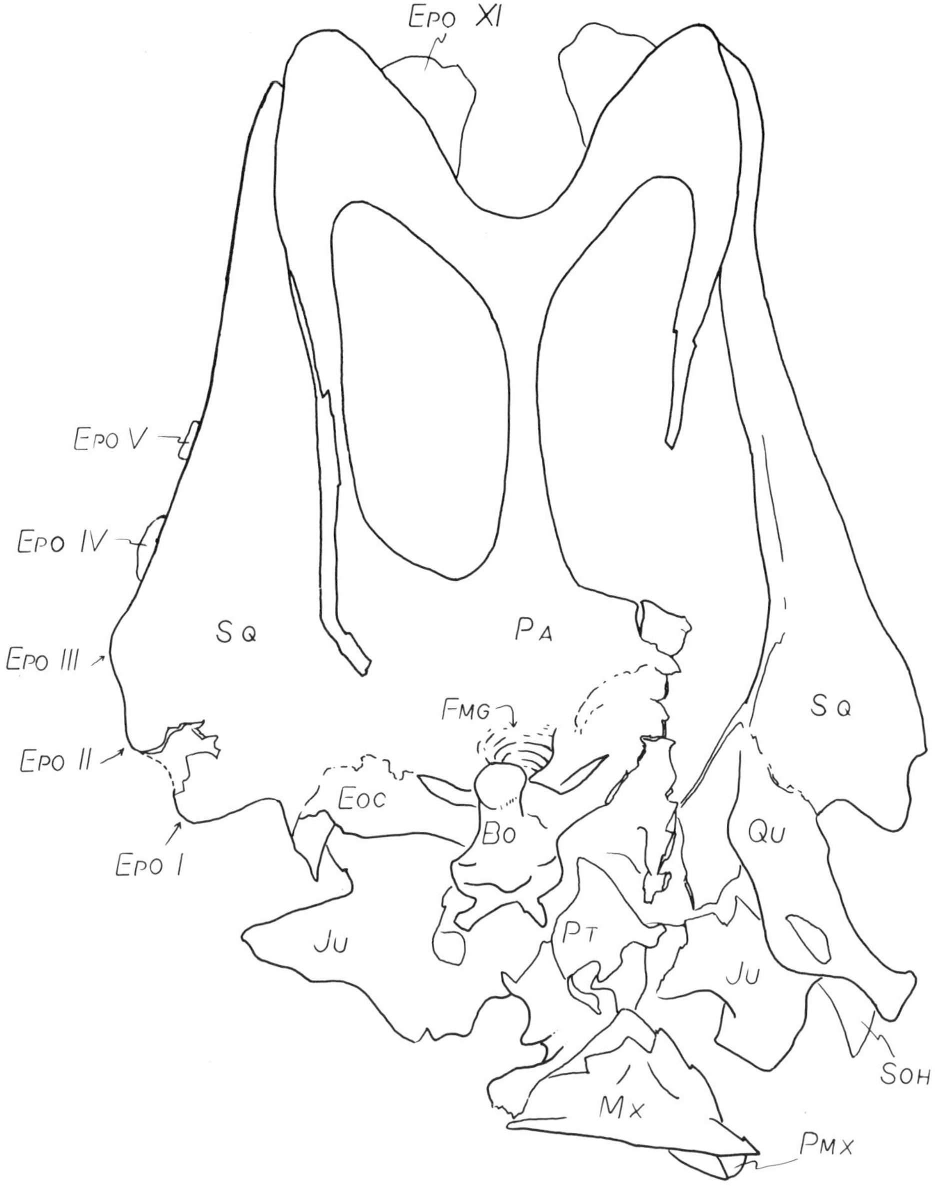

TEXT-FIG. 2.2—Key to skull features displayed in Text-fig. 2.1. *Bo*, basioccipital; *Eoc*, exoccipital; *Epo*, epoccipital; *Fmg*, foramen magnum; *Ju*, jugal; *Mx*, maxilla; *Qu*, quadrate; *Pa*, parietal; *Pmx*, premaxilla; *Pt*, pterygoid; *Sq*, squamosal.

in the American Museum of Natural History (AMNH 1624) has a frill in which, interestingly, almost the same parts are missing as in the type. It too has been restored, and the restoration follows the same pattern used in restoring the frill in the type specimen. Since there was no guide for such a restoration, the form of the reconstruction was shaped in a fashion that is now, on the basis of the Museum of Northern Arizona specimen (Text-figs. 2.1, 2.2), seen to be incorrect. A second *Pentaceratops* specimen with the rear part of the frill preserved (Text-fig. 2.3) is in the Museum of Natural History, University of Kansas (UKVP 16100), and agrees closely with MNA Pl.1747. A third specimen, consisting of the rear part of a frill referred to *Pentaceratops* (Lull, 1933), is reportedly housed in the American Museum collection (AMNH 1625), but at this date it has not been located.

The skull described by Wiman (1930) as *Pentaceratops fenestratus* is strongly compressed from side to side, and this distortion together with some breakage along the posterior margin of the frill has obscured the shape of this particular part of the skull (see Mateer, this volume). Wiman attempted a restoration of the skull as it might have appeared in an undistorted view, but one may question the form of the posterior border of the frill as restored.

The Museum of Northern Arizona *Pentaceratops* skull is characterized, as are other specimens of the same genus, by an elongated frill, thus placing it within the category of long-frilled ceratopsians as distinguished by Lull (1933). Other genera of long-frilled horned dinosaurs are *Chasmosaurus* from the Oldman Formation, *Anchiceratops* and *Arrhinoceratops* from the Edmonton Formation, and *Torosaurus* from the Lance Formation. All of these ceratopsians are characterized by large fenestrae within the frill.

The frill of the specimen under consideration is perhaps one and a half times the length of the anterior part of the skull as measured from the posterior edge of the horn base on the postorbital bone to the tip of the rostrum. Thus it closely resembles the type of *Pentaceratops sternbergii* and AMNH 1624 in this respect, as estimated from the contours of the portions of the frill that have been preserved. It is difficult to ascertain the precise length of the frill in relation to the anterior section of the skull with accuracy in MNA Pl.1747 because the premaxillary, nasal and maxillary bones have been separated and displaced. Similar proportions of frill length to anterior length are seen in the skull of *Pentaceratops fenestratus*.

The two large fenestrae of the frill of MNA Pl.1747 are elongated much as they are in the type of *Pentaceratops sternbergii*. In this latter specimen the length of each fenestra is about 60% of the total frill length. As in other ceratopsians, the fenestrae are included within the ex-

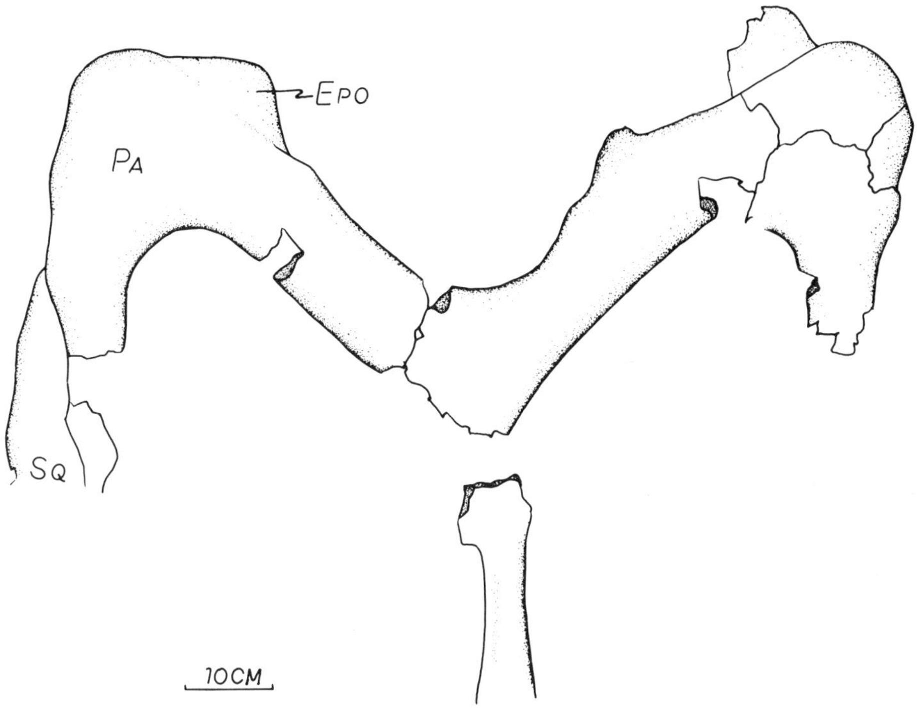

TEXT-FIG. 2.3—Posterior view of rear part of frill of *Pentaceratops* cf. *sternbergii* (UKVP 16100). (See Text-fig. 2.2 for key to abbreviations.)

panded parietal bones, which form the greater part of the frill. Elongated and expanded squamosals constitute the lateral borders of the frill.

These two fenestrae are the only openings in the frill, as is true for the type of *Pentaceratops sternbergii* and AMNH 1624. The skull of *Pentaceratops fenestratus* is unique in having a small fenestra on the left side, about midway along the frill, lateral to the parietal fenestra, and enclosed entirely within the expanded squamosal bone. Wiman (1930) designated this as the squamosal fenestra and considered it to be a valid character distinguishing the specimen specifically from *Pentaceratops sternbergii*, hence the name *Pentaceratops fenestratus*. In his restoration Wiman showed the squamosal fenestra on each side of the frill.

Lull (1933), however, considered this as a pathological feature present on the one side of the skull of *Pentaceratops fenestratus*. Dr. Niall J. Mateer (written commun., 1979; see also this volume) reexamined the specimen and concluded that this feature is not pathological:

> I tend to agree with Lull in that Wiman's choice of a specific name is not a good one, but the status of this specimen is good, i.e. it

> being a separate species. However I cannot agree with Lull's opinion on the nature of the squamosal fenestra; I cannot see that this could have been either a lesion or a puncture. Contrary to Wiman's reconstruction, this fenestra is quite uniformly elliptical with typical fenestra borders. Whether a similar fenestra existed on the right side can only be solved through finding a new specimen.

Much of the distinctive character of the frill in *Pentaceratops* is a result of the development of large epoccipital bones along the lateral borders of the squamosal and on the posterior edge of the parietal, thus giving a scalloped appearance to the outer border of the frill. In the type specimen there are 11 such scallops in a continuous series along the side and back of the frill, some of them restored in plaster. In his restoration of the crushed skull of *Pentaceratops fenestratus*, Wiman (1930) showed 15 scallops in a continuous series from the anterior edge of the squamosal around to the midline at the back. Eleven of the scallops are shown as forming the edge of the squamosal, four of the parietal.

Gilmore (1935) described and figured a left squamosal referred to *Pentaceratops sternbergii* (USNM 12002). There are from front to back seven large epoccipitals in a series along the edge of the bone, followed by a gap that would have accommodated an additional epoccipital and after that a large triangular epoccipital bone at the very end of the squamosal.

In MNA Pl. 1747 there are five anterior scallops on the left side of the frill, the first three being large and fused to the squamosal. The following two epoccipitals are distinctly separated from the squamosal by sutures. Behind the fifth epoccipital, which is incomplete, is the smooth edge of the squamosal with space for four more epoccipitals comparable in length to those present on the more anterior parts of the frill.

Four isolated epoccipital bones were found with the skull (Text-fig. 2.4), and these nicely fill the remaining space along the edge of the left squamosal. One of these epoccipitals is low and probably can be identified as number 6 of the series. The next two are higher, and are probably numbers 7 and 8, the eighth being somewhat higher than the seventh. The final epoccipital is a very large triangular bone, which may have occupied a terminal position along the edge of the squamosal. Interestingly, this bone is identical in shape to the terminal epoccipital in Gilmore's (1935) figure of the squamosal (USNM 12002) referred to *Pentaceratops sternbergii*. If the above identifications and interpretations of the epoccipitals in MNA Pl.1747 are correct, this skull had nine such bones along the edge of the squamosal, the ninth being large and of deeply triangular shape. The right squamosal in MNA Pl.1747 shows

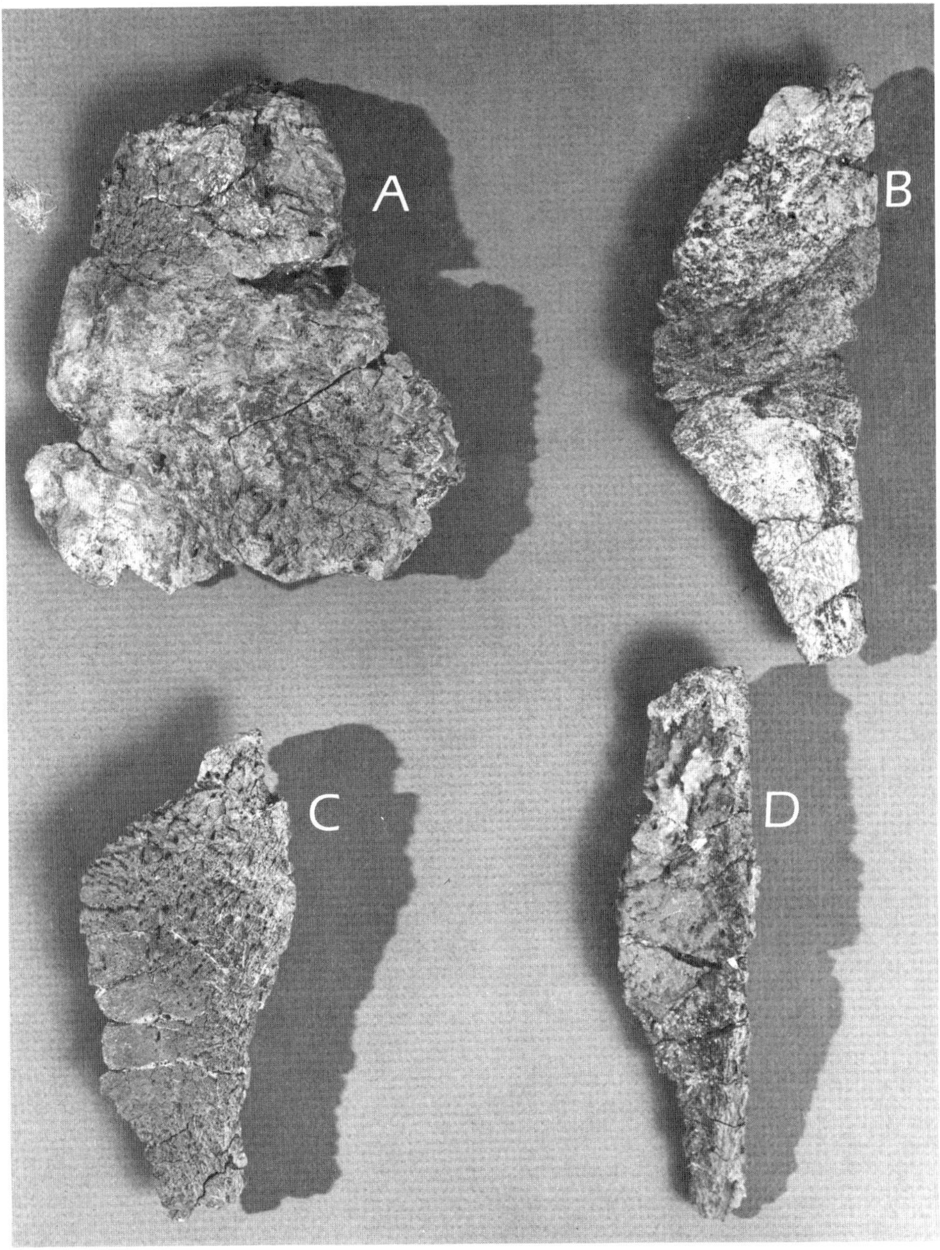

TEXT-FIG. 2.4—Isolated epoccipital bones found associated with the left side of the frill of MNA Pl. 1747. *A*, epoccipital number 9, which may have occupied a terminal position along the side of the squamosal; *B*, epoccipital number 8; *C*, epoccipital number 7; *D*, epoccipital number 6 (?). All × 0.53.

four anterior epoccipital scallops; the remaining edge of the bone lacks the epoccipitals.

The posterior edge of the parietals, forming that portion of the frill never before clearly visible, is quite distinctive and unlike any previous conception of this region. These edges of the two parietals together form a very deep U-shaped reentrant into the back of the frill. This large emargination is of such size as to extend anteriorly so that its anterior border, at the bottom of the U, is some 4 cm from the back borders of the parietal fenestrae. The width of the emargination is of such extent that between its border and the opposite border of the parietal fenestra the parietal bone is narrowed to form a flat bar. It is similar to, but proportionately different from, the same region in the frill of *Chasmosaurus.*

On the inner posterior border of this emargination there is on each side a very large triangular epoccipital, the point of which is directed medially. The points of these epoccipitals are separated from each other by a distance of about 12 cm. Moreover, the anteromedial borders of the epoccipitals are curved, extending the curve formed by the U-shaped emargination of the parietals so that it becomes, in effect, the greater part of a slightly elongated ellipse.

The posterior edge of each parietal is smooth, but it seems reasonable to suppose that each edge supported a large triangular epoccipital, one on each side between the terminal epoccipital of the squamosal and the large posteromedial epoccipital of the parietal just described. It should be noted that there is a large triangular epoccipital in this position on the parietal of *Chasmosaurus belli*, AMNH 5402.

The rear part of the frill in the University of Kansas specimen (UKVP 16100) is present, although not as well preserved as in MNA Pl.1747. Nevertheless, it closely resembles the Museum of Northern Arizona specimen. The posterior border of the parietals forms the deep U-shaped reentrant into the back of the frill and a large, seemingly triangular, epoccipital is present on each side within the reentrant.

It would therefore seem probable that there were 11 epoccipitals on each side of the frill in *Pentaceratops*, of which the last three, one on the squamosal and two on the parietal, are large and triangular in shape. The most anterior epoccipitals, while large, have laterally rounded borders, and the intermediate ones, on the more posterior part of the squamosal, are comparatively low.

FUNCTION OF THE FRILL IN *PENTACERATOPS*

The function of the ceratopsian frill has been the subject of much speculation, although the only functions given serious consideration in the literature are those of support for cranial musculature, sexual display, and protection.

It seems well established that the frill served as an expanded frame for the origin of large mandibular adductor muscles (Lull, 1908; Russell, 1935; Haas, 1955; Ostrom, 1964, 1966). Ostrom pointed out that only the *M. adductor mandibulae externus* was involved with the frill, and that this was the dominant jaw adductor. In an analysis of ceratopsian jaw mechanics, Ostrom (1966) found that the relative positions of the jaw articulation, height and position of the coronoid process, and length and position of the tooth row left the long-frilled forms, including *Pentaceratops*, mechanically inferior to the short-frilled ceratopsians. He suggested that the great relative lengthening of the frill, and the presumed concomitant increase in bulk of the *M. adductor mandibulae externus*, may have been a means of compensating for the relatively poor mechanical arrangement of the masticatory apparatus in long-frilled ceratopsians.

The wide morphological variability in the ceratopsian frill has been interpreted by Davitashvilli (1961), Farlow and Dodson (1975), Hopson (1977), and Molnar (1977) as indicating species—specific differences in agonistic and ritualized display behaviors. Dodson (1976) recognized sexual dimorphism in the frill of *Protoceratops*, of which a large sample is available. Frill dimorphism may have been characteristic of most ceratopsians, including *Pentaceratops*, but the small sample available for most genera renders its recognition difficult.

In models of ceratopsian behavior developed by Farlow and Dodson (1975) and Molnar (1977), the frill in long-frilled forms such as *Pentaceratops* is seen as being used as a frontal threat display, exhibited by inclining the head forwards, possibly in conjunction with side to side shaking movements. Intraspecific combat involved frontal fencing with the elongated supraorbital horns.

Molnar (1977) observed that the supraorbital horn cores of *Pentaceratops* are oriented in such a way that their long axes are directed nearly through the occipital condyles, serving to direct stresses from horn thrusts to the vertebral column. The lateral edges of the front surface of the *Pentaceratops* frill were not covered by the jaw adductor musculature, and may have served to catch blows from an opponent's horns during intraspecific combat. Nonetheless, the large jaw adductor muscles were quite exposed and vulnerable. In order to avoid grave injury, intraspecific battles using horn thrusts were probably rather strictly stereotyped. While horn thrusts may have been a part of the battle repertoire, frontal shoving and wrestling were probably the dominant modes of combat. In *Triceratops*, where the frill lacks fenestrae and only a small part was covered by the jaw adductor musculature, the frill may have performed a relatively more important protective role (Ostrom, 1966; Farlow & Dodson, 1975).

SUMMARY

The ceratopsian dinosaur *Pentaceratops* is known only from the southwestern edge of the San Juan Basin, in San Juan County, New Mexico. *Pentaceratops* extends through at least the upper two-thirds of the Fruitland Formation into the Kirtland Shale where it reaches the upper shale member, and has been tentatively identified from the overlying Naashoibito Member (Lehman, this volume). *Pentaceratops* is of Maastrichtian (probably Edmontonian to earliest Lancian) age.

Previous accounts of *Pentaceratops* are based on specimens in which the rear part of the frill is incomplete, or is crushed in such manner that its true shape is obscured. Although seemingly complete frills are figured in the literature, these figures were drawn from restorations based on other ceratopsian genera. The recent discovery by the Museum of Northern Arizona of a specimen with a complete frill revealed that these restorations are incorrect.

This specimen demonstrates that in *Pentaceratops* the posterior edge of the parietals forms a deep U-shaped reentrant into the back of the frill. There were evidently 11 epoccipitals on each side of the frill, with nine on the squamosal and two on the parietal. The posterior three epoccipitals, one on the squamosal and two on the parietal, are large and triangular in shape. The anterior eight epoccipitals, along the edge of the squamosal, are more rounded with relatively lower profiles.

ACKNOWLEDGMENTS

We gratefully acknowledge R. Cifelli (American Museum of Natural History), W. Clemens (University of California, Berkeley), N. Hotton (U.S. National Museum), L. Jacobs (Museum of Northern Arizona), L. Martin (University of Kansas), and N. Mateer (University of Uppsala) for useful comments and information on *Pentaceratops* specimens. We would like to thank E. Lindsay (University of Arizona) for making available to us an unpublished manuscript, and for information on the U. of A. specimen. Thanks are also due to Consolidation Coal Company and El Paso Natural Gas Company, who donated the funds, machinery and some of the manpower needed to collect and partially prepare the MNA specimen. Special thanks are due to Will Downs, who supervised the difficult task of collecting and preparing the MNA specimen.

REFERENCES

Alvarez, W., and D. W. Vann. 1979. Comment on "Biostratigraphy and magnetostratigraphy of Paleocene terrestrial deposits, San Juan Basin, New Mexico." Geology 7:66–67.

Anderson, R. Y. 1960. Cretaceous–Tertiary palynology, eastern side of the San Juan Basin, New Mexico. N. Mex. Bur. Mines & Mineral Resources Mem. 6:66 p.

Baltz, E. H., S. R. Ash, and R. Y. Anderson. 1966. History of nomenclature and stratigraphy of rocks adjacent to the Cretaceous–Tertiary boundary, western San Juan Basin, New Mexico. U.S. Geol. Survey Prof. Paper 534-D:D1–D23.

Bauer, C. M. 1916. Stratigraphy of a part of the Chaco River Valley. U.S. Geol. Survey Prof. Paper 98-P:271–278.

Butler, R. F., E. H. Lindsay, L. L. Jacobs, and N. M. Johnson. 1977. Magnetostratigraphy of the Cretaceous–Tertiary boundary in the San Juan Basin, New Mexico. Nature 267:318–323.

Clemens, W. A. 1973. The roles of fossil vertebrates in the interpretation of Late Cretaceous stratigraphy of the San Juan Basin, New Mexico. Four Corners Geol. Soc. Mem. p. 154–167.

Davitashvilli, L. S. 1961. Teoriya Polovovo Otbora. [The Theory of Sexual Selection]. Moscow Acad. Sci. [Izdatel'stvo Akad. Nauk]. 538 p.

Dodson, P. 1976. Quantitative aspects of relative growth and sexual dimorphism in *Protoceratops*. J. Paleontol. 50:929–940.

Farlow, J. O. and P. Dodson. 1975. The behavioral significance of frill and horn morphology in ceratopsian dinosaurs. Evol. 29:353–361.

Fassett, J. E. 1979. Comment on "Biostratigraphy and magnetostratigraphy of Paleocene terrestrial deposits, San Juan Basin, New Mexico." Geology 7:69–70.

Fassett, J. E. and J. S. Hinds. 1971. Geology and fuel resources of the Fruitland Formation and Kirtland Shale of the San Juan Basin, New Mexico and Colorado. U.S. Geol. Survey Prof. Paper 676:1–76.

Folinsbee, R. E., H. Baadsgaard, G. L. Cumming, J. Nascimbene, and M. Shafiqullah. 1965. Late Cretaceous radiometric dates from the Cypress Hills of western Canada, *in* Cypress Hills Plateau, Alberta and Saskatchewan. Alberta Soc. Petrol. Geology, 15th Field Conf. Guidebook, p. 162–174.

Gilmore, C. W. 1933. Two new dinosaurian reptiles from Mongolia with notes on some fragmentary specimens. Am. Mus. Novitates 679, 20 p.

————. 1935. On the Reptilia of the Kirtland Formation of New Mexico, with descriptions of new species of fossil turtles. Proc. U.S. Nat. Mus. 83:159–188.

Haas, G. 1955. The jaw musculature in *Protoceratops* and in other ceratopsians. Am. Mus. Novitates 1729, 24 p.

Hopson, J. A. 1977. Relative brain size and behavior in archosaurian reptiles. Ann. Rev. Ecol. Syst. 8:429–448.

Kues, B. S., J. W. Froehlich, J. A. Schiebout, and S. G. Lucas. 1977. Paleontological survey, resource assessment, and mitigation plan for the Bisti–Star Lake area, northwestern New Mexico. Unpub. report to Bureau of Land Management, Albuquerque, New Mexico, 1525 p.

LaBrecque, J. L., D. V. Kent, and S. C. Cande. 1977. Revised magnetic polarity time scale for Late Cretaceous and Cenozoic time. Geology 5:330–335.

Lambert, R. St. J. 1971. The pre-Pleistocene Phanerozoic timescale, a review. Geol. Soc. London Spec. Pub. 5:9–31.

Lindsay, E. H., L. L. Jacobs, and R. F. Butler, 1978. Biostratigraphy and magnetic polarity stratigraphy of Upper Cretaceous and Paleocene terrestrial deposits, San Juan Basin, New Mexico. Geology 6:425–429.

Lindsay, E. H., R. F. Butler, N. N. Johnson, and L. L. Jacobs. 1979a. Reply to comment by Alvarez and Vann. Geology 7:68–69.

————. 1979b. Reply to comment by Fassett. Geology 7:69–71.

Lindsay, E. H., R. F. Butler, and N. M. Johnson. 1981. Magnetic polarity zonation and biostratigraphy of Late Cretaceous and Paleocene continental deposits, San Juan Basin, New Mexico. Am. J. Sci. 281:390–435.

Lucas, S. G. and J. K. Rigby, Jr. 1979. Comment on "Biostratigraphy and magnetostratigraphy of Paleocene terrestrial deposits, San Juan Basin, New Mexico." Geology 7:323–325.

Lull, R. S. 1908. The cranial musculature and the origin of the frill in the ceratopsian dinosaurs. Am. J. Sci. Ser. 4, 25:387–399.

————. 1933. A revision of the Ceratopsia or horned dinosaurs. Peabody Mus. Nat. Hist. Mem. 3(3):175 p.

Molnar, R. E. 1977. Analogies in the evolution of combat and display structures in ornithopods and ungulates. Evol. Theory 3:165–190.

Osborn, H. F. 1923. A new genus of Ceratopsia from New Mexico, *Pentaceratops sternbergii*. Am. Mus. Novitates 93, 3 p.

Ostrom, J. H. 1964. A functional analysis of jaw mechanics in the dinosaur *Triceratops*. Peabody Mus. Nat. Hist. Postilla 88, 35 p.

————. 1966. Functional morphology and evolution of the ceratopsian dinosaurs. Evol. 20:290–308.

O'Sullivan, R. B., C. A. Repenning, E. C. Beaumont, and H. G. Page. 1972. Stratigraphy of the Cretaceous rocks and the Tertiary Ojo Alamo Sandstone, Navajo and Hopi Indian Reservations, Arizona, New Mexico, and Utah. U.S. Geol. Survey Prof. Paper 521-E: E1–E65.

Powell, J. S. 1973. Paleontology and sedimentation of the Kimbeto Member of the Ojo Alamo Sandstone. Four Corners Geol. Soc. Mem., p. 111–122.

Russell, L. S. 1935. Musculature and function in the Ceratopsia. Canada Dept. of Mines Bull. 77 (Geol. Ser. 52):39–48.

————. 1975. Mammalian faunal succession in the Cretaceous System of western North America. Geol. Assoc. Canada Spec. Paper 13:137–161.

Stanton, T. W. 1916. Nonmarine Cretaceous invertebrates of the San Juan Basin. U.S. Geol. Survey Prof. Paper 98-R:309–326.

Tschudy, R. H. 1973. The Gasbuggy core—a palynological appraisal. Four Corners Geol. Soc. Mem:131–143.

Wiman, C. 1930. Über Ceratopsia aus der Oberen Kreide in New Mexico. Nova Acta Regiae Soc. Scient. Upsaliensis, Ser. 4, 7(2):1–19.

Zavada, M. S. 1976. Palynology of the Upper Cretaceous Fruitland Formation, San Juan Basin, New Mexico. Unpublished MA Thesis, Ariz. State Univ., 157 p.

3 The reptilian megafauna from the Kirtland Shale (Late Cretaceous) of the San Juan Basin, New Mexico

Niall J. Mateer
Uppsala, Sweden

INTRODUCTION

This paper describes the collection of fossil reptiles from the San Juan Basin now housed at the Paleontological Institute at Uppsala University in Sweden. During 1920 and 1921, C. H. Sternberg was employed by Carl Wiman of Uppsala to collect fossil vertebrates from the San Juan Basin. The field work of this period is partially described by Sternberg (1932) as well as in his private correspondence with Wiman.

The San Juan Basin provided Sternberg with rich finds, primarily of turtles, ornithischians, one sauropod and a crocodile. These were subsequently described by Wiman (1930, 1931, 1932, 1933) though few of these specimens were illustrated by photographs. Mateer (1976) described the sauropod remains. Owing to advances in the systematics of the groups discussed here during the last 50 years, a reexamination of this collection is necessary. Since this is the only collection of reptiles from the San Juan Basin that is now outside the United States, many workers on these groups have not been able to study this material, thus creating a vacuum in the knowledge of these groups. Many questions have been raised regarding certain specimens in this collection, and it is hoped that this paper will answer them where possible.

The stratigraphy of the San Juan Basin has been, and still is, somewhat difficult to understand fully. However, in this paper, the terminology of Baltz et al. (1966) will be followed. All material in the Uppsala collection has come from the Kirtland Shale. Sternberg used the terminology of Gilmore (1916) and Bauer (1916) in his correspondence to Wiman. The type specimen of *Parasaurolophus tubicen* and the topotypes of *Alamosaurus sanjuanensis* both came from what Sternberg called the Ojo Alamo Sandstone, which is now correlated with the Naashoibito Member of the Kirtland Shale (Baltz et al., 1966; Clemens, 1973). Specimens that Wiman referred to as having come from the Fruitland Formation are either misinterpretations of Sternberg's letters or, in some cases, Sternberg's own muddling of his stratigraphy. He generally collected in areas where the Fruitland is not exposed. Because of this confusion, it is sometimes difficult to be quite sure about the stratigraphic provenance of some of these specimens. However, a recent visit to the San Juan Basin has cleared the element of doubt in most cases.

The collection discussed in this paper comprises the following:

Testudines:	*"Baena" nodosa* Gilmore, 1916
	Neurankylus eximius (= *baueri*) Lambe, 1902
	Thescelus insiliens (= *rapiens*) Hay, 1908
	Basilemys nobilis Hay, 1910
	Plastomenus sp.
	Platypeltis sternbergi Wiman, 1933
Crocodilia:	*Goniopholis kirtlandicus* Wiman, 1932
Ornithischia:	*Pentaceratops fenestratus* Wiman, 1930
	Parasaurolophus tubicen Wiman, 1931
	Kritosaurus cf. *navajovius* Brown, 1910
Saurischia:	*Alamosaurus sanjuanensis* Gilmore, 1922

Numerous fragments of dinosaurian limb bones are present in the collection but are poorly preserved and are not identifiable. In this paper, PMU.R refers to the reptile collection at the Paleontological Museum in Uppsala, Sweden.

SYSTEMATIC PALEONTOLOGY

Order ORNITHISCHIA Seeley, 1888
Suborder CERATOPSIA Marsh, 1890
Genus PENTACERATOPS Osborn, 1923
PENTACERATOPS FENESTRATUS Wiman, 1930
Pl. 3.1

Holotype.—A nearly complete cranium. PMU.R200.

Locality.—South flank of Kimbeto Wash, San Juan County, New Mexico, (T22N, R10W).

Horizon.—Kirtland Shale, below the Naashoibito Member.

Comments.—The cranium is virtually complete but without the jaws and is badly compressed laterally. Wiman (1930) described this specimen briefly, an account that Lull (1933) based his description on. A nearly complete postcranial skeleton of what was presumed to be a *Pentaceratops* was also described by Wiman (1930), and he allocated it to the same species as the cranium, *P. fenestratus.* Lull (1933) quoted these two specimens as being cotypes. This postcranial skeleton (PMU.R268) is the only known ceratopsian skeleton from the San Juan Basin.

There is no direct evidence that this postcranial skeleton material belongs to *Pentaceratops*, only that *Pentaceratops* is known to have existed in the area at the same time. The two specimens were not in association with each other; the postcranial skeleton was found at what is now called Ah-shi-sle-pah Wash (T22N, R11W), which is a considerable distance from where the cranium was found in Kimbeto Wash. There is nothing remarkable about this specimen that precludes its allocation within the genus *Chasmosaurus* or *Monoclonius.* The differences that do exist can be accounted for by individual variation. The main character that Lull (1933) found unusual was the "bowleggedness" of the hindlimbs. Wiman's (1930) drawing of this condition is somewhat misleading in that the contours of the femora are given in detail but not the crushed nature of the shaft. The femora, particularly on the left side, have been deformed, giving the impression that they are bowed, but this is not the original state.

The cranium, which is the holotype, is not as well preserved as could be wished (Pl. 3.1, fig. 2). Wiman (1930) named the new species primarily on the basis of a squamosal fenestra on the left side (Pl. 3.1, fig. 3); in the absence of the right squamosal, he assumed that such a fenestra was present on the right side. Lull (1933) suggested that this attribute was pathological, and thus not a valid character by comparison with other single fenestrated ceratopsians (for example *Arrhinoceratops brachyops*). Whether the Uppsala specimen was symmetrically fenestrated can, of course, not be shown. However, the nature of the squamosal fenestra is roughly elliptical with a smooth fenestral border (Pl. 3.1, fig. 3; the irregularity on the right side is an erroneous reconstruction). It in no way resembles a puncture from another ceratopsian or an injury through infection or disease. A puncture would seem unlikely since aggressive ceratopsians would surely have faced each other head-on (Farlow & Dodson, 1975). The sheer mechanics of horn fighting would make such a puncture high up on the left side extremely improbable. Such a puncture would have shown signs of repair, unless of course, the animal died soon after these wounds. If death did occur, the postcranial skeleton seems to have disappeared very quickly in the coal swamp environment. The improbability of this sequence of events, together with the even outline of the fenestra, strongly suggests that this was not a puncture. Impressions from the blood vessels lead right up to the edge of the fenestra, which would otherwise be very difficult to discern if there were a lesion or infection in this area.

While a case could be made for this fenestra being congenital, an equally strong case could be made for such a deformation being a specific attribute. Until further material is found, this will remain an open question. The *Pentaceratops* specimen described by Rowe et al. (this volume) has no fenestral component and thus does not help resolve this issue.

Apart from the fenestra, there are other characters of this cranium that show significant differences from the type species of this genus, *P. sternbergii* (Lull, 1933). However, some characters still remain unclear. The apparently posteriorly directed nasal horn of *P. fenestratus* is due to the fracturing of the skull in such a way that the premaxillae and the nasals have, in effect, been pushed up. The orientation of this horn is actually identical with that of *P. sternbergii.* The narial opening in *P. sternbergii* has a large projection from the ventral border of the premaxillae, which is not present in *P. fenestratus.* The epijugals differ from those of *P. sternbergii* in that they indeed do have a more forward aspect. There is no sign of a preorbital foramen, although the poor preservation makes this difficult to confirm (Pl. 3.1, fig. 1).

Whether *P. fenestratus* and *P. sternbergii* are taxonomically separate and whether the stratigraphic separation of the types is significant, is still somewhat uncertain. However, with the rapid rate of diversification of ceratopsians during the Late Cretaceous, the presence of two species in the San Juan Basin separated stratigraphically may be real.

Suborder ORNITHOPODA Marsh, 1881
Genus PARASAUROLOPHUS Parks, 1922
PARASAUROLOPHUS TUBICEN Wiman, 1931
Text-fig. 3.1

Holotype.—Nasal and premaxilla crest, jugal, maxilla, postorbital and endocranium, PMU.R222.

Locality.—De-na-zin Wash (Barrel Springs area), San Juan County, New Mexico (T24N, R11W).

Horizon.—Naashoibito Member, Kirtland Shale.

Comments.—The three described species of *Parasaurolophus* were collected within three years of each other (1920–23): *P. walkeri* (Parks, 1922) was found in the Oldman Formation of Alberta; *P. tubicen* (Wiman, 1931) was collected from the Naashoibito Member of the Kirt-

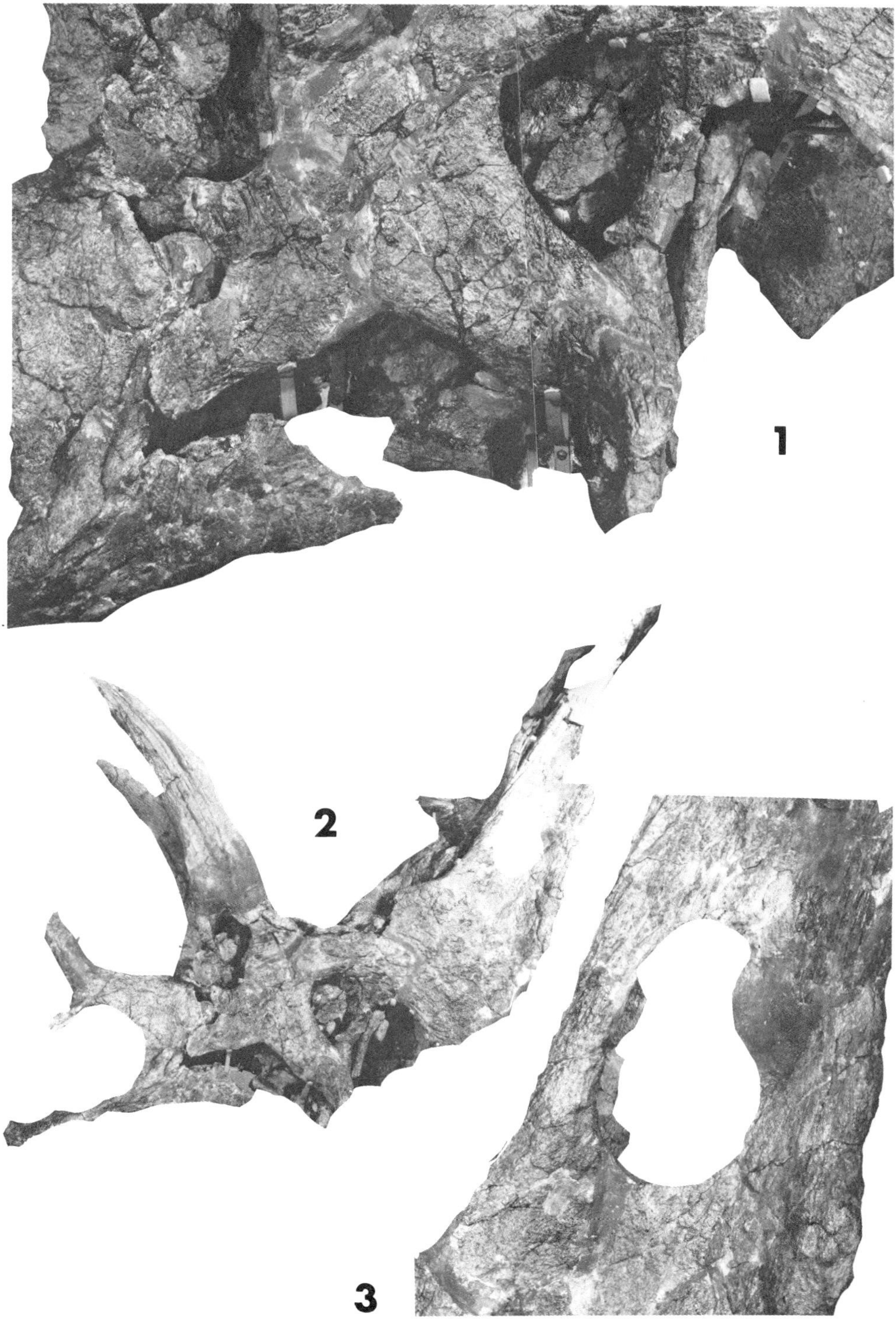

EXPLANATION OF PLATE 3.1

FIGS. *1–3*—*Pentaceratops fenestratus. 1*, Holotype (PMU.R200), detail of the left orbit and infratemporal region, × 0.17; *2*, Holotype, complete cranium, left side, × 0.05 (Note: because of restrictive museum space, a 28 mm wide angle lens was used, thus causing some distortion); *3*, Holotype, detail of squamosal fenestra, left side, × 0.17.

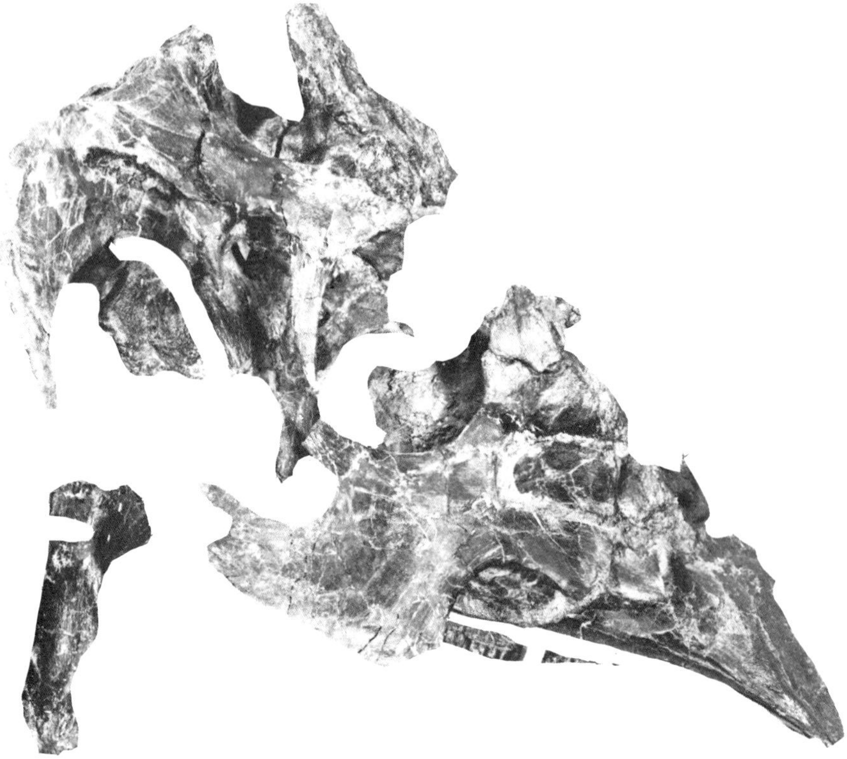

TEXT-FIG. 3.1—*Parasaurolophus tubicen* holotype (PMU.R222). Lateral view of right side of the skull, excluding the premaxillae crest, ×0.22.

land Shale in New Mexico; and *P. cyrtocristatus* (Ostrom, 1961, 1963) is from the Fruitland Formation of the San Juan Basin. There is an undescribed specimen of *Parasaurolophus* from the Fruitland Formation presently housed at the Museum of Northern Arizona (MNA 230). Gilmore (1935) mentioned a *Parasaurolophus* caudal series, a left femur, and part of an ilium and maxillae collected north of Hunter's Store, San Juan County in 1929. This material, at the United States National Museum, is probably correctly identified, though none of these elements are good generic indicators.

The Uppsala material (Text-fig. 3.1), although fragmentary, is unique in the nature of the posterior projection of the frontals and premaxillae. The form of the jugal also differs between *P. walkeri* and *P. tubicen*. The jugal of the specimen from the Museum of Northern Arizona shows a slightly different form, and might, on the basis of its having come from the same horizon as *P. cyrtocristatus*, represent this species. The holotype of *P. cyrtocristatus* does not have a jugal or maxilla preserved.

The interior of the premaxilla crest was shown by Wiman to comprise two passages when seen in cross section. It appears that this is an assumption only, since the crest has never been sectioned; presumably, the models of Wiman's cross sections were obtained by making a mold from the crest, and the resulting cast showed an hourglass contour. Hence Wiman assumed that this contour showed only two passages, a dorsal and a ventral. This is not so in *P. walkeri* and *P. cyrtocristatus* (Lull & Wright, 1942; Ostrom, 1963) where four passages are present. A forthcoming redescription of *P. tubicen* will evaluate the nature of this character more definitely.

A new specimen of *Parasaurolophus* has been described by Weishampel & Jensen (1979). This fragmentary specimen of the crest is from the Kaiparowits Formation (upper Maastrichtian) of Utah. This material indicates that the crest was four-chambered.

Genus KRITOSAURUS[1] Brown, 1910.
KRITOSAURUS cf. NAVAJOVIUS Brown, 1910.
Pl. 3.2.

Material.—Left ilium, PMU.R231; left pubis, PMU.R224; left humerus, PMU.R223; left ischium, PMU.R236; sternum, PMU.R245.

Locality.—South flank of Kimbeto Wash, San Juan County, New Mexico (T22N, R9W).

Horizon.—Kirtland Shale, below the Naashoibito Member.

Comments.—Sternberg collected numerous skeletal elements of *Kritosaurus* for Wiman. All identifiable specimens are referred to *Kritosaurus*; however, since the only species of this genus known from the San Juan Basin is *K. navajovius*, it has become something of a dustbin species. (The type is a cranium.) Postcranial elements of hadrosaurs are generically determinable, but not to a specific level (Brett-Surman, 1975, 1978).

The humerus (Pl. 3.2, fig. 4), although not complete at the proximal end, does show the typical proportions of kritosaurs. The length of the deltoid blade is about half the length of the bone. The prominence of the crest running almost immediately through the deltoid blade is a further typical kritosaur condition. The posterior median angle of the deltoid blade is somewhat larger than many kritosaurs, but this is a variable feature throughout the group.

The ilium (Pl. 3.2, fig. 3) is sufficiently complete to enable a secure allocation to *Kritosaurus*. The postacetabular process is typically narrow and somewhat elongated. The antitrochanter, while prominent, has

---

1. Baird and Horner (1977) and Horner (1979) regard *Kritosaurus* as a junior synonym of *Hadrosaurus*. At this time, no complete evidence for this has been published, and thus the name *Kritosaurus* is retained here.

no lip extension typical of the Lambeosaurinae. The pubic peduncle is short and bluntly triangular, bordering the gently curved acetabulum.

The pubis (Pl. 3.2, fig. 1) also is not quite complete. The iliac peduncle is broad and prominent; the ischial peduncle, on the other hand, is small and fluted with the postpubis. The angle of the acetabular border is sharply right-angled and smooth. These are all characters typical of *Kritosaurus.*

The ischium (Pl. 3.2, fig. 2) is well preserved, although the foot and most of the shaft are missing. The obturator notch is well developed and is partially closed, indicating that this was a mature individual. The pubic peduncle is long and narrow. The angle between the iliac and pubic peduncles is approximately 90°, the former pointing along the same axis as the shaft in the opposite direction. This specimen is almost identical with that of *K. notabilis* (Lull & Wright, 1942) except for the less well developed obturator foramen in the latter specimen.

As noted by Brett-Surman (1975), the hadrosaur sternum is a difficult bone to study owing to the lack of comparative material. The sternum collected by Sternberg (Pl. 3.2, fig. 5) is nearly complete, thus worth noting here. Whether or not it is kritosaurian is not possible to say, although the probability is high. The sternum is 45 cm long and 16 cm wide at its greatest breadth. The shaft is small and rather long, but quickly expanding to the main blade. The surface is smooth except for some elongated grooves at the distal end. There is a continuation of the shaft ridge into the proximal blade in the form of a prominent ridge. The lateral dilation is somewhat more pointed than that of *Anatosaurus* (Lull & Wright, 1942).

Numerous other fragments, primarily of limb bones, were collected by Sternberg, but their identification to a generic level is not possible.

Infraorder SAUROPODA Marsh, 1878
Family TITANOSAURIDAE Lydekker, 1877
Genus ALAMOSAURUS Gilmore, 1922.
ALAMOSAURUS SANJUANENSIS Gilmore, 1922
Text-fig. 3.2

Topotype.—The first three sacral vertebrae, PMU.R280; left ilium, PMU.R281.

Locality.—De-na-zin Wash (Barrel Springs area), San Juan County, New Mexico (T24N, R11W).

Horizon.—Naashoibito Member, Kirtland Shale.

Comments.—Remains of large sauropods are infrequent in the Upper Cretaceous of North America, having been found only in Utah, New Mexico and Texas, and possibly in Wyoming and Montana. The first specimen to be found was from approximately the same locality as the Uppsala specimen. This was described by Gilmore (1922) and given the

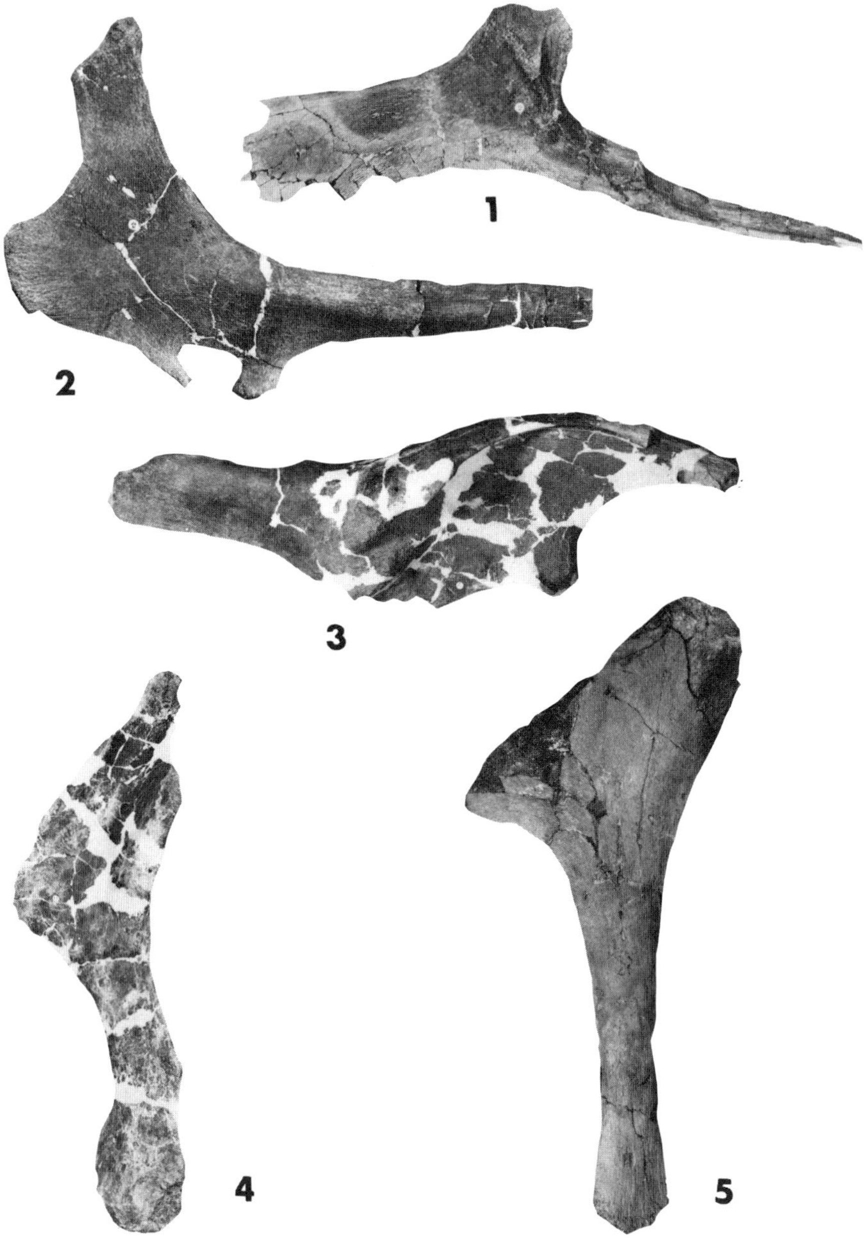

Explanation of Plate 3.2

Figs. *1–5*—*Kritosaurus* cf. *navajovius*. *1*, Left pubis (PMU.R224), external aspect, ×0.14; 2, Left ischium (PMU.R236), external aspect, ×0.19; *3*, Left ilium (PMU.R231), external aspect, ×0.11; *4*, Left humerus (PMU.R223), external aspect, ×0.12; 5, Sternum (PMU.R245), ventral aspect, ×0.20.

name *Alamosaurus sanjuanensis.* The Uppsala specimen was the next to be collected, but owing to the poor state of preservation, Wiman presumably deemed it not worth describing. Mateer (1976) briefly described this material. Gilmore (1946) described another specimen from the North Horn Formation of Utah. Additional material has since been described from the Javelina Formation of Big Bend National Park, Texas (Lawson, 1972).

Few elements described from these various finds are duplicated, hence giving very little comparative material. Generally, the material is titanosaurid and has been conveniently assigned to the form genus *Alamosaurus.* In most cases it appears that this is a sound taxonomic decision.

The Uppsala material (Text-fig. 3.2) comprises an ilium, the first three sacral vertebrae (the first with a lateral zygopophysis), and some cervical vertebrae. Since the description of this material (Mateer, 1976), W. Langston (written commun., 1976) has suggested that the first caudal is not present, and that it is the third sacral, contrary to my previous statement. The first vertebra in this sequence is characteristically greatly expanded anteriorly and has a concave anterior centra. The posterior surface is not visible.

The ilium is almost identical to the specimen from Texas (Lawson, 1972, Pl. 10), indicating a close relationship of these two specimens.

TEXT-FIG. 3.2—*Alamosaurus sanjuanensis.* Left, first three sacral vertebrae (PMU.R280); Right, left ilium, internal view (PMU.R281), ×0.15.

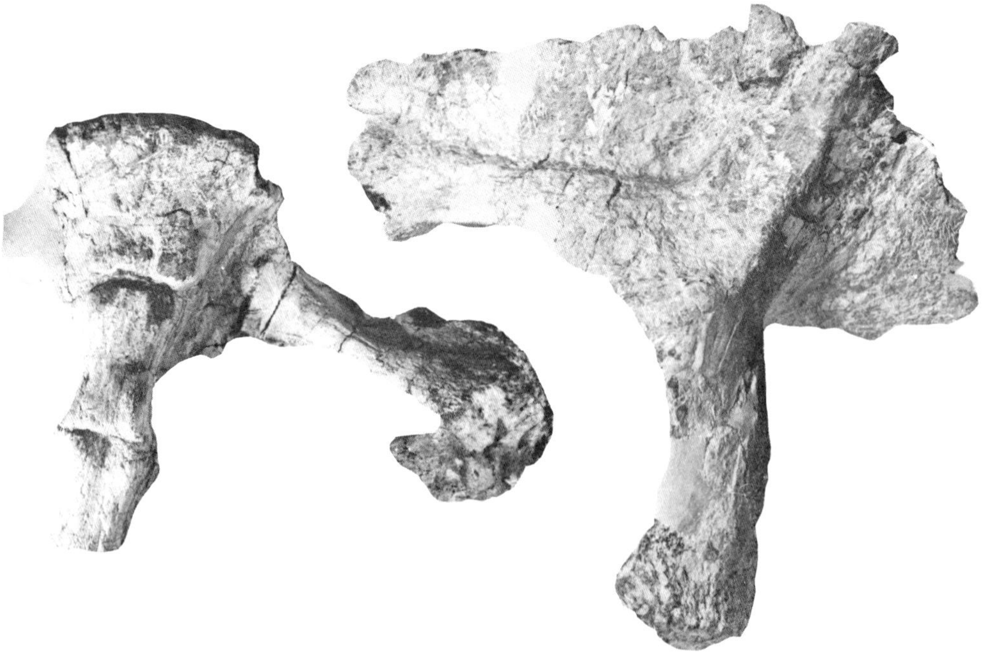

Order CROCODILIA Gmelin, 1788
Suborder MESOSUCHIA Huxley, 1875
Family GONIOPHOLIDAE Owen, 1841
Genus GONIOPHOLIS Owen, 1841
GONIOPHOLIS KIRTLANDICUS Wiman, 1932

Holotype.—A fragmentary skull, PMU.R230; a scute, PMU.R232; and a tooth, PMU.R231.

Locality.—De-na-zin Wash (Barrel Springs area), San Juan County, New Mexico (T24N, R11W).

Horizon.—Kirtland Shale, below Naashoibito Member.

Comments.—The material collected by Sternberg is somewhat fragmentary (Wiman, 1932, Pl. 3,4,5), creating difficulties in securing its systematic affinity. *Goniopholis* specimens from North America are not quite the same as those from Europe and would appear to belong to the genus *Eutretaurasuchus* (W. Langston, written commun., 1975). A characteristic feature of goniopholids is a depressed area immediately anterior to the orbit, a feature which is not clearly present in the Uppsala specimen, since the preservation is not good. The high stratigraphic position of the specimen adds to its uniqueness, if it is *Goniopholis*; goniopholids are otherwise found in North America only in the Upper Jurassic. Until more is known about the North American goniopholids, Wiman's (1932) designation remains somewhat uncertain due to a lack of a definite antorbital depression.

Order TESTUDINES Linnaeus, 1758
Suborder CRYPTODIRA Cope, 1870
Family BAENIDAE (Cope), 1872 (*sensu* Gaffney, 1972)
Genus BAENA Leidy, 1870
"BAENA" NODOSA Gilmore, 1916
Pls. 3.3–3.5

Material.—Eleven carapaces: PMU.R11–14, 16–19, 21. (Wiman's (1933) specimens 1–4, 7, 9–13, 15). PMU.R15, 20 are missing.

Locality.—Head of Ah-shi-sle-pah Wash (T22N, R10W): PMU.R11, 13, 18 and 19; De-na-zin Wash (T24N, R11W): PMU.R12; South branch of Hunter Wash (T24N, R13W): PMU.R14, 16, 17, 21.

Horizon.—All specimens are from the Kirtland Shale below the Naashoibito Member.

Comments.—Many turtles collected from the Upper Cretaceous of the Western Interior have been assigned to the genus *Baena* on the basis of the carapace. Gaffney (1972) regarded many of these specimens as being systematically indeterminate and left them in a form genus "*Baena.*"

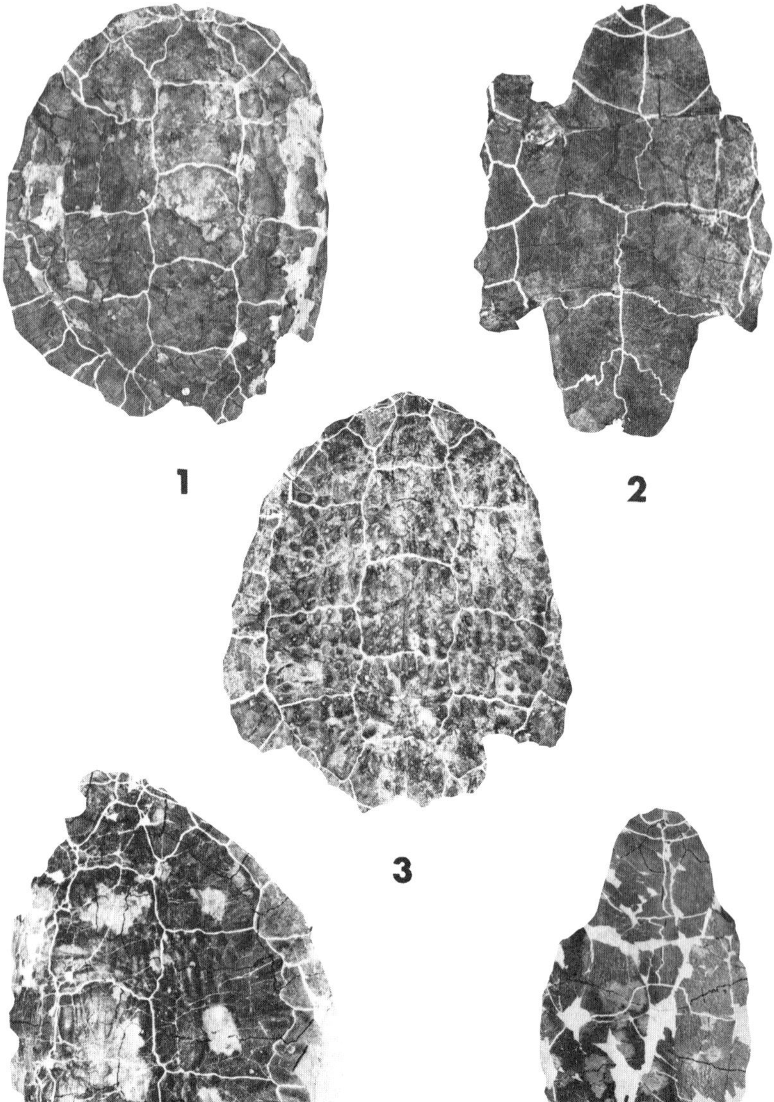

EXPLANATION OF PLATE 3.3

FIGS. *1–5*—"*Baena*" *nodosa*. *1*, Carapace (PMU.R11a) in dorsal view (cf. Wiman, 1933, Pl. 1, fig. 1), ×0.15; *2*, Plastron (PMU.R11b) in ventral view (cf. Wiman, 1933, Pl. 1, fig. 1a), ×0.17; *3*, Carapace (PMU.R12) in dorsal view (cf. Wiman, 1933, Pl. 1, fig. 2), ×0.17; *4*, Carapace (PMU.R13a) in dorsal view (cf. Wiman, 1933, Pl. 1, fig. 3), ×0.21; *5*, Plastron (PMU.R13b) in ventral view (cf. Wiman, 1933, Pl. 1, fig. 3a), ×0.20.

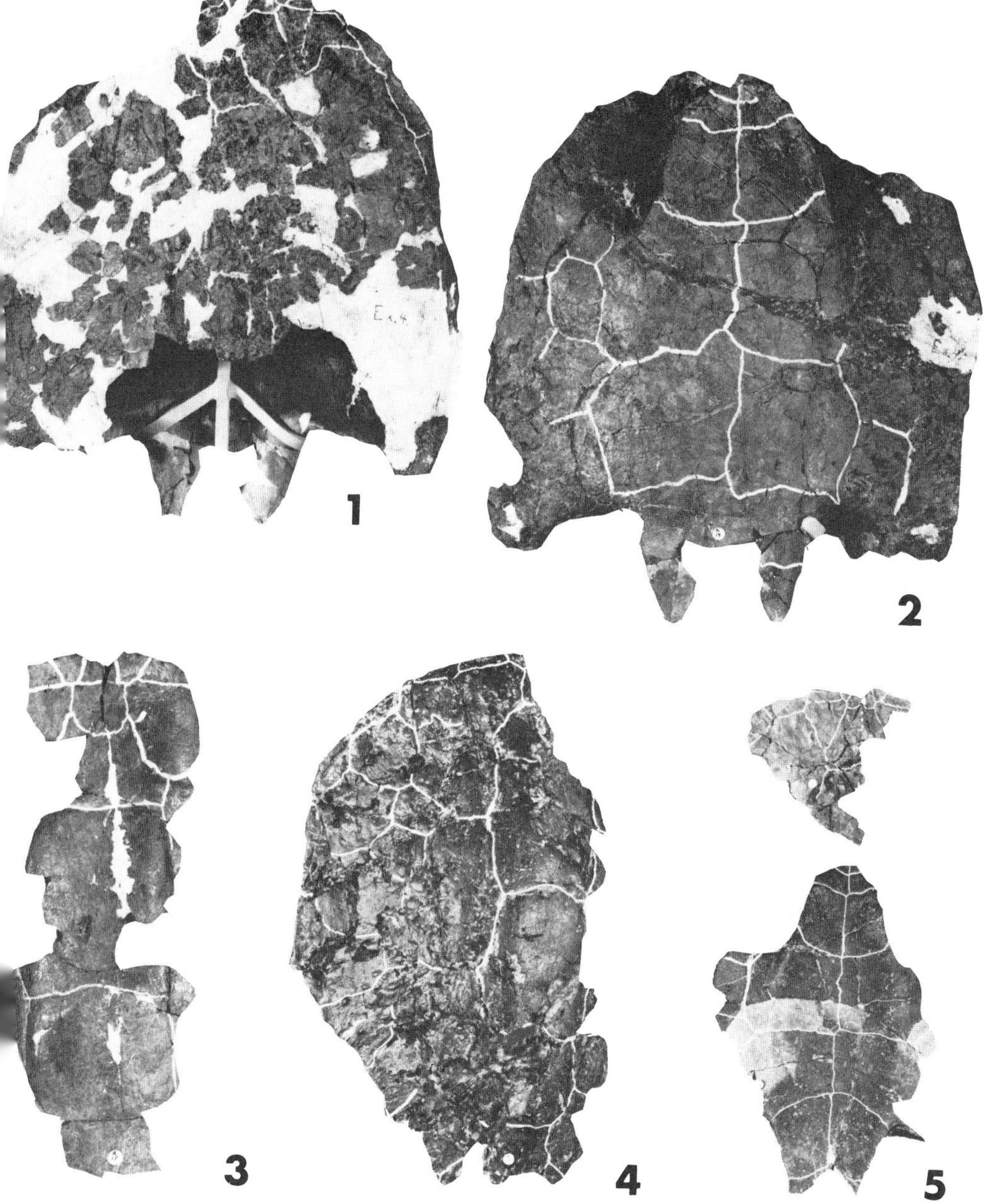

EXPLANATION OF PLATE 3.4

FIGS. *1–5*—*"Baena" nodosa*. *1*, Carapace (PMU.R14b) in dorsal view (cf. Wiman, 1933, Pl. 1, fig. 4), ×0.25; 2, Plastron (PMU.R14a) in ventral view (cf. Wiman, 1933, Pl. 1, fig. 4a), ×0.24; *3*, Midline (PMU.R15) of carapace in dorsal view (cf. Wiman, 1933, Pl. 1, fig. 5), ×0.29; *4*, Carapace (PMU.R16) in dorsal view (cf. Wiman, 1933, Pl. 1, fig. 6), ×0.22; 5, Fragments of carapace (above, PMU.R17a) and plastron (below, PMU.R17b) (cf. Wiman, 1933, Pl. 1, fig. 7a), ×0.17.

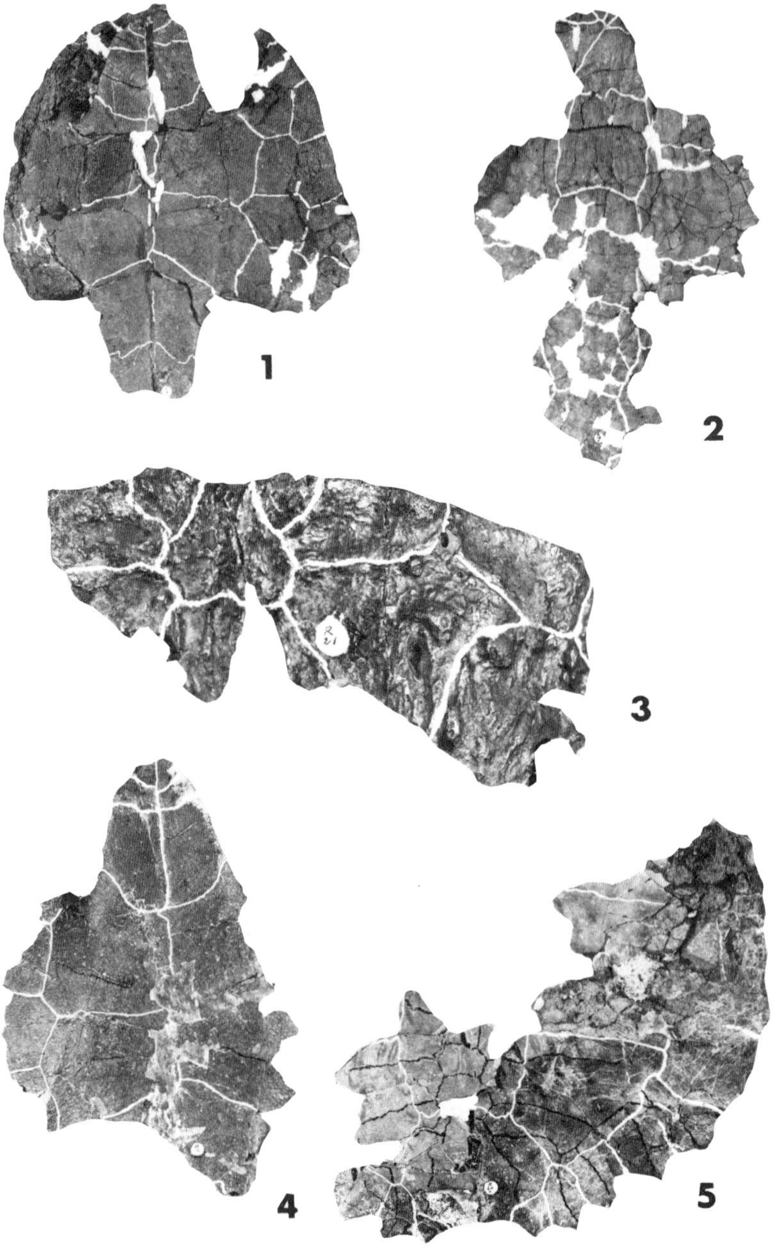

EXPLANATION OF PLATE 3.5

FIGS. *1–5*—"*Baena*" *nodosa*. *1*, Plastron (PMU.R18) in dorsal view (cf. Wiman, 1933, Pl. 1, fig. 8a), ×0.19; *2*, Carapace (PMU.R18a) in ventral view (cf. Wiman, 1933, Pl. 1, fig. 8), ×0.14; *3*, Fragment of carapace (PMU.R21) in dorsal view (cf. Wiman, 1933, Pl. 1, fig. 11), ×0.66; *4*, Partial plastron (PMU.R19) in ventral view (cf. Wiman, 1933, Pl. 1, fig. 9), ×0.18; *5*, Partial plastron (PMU.R20a) in dorsal view (cf. Wiman, 1933, Pl. 1, fig. 10), ×0.23.

"Baena" nodosa is one of five species belonging to this form genus, which represents the great deal of shell variation within the Baenidae. All of the Uppsala specimens fall within this variational range except for Wiman's specimen 9 (PMU.R16), which Gilmore (1935) regarded as being *Boremys pulchra.* This change in designation is followed by Gaffney (1972). Contrary to Wiman (1933) and Gaffney (1972), all specimens collected by Sternberg for Uppsala University were from the Kirtland Shale, somewhere below the Naashoibito Member. Although Canadian specimens collected from the Oldman Formation may be slightly older than the Kirtland specimens, there are no older specimens known from the San Juan Basin.

Genus NEURANKYLUS Lambe, 1902
NEURANKYLUS EXIMIUS Lambe, 1902
Pls. 3.6–3.8. Text-fig. 3.3.

Material.—A plastron and carapace, PMU.R24a,b; cervical vertebrae 2 through 8, PMU.R24c–i; caudal vertebrae series, PMU.R24j; pelvic girdle, PMU.R24q; and a right femur, PMU.R24n; two carapaces and plastrons, PMU.R25a, 26a, and PMU.R25b, 26b; a single carapace, PMU.R27 and a single plastron, PMU.R28.

Locality.—Head of Ah-shi-sle-pah Wash, San Juan County, New Mexico, (T22N, R10W).

Horizon.—Kirtland Shale, below the Naashoibito Member.

Comments.—*Neurankylus* is a baenid but with characters that separate it from other subfamilies of this group (Gaffney, 1972). Here, also, there is considerable confusion regarding the status of the species assigned to this genus. Although all the specimens of *Neurankylus* described by Wiman (1933) were referred by him to Gilmore's (1919) species *N. baueri*, Gaffney (1972) suggested that all specimens pertaining to this genus from the San Juan Basin should be regarded as being monotypic, *N. eximius.* The characters that have been used to distinguish these species of this genus (for example, the presence or absence of a keel on the neurals) are not regarded as being good systematic characters and reflect only the extensive variation within the group (Gaffney, 1972).

Wiman's specimen 1 (PMU.R24) is of considerable importance in that the cervical vertebral series is completely preserved. Doubt was raised about the correctness of Wiman's reconstruction of the vertebrae by Williams (1950) because of their unique cervical central formula. Inspection of this specimen in the light of Williams' comments shows that Wiman's description was quite correct, and that the vertebrae were found in close association with the carapace and other postcranial elements. The cervical formula is as follows: (2((3((4( )5))6((7((8). (The parentheses represent the concavity/convexity of each particular cervical central articulation.) As far as is known, this pattern is unique, although somewhat similar to some modern cryptodires (e.g. *Carrettochelys* and *Testudo giganteus*). Williams (1950) suggested, in his analysis of the variation and selection of turtle cervical vertebrae, that this unique cervical formula may reflect either a higher frequency of intraspecific variation in these turtles or a selective strategy that since the Late Cretaceous has proved to be non-optimal and has thus disappeared. The former situation suggests that the above formula would have been abnormal; if, however, it were normal, the latter case would have been more likely. Since only one specimen is known, this suggestion cannot be tested. However, judging by the extensive variation within the contemporary Trionychidae, one might be tempted to suspect that the Baenidae might also have been highly variable.

A further point of interest in this specimen is the nature of the articulation between the fifth and sixth cervical vertebrae. The posterior centra of the fifth vertebra has an ossified epiphysis partially preserved, accentuating the convex form of this surface. The anterior surface of the sixth centra is deeply concave for the accommodation of the epiphysis. This interface is one of the two joints used in the lateral retraction of the neck typical in cryptodires.

Genus THESCELUS Hay, 1908
THESCELUS INSILIENS Hay, 1908
Pl. 3.9.

Material.—Carapace and plastron from two specimens, PMU.R22, 23.

Horizon.—Kirtland Shale, below the Naashoibito Member.

Locality.—South branch of Hunter Wash, San Juan County, New Mexico (T24N, R13W).

Comments.—One of these specimens (PMU.R23) was described by Wiman (1933) as belonging to the species *T. insiliens*, but was referred by Gilmore (1935) to *T. hemisphaericus*. Gilmore thus proceeded to base his own incompletely known species *T. hemisphaericus* on the more complete Uppsala specimen. Gaffney (1972) found no reliable characters for differentiating species within this genus and thus regarded it as being monotypic, with *T. insiliens* the sole valid species. Variation, as with other Late Cretaceous turtles, is great, accounting for the proliferation of specific determinations by early workers. Both the Uppsala specimens are hereafter regarded as being *T. insiliens*. Wiman's other specimen (PMU.R22) was referred by him to *T. rapiens*.

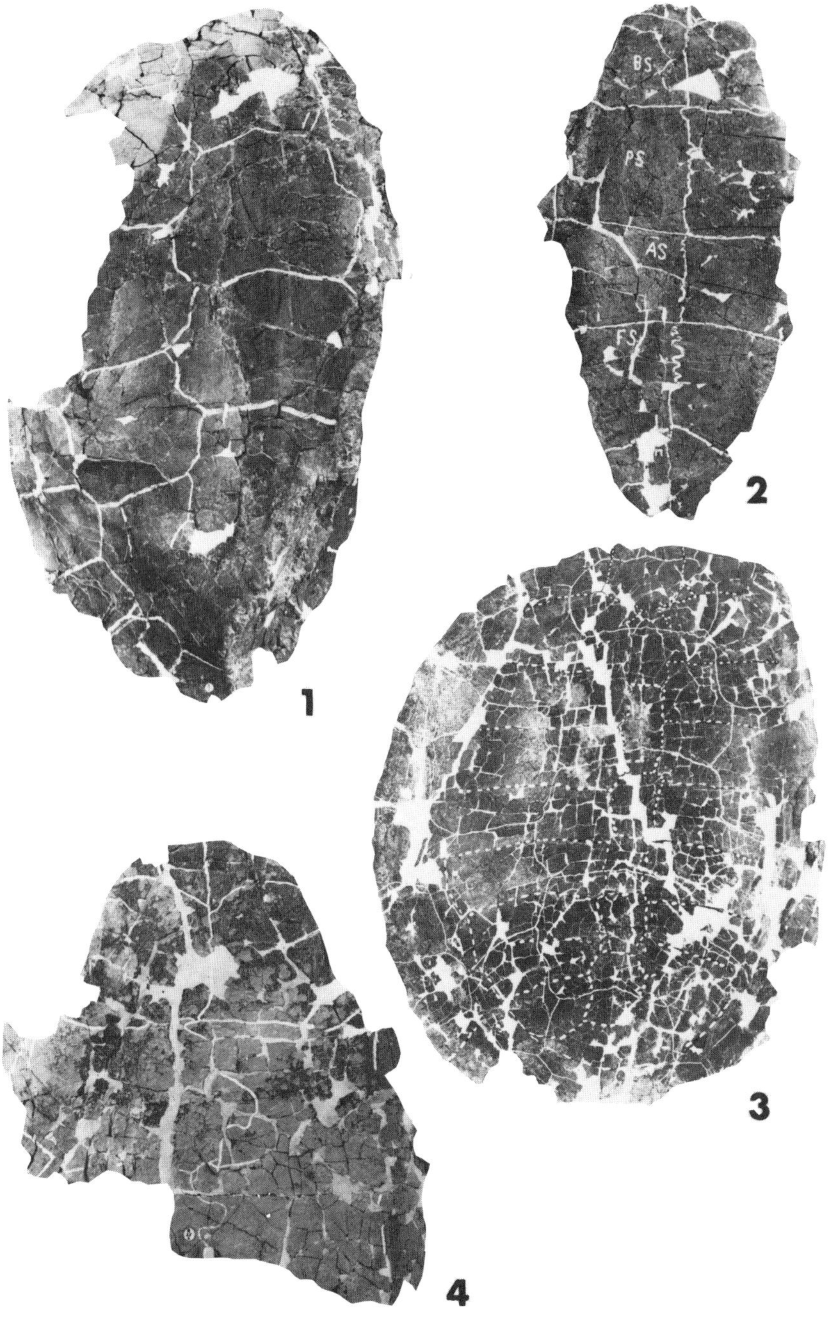

EXPLANATION OF PLATE 3.6

FIGS. *1–4*—*Neurankylus eximius. 1*, Carapace (PMU.R24a) in dorsal view (cf. Wiman, 1933, Pl. 2, fig. 1), ×0.20; *2*, Plastron (PMU.R24b) in ventral view (cf. Wiman, 1933, Pl. 2, fig. 1a), ×0.19; *3*, Carapace (PMU.R25a) in dorsal view (cf. Wiman, 1933, Pl. 2, fig. 2), ×0.17; *4*, Partial plastron (PMU.R25b) in ventral view (cf. Wiman, 1933, Pl. 2, fig. 2a), ×0.28.

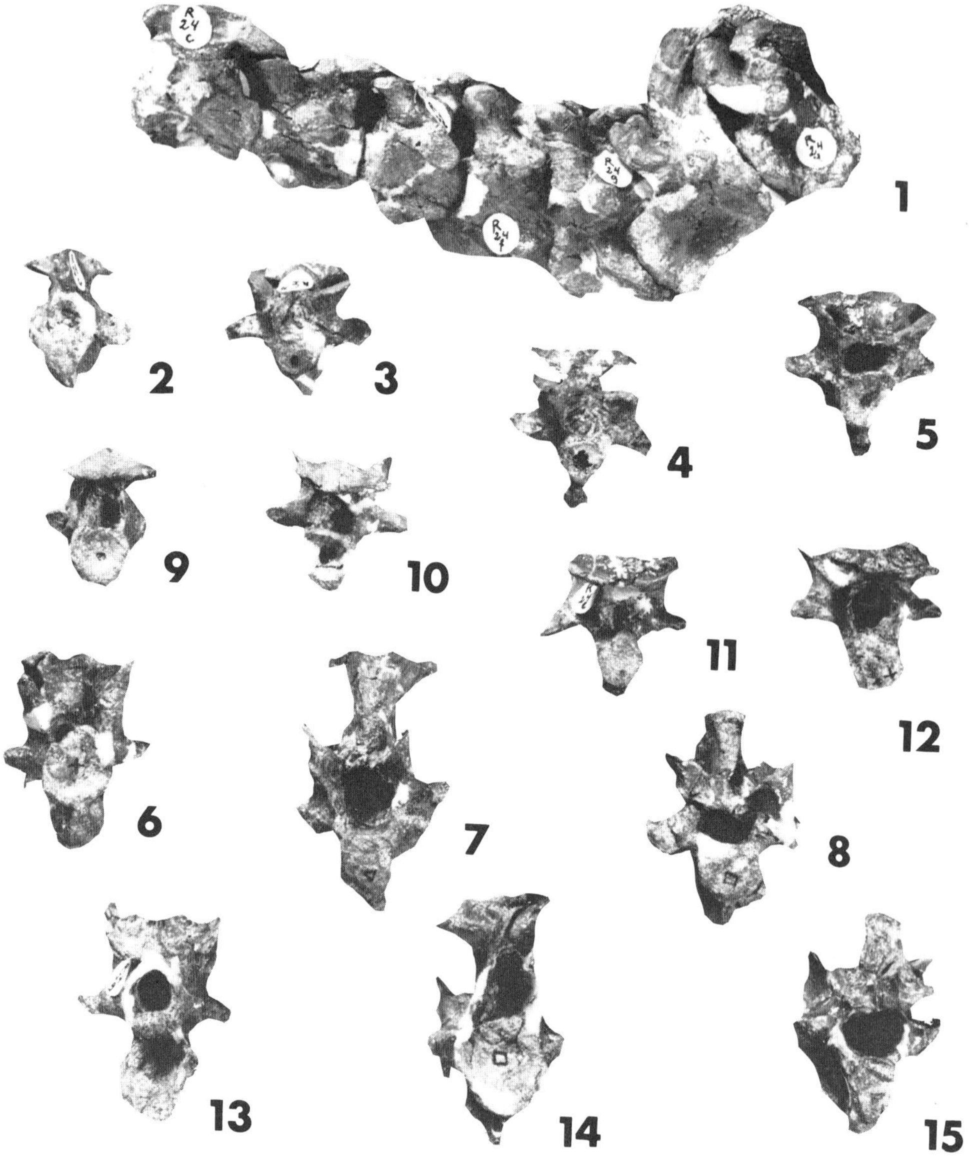

EXPLANATION OF PLATE 3.7

FIGS. *1–15*—*Neurankylus eximius*. *1*, Cervical vertebrae in series (PMU.24c–j), from the left side (cf. Wiman, 1933, Text-fig. 4), ×0.88; *2–8*, Cervical vertebrae (PMU.R24c–j), anterior view (cf. Wiman, 1933, Pl. 3, figs. 2a–8a), ×0.88; *9–15*, Cervical vertebrae (PMU.R24c–j), posterior view (cf. Wiman, 1933, Pl. 3, figs. 2b–8b), ×0.88.

2

1

3

4

EXPLANATION OF PLATE 3.8

FIGS. *1–3—Neurankylus eximius*. *1*, Partial plastron (PMU.R26b) in ventral view (cf. Wiman, 1933, Pl. 2, fig. 3a). ×0.15; *2*, Partial carapace (PMU.R26a) in dorsal view (cf. Wiman, 1933, Pl. 2, fig. 3), ×0.17; *3*, Partial carapace (PMU.R27) in dorsal view (cf. Wiman, 1933, Pl. 2, fig. 4), ×0.23.

4—Neurankylus sp. Partial plastron (PMU.R28) in ventral view (cf. Wiman, 1933, Pl. 2, fig. 5a), ×0.20.

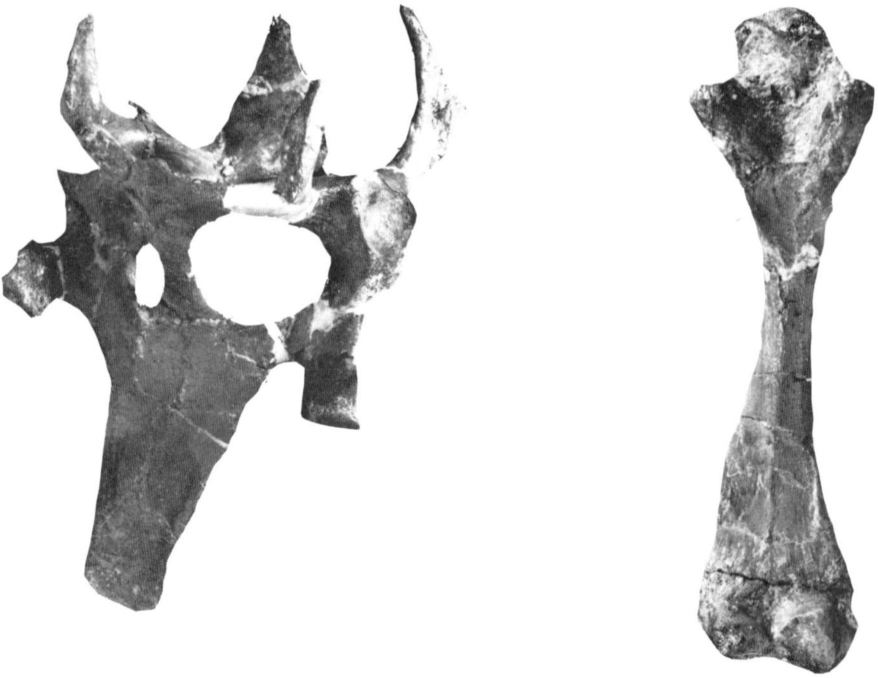

TEXT-FIG. 3.3—*Neurankylus eximius*. Left, pelvic girdle (PMU.R24q) (cf. Wiman, 1933, Text-fig. 6), ×0.67; Right, right femur (PMU.R24r), ×0.76.

Both these specimens come from the Kirtland Shale, and not the Fruitland Formation as previously reported (Wiman, 1933; Gaffney, 1972).

Suborder CRYPTODIRA Cope, 1870
Family DERMATEMYDIDAE Smith and Taylor, 1950
Genus BASILEMYS Hay, 1896
BASILEMYS NOBILIS Hay, 1911
Text-fig. 3.4.

Material.—A carapace and plastron (PMU.R29a,b)

Horizon.—Kirtland Shale, below Naashoibito Member.

Locality.—South branch of Hunter Wash, San Juan County, New Mexico (T24N, R13W).

Comments.—The Uppsala specimen differs from that described by Gilmore (1935) by the absence of an anteriorly expanded first central. The specimen figured by Gilmore shows many irregularities, presumably parts that are not preserved in the carapace, and shows little similarity to the Uppsala specimen. Gilmore suggested that the large first central may be the result of sexual dimorphism, the Uppsala specimen being female, rather unjustifiably on the basis of its small size. Gilmore's (1935) specimen came from the same general area as the Upp-

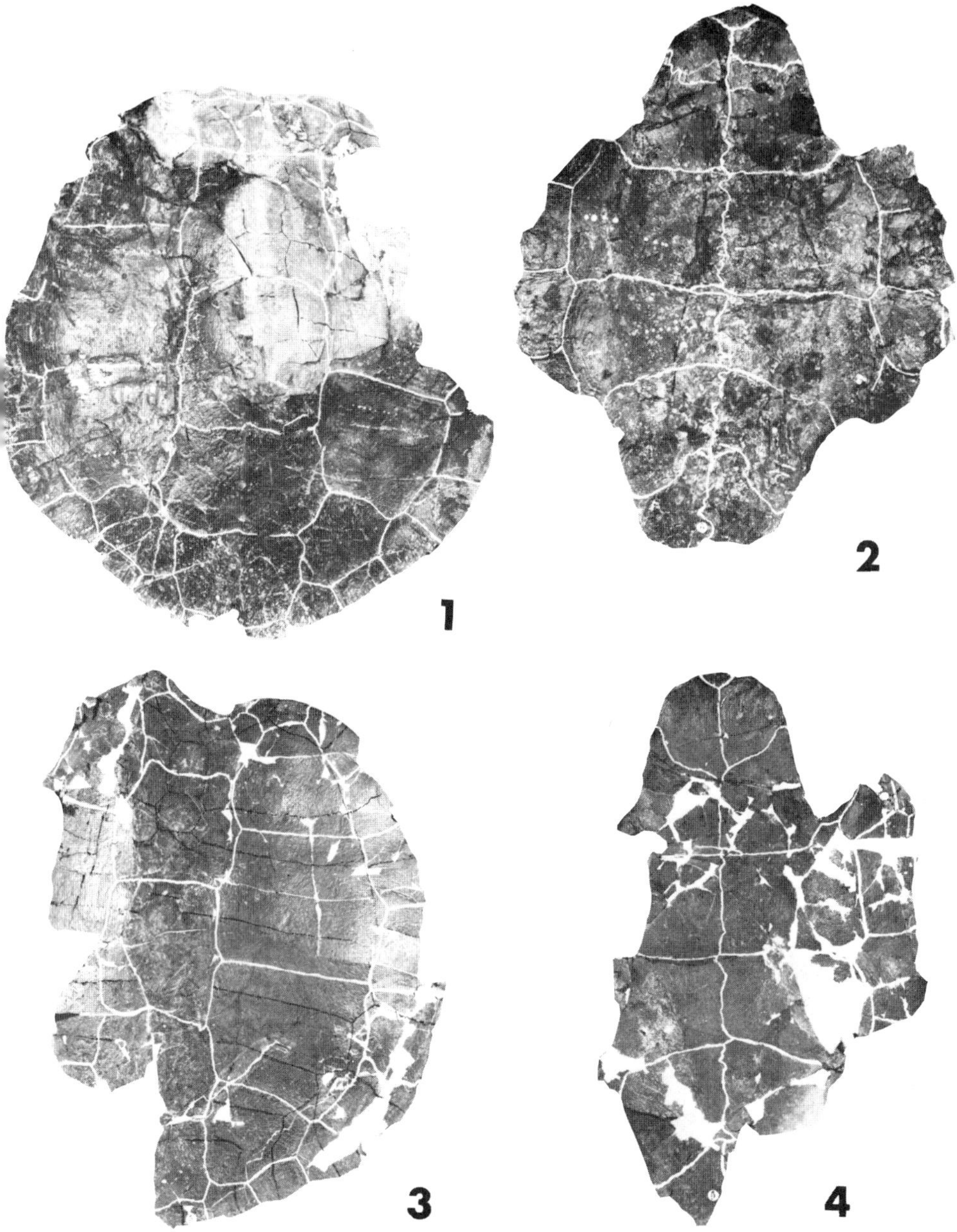

EXPLANATION OF PLATE 3.9

FIGS. *1–4*—*Thescelus insiliens*. *1*, Carapace (PMU.R22) in dorsal view (cf. Wiman, 1933, Pl. 2, fig. 6), ×0.24; *2*, Plastron (PMU.R22b) in ventral view (cf. Wiman, 1933, Pl. 2, fig. 6a), ×0.24; *3*, Carapace (PMU.R23a) in dorsal view (cf. Wiman, 1933, Pl. 2, fig. 7), ×0.25; *4*, Plastron (PMU.R23b) in ventral view (cf. Wiman, 1933, Pl. 2, fig. 7a), ×0.25.

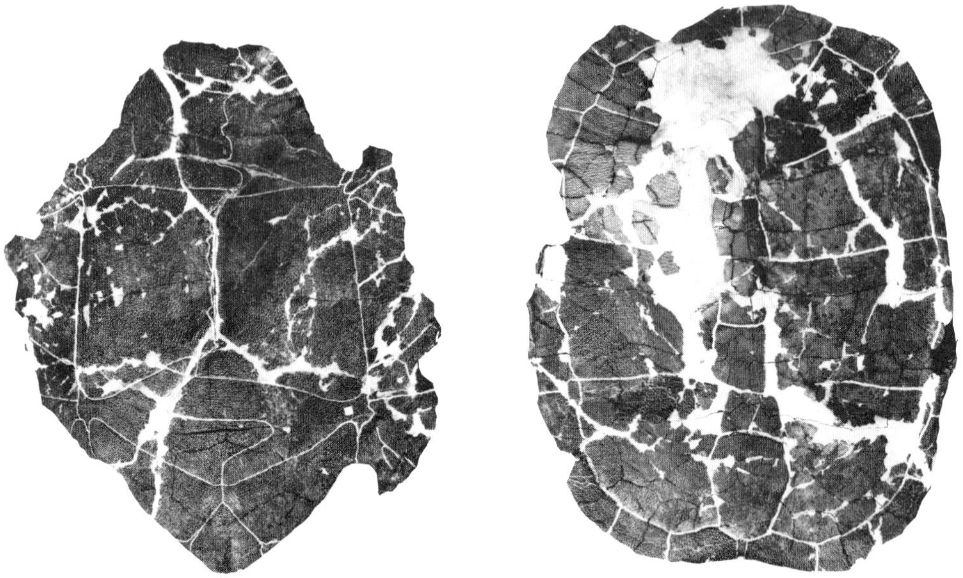

TEXT-FIG. 3.4—*Basilemys nobilis* (PMU.R29a, b). Left, carapace in dorsal view; Right, plastron in ventral view (cf. Wiman, 1933, Pl. 4, figs. 1–1a), ×0.1.

sala specimen and was also collected by C. H. Sternberg. The type specimen is from the Naashoibito Member of the Kirtland Shale.

In many attributes, the genera *Adocus* and *Basilemys* are very similar, the principal difference being an increase in number of the posteriormost neurals in some species of *Basilemys*. A review of these genera, preferably with crania, might streamline their systematics.

Family TRIONYCHIDAE Bell, 1828
Genera PLATYPELTIS Hay, 1908 and PLASTOMENUS Cope, 1873
PLATYPELTIS STERNBERGI Wiman, 1933 and PLASTOMENUS sp. Cope, 1873

Text-figs. 3.5, 3.6.

Material.—Carapace, PMU.R30 and plastron, PMU.R31, of *Plastomenus* sp.; two carapaces, one complete, PMU.R32, and one fragmentary, PMU.R33, of *Platypeltis sternbergi*.

Horizon.—Kirtland Shale, below Naashoibito Member.

Locality.—North flank of Ah-shi-sle-pah Wash, San Juan County, New Mexico (T22N, R11W).

Comments.—Romer (1966) included the genera *Plastomenus*, *Platypeltis*, and *Aspideretes* in the genus *Trionyx*, with some justification. While there is a certain amount of variation in the surface relief of the carapace, there is little to separate these genera. However, in the absence

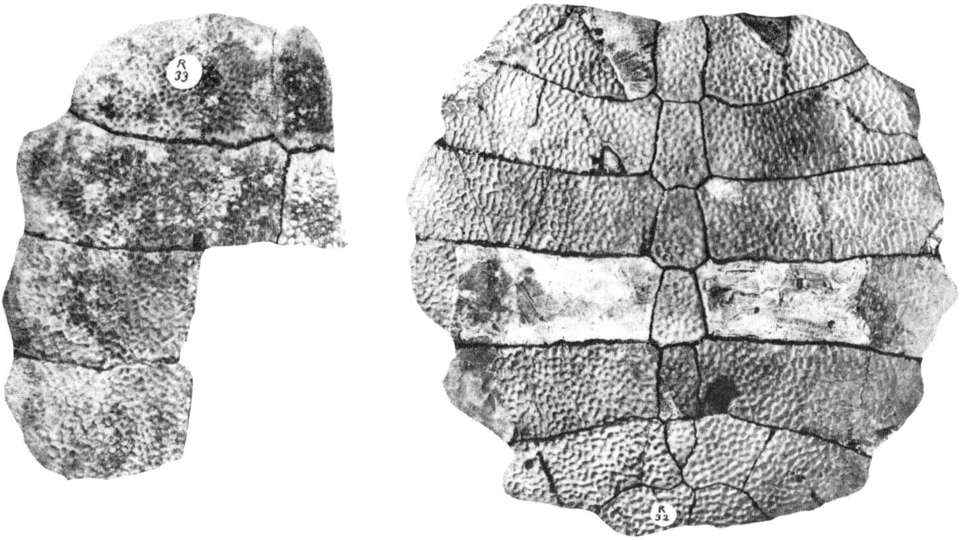

TEXT-FIG. 3.5—*Platypeltis sternbergi*. Left, a fragment of carapace, dorsal view (PMU.R33), ×0.70; Right, a nearly complete carapace, dorsal surface (PMU.R32, the holotype) (cf. Wiman, 1933, Pl. 6), ×0.51.

TEXT-FIG. 3.6—*Plastomenus* sp. Left, carapace in dorsal view (PMU. R30), ×0.22; Right, plastron in ventral view (PMU.R31) (cf. Wiman, 1933, Pl. 5, figs. 1–2), ×0.26.

of a full revision of the family, the original names are kept here, although almost certainly they will prove to be synonymous.

DISCUSSION

The specimens collected by Sternberg mentioned here reflect the taxonomic diversity of the reptilian megafauna from the San Juan Basin fairly accurately. Carnivorous dinosaurs, however, are the major exception. These, as to be expected, are not common; only a very few teeth have been found. Sternberg was primarily interested in sending exhibition specimens to Uppsala, thus smaller and fragmentary specimens (then seemingly insignificant) were not collected. Other groups not collected by Sternberg are the ankylosaurs and *Torosaurus*. The microvertebrate fauna from the Fruitland Formation and the Kirtland Shale is poorly known, although some collecting has been done (Clemens, 1973; Armstrong-Ziegler, 1978; Lehman, Lucas, this volume, for fuller accounts).

Perhaps the most neglected element of the Late Cretaceous Western Interior faunas are the turtles. As Williams indicated (1950), the primitive cryptodires experienced a period of high evolutionary activity during the Late Cretaceous involving a large degree of morphological experimentation as an adaptive strategy in the colonization of a new environment. The San Juan Basin Late Cretaceous turtle fauna, comprising shallow and deeper water forms, shows close affinities to faunas from other parts of the Western Interior, and from Mongolia (Gilmore, 1931, 1934; Khozatsky & Mlynarski, 1971) which may well have been the place of origin of the evolutionary radiation of these turtles. Parallels may be drawn with the acme of hadrosaurian evolution, a group also thought to have originated in Central Asia (Riabinin, 1939; Brett-Surman, 1975, 1978) and to have evolved en route through the Western Interior down to South America. Sauropods, although not common, are also found within this Asia–Western Interior–South America region; however, there is some evidence that the Late Cretaceous sauropods of North America are remnant stocks from the Late Jurassic.

A plausible ecological picture of the dinosaurian communities in the San Juan Basin during the Late Cretaceous is drawn by Lucas (this volume); here, ecological and spatial factors are primarily considered, although biogeographical and temporal effects were presumably in operation, including both movements of faunas and their evolutionary activity. Both the turtles and hadrosaurs were experiencing a period of maximum evolutionary expansion into environments that were not previously saturated or were newly created at this time. While a shift in environments from the Fruitland coastal swamps to the Kirtland hinterland in what is now northwestern New Mexico is recorded, a shift in dinosaurian movements in the Western Interior is also apparent. *Parasaurolophus*, for example, is known from the Oldman Formation of Alberta, which is generally regarded as being somewhat older than both the Fruitland and Kirtland of New Mexico. A migration of the *Parasaurolophus* stock from north to south can be envisioned, with some degree of modification on the way. The recently described *Parasaurolophus* from the Upper Maastrichtian of Utah (Weishampel & Jensen, 1979) is not discordant with this picture. Further, there is a modification of this genus between the Fruitland and Kirtland (*P. cyrtocristatus* and *P. tubicen* respectively). Whether this reflects *in situ* evolution, or the remains of migrating pulses through the San Juan Basin area is not possible to say. The reasons for such movements are not clear, although climatic change and competitive pressure from the north possibly played a part. Such pressures, and undoubtedly many others, may well have caused many dinosaurs to migrate in search of environments that, for them, were unexploited and that induced rapid development and ecological saturation. The turtles appeared to have adopted a strategy of morphological experimentation in their expansion and colonization, a rather typical r-selective strategy. Having already evolved a successful strategy, the hadrosaurs, ceratopsians, and sauropods may have merely evolved in tune with shifting local environments. Undoubtedly, further field work and perhaps inferences drawn from other comparable areas might help define the environmental and biogeographical roles at work and lead to a fuller understanding of the dinosaurian communities of the San Juan Basin Upper Cretaceous.

ACKNOWLEDGMENTS

Although C. H. Sternberg has long since gone the way of the dinosaurs, it is quite clear that this paper, and many other like it, would not have been possible without his continued efforts and hardships in the exploration of the San Juan Basin and other regions. Sternberg, aware of his contributions to paleontology, wrote to Wiman in 1919: "I feel with you I can look back on my life with gratitude to God that has chosen me, a small insignificant crippled man to add to His glory, by adding to human knowledge the wonderful story of His buried dead. With that sentiment I have been able to endure contempt, lack of support and a thousand other things that irritate a man, who works with people so wrapped up in the pursuit of the Almighty Dollar that they see nothing else." While his style may seem somewhat incongruous today, the problems of the paleontologist still remain.

More directly, I am indebted to Thomas Lehman and Spencer Lucas for their help in elucidating Sternberg's old localities, their stimulating discussions and critical reviewing of the manuscript. I acknowledge the help of the Bureau of Land Management (Albuquerque) for arranging logistics in the field. I wish to thank Richard Reyment for permitting me to study the Sternberg Collection in his care and Gustav Andersson for helping with the photography.

Financial assistance was provided by the Swedish Natural Science Research Council (Grant R 4001-101) and from the Liljewalch resestipendiefond of Uppsala University, enabling me to examine the Sternberg localities first hand.

REFERENCES

Armstrong-Ziegler, J. 1978. An aniliid snake and associated vertebrates from the Campanian of New Mexico. J. Paleontol. 52:480–483.

Baird, D. and J. R. Horner. 1977. A fresh look at the dinosaurs of New Jersey and Delaware. New Jersey Acad. Sci. Bull. 22:50.

Baltz, E. H., S. R. Ash and R. Y. Anderson. 1966. History of nomenclature and stratigraphy of rocks adjacent to the Cretaceous–Tertiary Boundary western San Juan Basin, New Mexico, U.S. Geol. Surv. Prof. Paper 524D:1–23.

Bauer, C. M. 1916. Stratigraphy of a part of the Chaco River Valley. U.S. Geol. Surv. Prof. Paper 98P:271–278.

Brett-Surman, M. K. 1975. The appendicular anatomy of hadrosaurian dinosaurs. Unpublished MA thesis, Univ. Calif., Berkeley. 70 p.

————. 1978. Phylogeny and paleobiogeography of hadrosaurian dinosaurs. Nature 277:560–562.

Clemens, W. A. 1973. The roles of fossil vertebrates in the interpretation of Late Cretaceous stratigraphy in the San Juan Basin, New Mexico. Four Corners Geol. Soc. Mem.: 154–167.

Farlow, J. O. and P. Dodson. 1975. The behavioral significance of frill and horn morphology in ceratopsian dinosaurs. Evol. 29:353–361.

Gaffney, E. S. 1972. The systematics of the North American family Baenidae (Reptilia, Cryptodira). Bull. Am. Mus. Nat. Hist. 147:243–319.

Gilmore, C. W. 1916. Vertebrate faunas of the Ojo Alamo, Kirtland and Fruitland Formations. U.S. Geol. Surv. Prof. Paper 98Q:279–308.

————. 1919. Reptilian faunas of the Torrejon, Puerco and underlying Upper Cretaceous formations of San Juan County, New Mexico, U.S. Geol. Surv. Prof. Paper 119:1–68.

————. 1922. A new sauropod dinosaur from the Ojo Alamo formation of New Mexico. Smithsonian Misc. Coll. 72:1–9.

————. 1931. Fossil turtles of Mongolia. Bull. Am. Mus. Nat. Hist. 59:213–257.

————. 1934. Fossil turtles of Mongolia: second contribution. Am. Mus. Novitates 689, 14 p.

————. 1935. On the Reptilia of the Kirtland Formation of New Mexico, with descriptions of new species of fossil turtles. Proc. U.S. Natl. Mus. 83:159–188.

————. 1946. Reptilian fauna of the North Horn Formation of central Utah. U.S. Geol. Surv. Prof. Paper 210C:29–53.

Horner, J. R. 1979. Upper Cretaceous dinosaurs from the Bearpaw Shale (marine) of south central Montana with a checklist of Upper Cretaceous dinosaur remains from the marine sediments in North America. J. Paleontol. 53:566–577.

Khozatsky, L. I. and M. Mlynarski. 1971. Chelonians from the Upper Cretaceous of the Gobi Desert, Mongolia. Palaeontol. Polonica. 25:131–144.

Lawson, D. A. 1972. Paleoecology of the Tornillo Formation, Big Bend National Park, Brewster County, Texas. Unpublished MA thesis, Univ. Texas, Austin. 182 p.

Lull, R. S. 1933. Revision of the Ceratopsia or horned dinosaurs. Memoirs of the Peabody Mus. Nat. His. 3:1–135.

Lull, R. S. and N. E. Wright. 1942. Hadrosaurian dinosaurs of North America. Geol. Soc. Am. Spec. Paper 40:1–242.

Mateer, N. J. 1976. New topotypes of *Alamosaurus sanjuanensis* Gilmore (Reptilia: Sauropoda). Bull. Geol. Instn. Univ. Uppsala. n.s. 6:93–95.

Ostrom, J. H. 1961. A new species of hadrosaurian dinosaur from the Cretaceous of New Mexico. J. Paleontol. 35:575–577.

———. 1963. *Parasaurolophus cyrtocristatus*, a crested hadrosaurian dinosaur from New Mexico. Fieldiana: Geology 14:143–168.

Parks, W. A. 1922. *Parasaurolophus walkeri*, a new genus and species of crested trachodont dinosaur. Univ. Toronto Stud. 13:1–32.

Riabinin, A. N. 1939. The Upper Cretaceous vertebrate faunas of south Kazakhstan. 1 Reptilia, Part 1. Ornthischia. Trans. Central Geol. and Prospecting Inst. Moscow. 118:1–39.

Romer, A. S. 1966. Vertebrate Paleontology. Univ. Chicago Press. Chicago. 468 p.

Sternberg, C. H. 1932. Hunting dinosaurs in the badlands of the Red Deer River, Alberta, Canada. (Published by the author) San Diego. 261 p.

Weishampel, D. B. and J. A. Jensen. 1979. *Parasaurolophus* (Reptilia: Hadrosauridae) from Utah. J. Paleontol. 53:422–1427.

Williams, E. E. 1950. Variation and selection in the cervical central articulations of living turtles. Bull. Am. Mus. Nat. Hist. 94:509–561.

Wiman, C. 1930. Über Ceratopsia aus der Oberen Kreide in New Mexico. Nov. Acta Reg. Soc. Scient. Upsaliensis. ser. 4. 7:1–19.

———. 1931. *Parasaurolophus tubicen* n.sp. aus der Kreide in New Mexico. Nov. Acta Reg. Soc. Scient. Upsaliensis. ser.4. 9:1–34.

———. 1932. *Goniopholis kirtlandicus* n.sp. aus der Oberen Kreide in New Mexico. Bull. Geol. Instn. Univ. Uppsala. 23:181–190.

———. 1933. Über Schildkröten aus der Oberen Kreide in New Mexico. Nov. Acta Reg. Soc. Scient. Upsaliensis. ser.4. 9:1–34.

4 Fossil lizards from the San Juan Basin, New Mexico

Robert M. Sullivan
Geology Section
University of Colorado Museum
Boulder, Colorado

INTRODUCTION

The San Juan Basin, long noted for its rich and diverse Late Cretaceous dinosaur and Paleogene mammal faunas, has also yielded some of the most important lacertilian remains known from the North American Paleogene. So far the rarity of these fossil lizards has been surpassed only by their importance to our understanding of the early Tertiary evolution of modern day lizard families.

Fossil lizards have been known from the San Juan Basin for over a century, and were first recorded by Cope (1877). Gilmore (1928), however, was the first to describe fossil lizards from this region in any comprehensive way. Some of Gilmore's descriptions were later reprinted in Matthew's (1937) classic monograph, *Paleocene Faunas of the San Juan Basin, New Mexico*. A few of Gilmore's descriptions are now outdated and need to be reevaluated in the light of our present knowledge. There have been relatively few studies, however, concerning fossil lizards from the San Juan Basin since Gilmore's work. Some fossil lizards from this area have been listed by Armstrong-Ziegler (1978) and redescribed or discussed in papers by Meszoely (1970), Meszoely et al. (1978) and Sullivan (1979). Other specimens have remained undescribed for some years while additional material continues to be discovered.

The purpose of this paper is to bring together for the first time a stratigraphically arranged account of some of the fossil lizards of the San Juan Basin, to redescribe the type of *Machaerosaurus torrejonensis* Gilmore (1928) and to report on some hitherto undescribed specimens. Owing to space limitations and current studies in progress on San Juan Basin fossil lizards, this paper cannot be comprehensive, but rather will serve as a reference point for future work.

INSTITUTIONAL ABBREVIATIONS

AMNH American Museum of Natural History, New York.
KU Museum of Natural History, University of Kansas, Lawrence.
UNM Department of Geology, University of New Mexico, Albuquerque.

SYSTEMATIC PALEONTOLOGY

Class REPTILIA
Subclass LEPIDOSAURIA Romer, 1945
Order SQUAMATA Oppel, 1811
Suborder SAURIA MacCartney, 1802
Infraorder SCINCOMORPHA Camp, 1923
Family TEIIDAE Gray, 1827
Genus CHAMOPS Marsh, 1892
cf. CHAMOPS SEGNIS Marsh, 1892
Text-fig. 4.1

Referred material.—UNM FKK-038a, fragmentary right dentary.

Locality.—UNM V-072.

Horizon.—Fruitland Formation.

Age.—Late Cretaceous (late Campanian).

Collector.—Thomas Lehman, July, 1979.

Description.—This dentary (Text-fig. 4.1) measures 24 mm in length and is fractured in several places. Although the posteriad ventral border of the dentary is missing, most of the tooth row is intact. Twelve teeth are present with spaces for at least four others. The teeth are moderately heterodont with expanded bases. Posteriad teeth are tricuspid; anteriad teeth are pointed and slightly recurved.

Discussion.—Estes (1964) provided a thorough discussion, description and diagnosis of this Late Cretaceous lizard. This specimen, although somewhat larger than previously described specimens (Estes, 1964), agrees in every respect with the species *Chamops segnis*, where heterodonty is little pronounced and tooth crowns are not swollen. *Chamops segnis* has previously been reported from the San Juan Basin by Armstrong-Ziegler (1978)

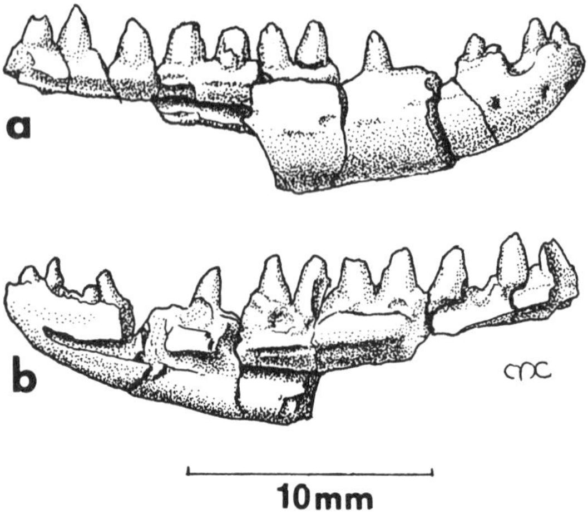

TEXT-FIG. *4.1*—cf. *Chamops segnis* Marsh. UNM FKK-038a, right dentary. *a*, labial view; *b*, lingual view.

Family ANGUIDAE Cope, 1864
Subfamily INCERTAE SEDIS
Genus ODAXOSAURUS Gilmore, 1928
ODAXOSAURUS cf. PIGER (Gilmore, 1928)
Text-fig. 4.2

Referred material.—KU 7897, crushed and weathered anterior part of skull; and KU 9734, anterior part of the frontal complex, posterior part of the right dentary and splenial, right supratemporal process, ilium fragment, with associated unidentifiable concretionary fragments and a fragment of a crocodilian scute.

Locality.—Kutz Canyon (KU New Mexico locality 13).

Horizon.—Nacimiento Formation.

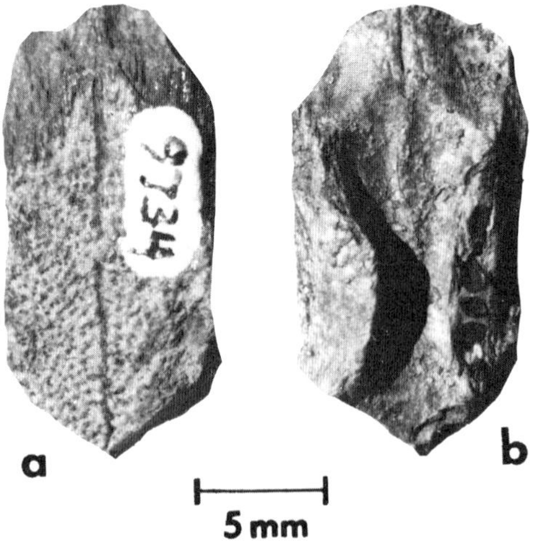

TEXT-FIG. *4.2*—*Odaxosaurus* cf. *piger* (Gilmore) KU 9734, anterior portion of frontal complex. *a*, dorsal view; *b*, ventral view. Anterior direction is up.

Age.—Middle Paleocene (Torrejonian).

Collectors.—For KU 7897: Robert W. Wilson, 1948; for KU 9734: R. W. Wilson and R. Rinker, 16 June 1950.

Description.—KU 7897, a crushed, weathered anterior part of a skull measuring approximately 40 mm in length is totally encrusted by matrix. Both maxillae and dentaries with teeth are present as well as the frontal and nasal regions of the skull, the latter of which is severely crushed. The osteoderms are encrusted by matrix, but appear to be of the pit and ridge (vermiculate) type. The tooth crowns are blunt and are not striated.

The other specimen (KU 9734) consists of a 16 mm anterior portion of an elongated frontal complex (Text-fig. 4.2). The frontals are partially fused along the metopic suture which is distinct dorsally but is nearly obliterated ventrally towards the posterior end of the frontals. The dermal armor sculpturing consists principally of a pit and ridge morphology; however, there are a number of small tubercular mounds on the dermal armor. A medial part of the right mandible is represented by the posterior part of the dentary with one tooth and the anterior part of the splenial. The tooth is compressed anterior-posteriorly, has a blunt crown with striations, and a moderately swollen base. Other elements include the right supratemporal process of the parietal and ilium fragment.

Discussion.—Based on tooth structure and osteoderm morphology, both KU 7897 and 9734 are tentatively referred to *Odaxosaurus* cf. *piger*. The skull (KU 7897) is encrusted with a hard matrix that obliterates most of its detail. There are a number of fractures on this specimen which also make study of the specimen difficult. The teeth appear smooth, almost polished, suggesting that weathering has eliminated the striations normally seen on the tooth crowns. KU 9734, represented by an anterior portion of frontal complex whose dermal armor morphology is that of a combined pit and ridge and tuberculate type, is significant in that it suggests a possible transitional form between an *Odaxosaurus* and glyptosaurines; thus the species "*Odaxosaurus*" *jepseni* may be represented by the latter specimen.

Although these two specimens are somewhat fragmentary, they clearly establish the presence of *Odaxosaurus* in the Torrejonian horizons of the Nacimiento Formation. The specific and generic designation is tentative due to the nature of the specimens.

?ODAXOSAURUS sp.

Text-fig. 4.3

Referred material.—UNM NP-334: part of left opisthotic, part of left splenial and 23 body osteoderms.

Locality.—Betonnie Tsosie Wash, 0.9 km northwest of Mammalon Hill, Kimbeto Quadrangle (7½ min.), SW¼SE¼ Sec. 34, T23N, R9W.

Horizon.—Nacimiento Formation (lower part).

Age.—Early Paleocene (Puercan).

Collector.—Robert M. Sullivan, 8 July 1979.

Description.—The left opisthotic process (Text-fig. 4.3a,b) is represented only by the distal end and thus is relatively incomplete. This bone appears slightly more robust and abbreviated when compared to this element in Recent anguines of comparable size like *Pseudopus* (= *Ophisaurus*) *apodus*.

TEXT-FIG. 4.3—?*Odaxosaurus* sp, UNM NP-334. *a* and *b*, anterior and posterior views of left opisthotic process; *c* and *d*, lingual and labial views of splenial; and *e* and *f*, dorsal body osteoderms.

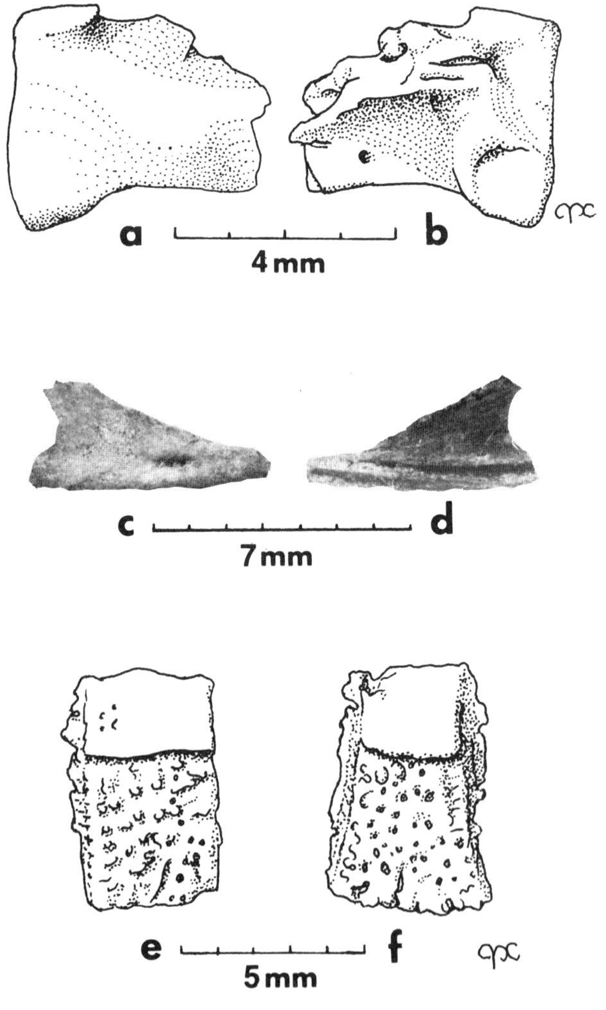

The preserved left splenial (Text-fig. 4.3c,d) represents the posterior part of this bone that articulates with the posterior part of the left dentary. The splenial is thicker than that of *Pseudopus apodus* of comparable size especially on the dorsal edge where it joins the dentary. This region is flared and grooved, indicating that the dentary probably possessed a ventral projection below the dentary shelf that fitted into the splenial. Lingually the bone is smooth with a well developed anterior mylohyoid foramen that is near the posterior part of the anterior splenial projection (Text-fig. 4.3c). The anterior inferior alveolar foramen was apparently formed by the ascending edge of the splenial, which is missing in part, and was primarily situated on the dentary, not unlike the situation seen in the glyptosaurines and some anguines. Labially, the ventral groove that articulated with the dentary is well developed and possessed a dorso-lateral flange that is directed obliquely downward (Text-fig. 4.3d), unlike the condition in the anguines *Pseudopus apodus* and *Ophisaurus ventralis* in which this lateral flange is directed obliquely upward. The condition of the flange in glyptosaurines is not known due to the fact that this articulating region is rarely observed. The posteriormost extent of the splenial is incomplete, giving the preserved bone a subtriangular appearance.

The body osteoderms (Text-fig. 4.3e,f) morphologically range from the strict pit and ridge (vermiculate) type characteristic of the genus *Odaxosaurus* and some anguines, to that of a pit and ridge-tuberculate pattern. These osteoderms also display lateral beveling with sutured surfaces, conditions that are also present in the body osteoderms of later glyptosaurine lizards. The scutes mainly represent dorsal osteoderms from the right, left and midline regions. All well preserved osteoderms display a well developed anterior gliding surface.

Discussion.—Based on body osteoderms and the size of the material, this specimen is questionably referred to *Odaxosaurus*. The specimen was surface picked from within a 1 m^2 area. Screening of this site may have yielded additional osteological elements of this specimen. The occurrence of so many osteoderms from within this small area strongly suggests that this material is from a single individual.

Subfamily ANGUINAE McDowell and Bogert, 1954
Genus MACHAEROSAURUS Gilmore, 1928
MACHAEROSAURUS TORREJONENSIS Gilmore, 1928
Pl. 4.1

Holotype.—AMNH 5184, incomplete skull and lower jaws consisting of: peripheral region of the left side of left frontal and parietal; medial part of the left orbital osteoderm series; right and left prefrontals, both

1

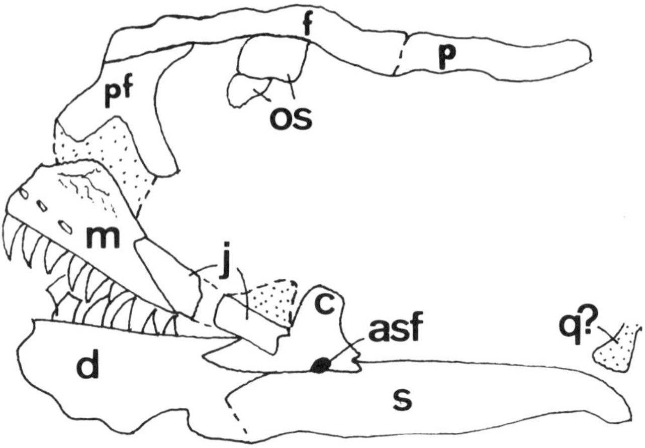

2

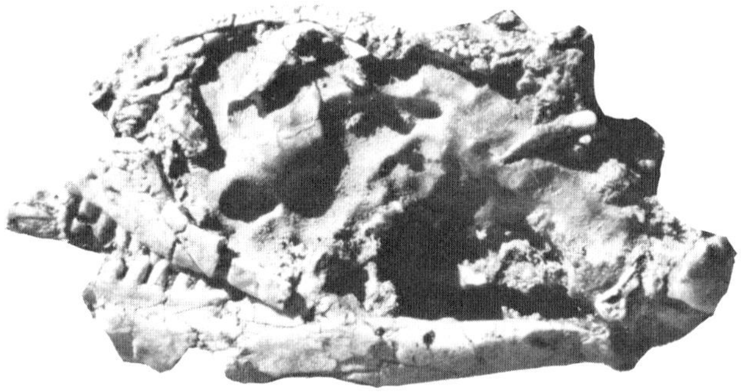

5 mm

3

EXPLANATION OF PLATE 4.1

FIGS. *1–3*—*Machaerosaurus torrejonensis* Gilmore. AMNH 5184 (type), incomplete skull and lower jaws. *1*, left lateral view, osteological key. Abbreviations: asf, anterior supra-angular foramen; c, coronoid; d, dentary; f, frontal; j, jugal; m, maxilla; os, orbital osteoderms; p. parietal; pf, prefrontal; ?q, ?quadrate; and s, surangular. Anterior direction is to the left; *2*, left lateral view, and *3*, right lateral view, of skull showing the dentary and posterior part of the right maxilla.

coronoids, anterior part of left jugal; fragment of right maxilla; greater posterior part of left maxilla with teeth; posterior parts of both dentaries with teeth; nearly complete left surangular and anterior part of right surangular; right splenial; and fragment of left quadrate and left epipterygoid.

Type locality.—East fork of "Torrejon Arroyo" (=Torreon Wash), San Juan Basin, New Mexico.

Horizon.—Nacimiento Formation.

Age.—Middle Paleocene (Torrejonian)

Distribution.—Middle Paleocene of North America.

Collector.—William J. Sinclair, 30 August 1913.

Revised description.—An anguine exhibiting the following osteological characters:

1) Dermal armor sculpturing: The dermal armor sculpturing of *Machaerosaurus torrejonensis* differs from the gracile pit and ridge (vermiculate) type seen in the late Cretaceous *Odaxosaurus piger*, the early Oligocene "*Machaerosaurus*" *pawneensis*, and the Recent New World *Ophisaurus ventralis*. The sculpturing on *M. torrejonensis* is somewhat rugose and is similar to the kind seen on the Recent Old World *Pseudopus apodus*. The upper part of the maxilla exhibits crenulated sculpturing for osteoderm attachment as seen in *Odaxosaurus piger* as well as some fossil and extant anguines and is not part of the dermal armor as implied by Gilmore (1928).

2) Teeth: The teeth are simple, laterally compressed, nonstriated, and recurved. They differ significantly from the inflated conical type seen in *Ophisaurus ventralis* and "*Machaerosaurus*" *pawneensis* and differ from the bulbous, striated heterodont type seen in the Recent *Pseudopus apodus*. They also differ from the tooth type of *Odaxosaurus*, in which the teeth are obtuse, the tooth shafts are expanded and striations are confined to the tooth crown.

3) Anterior supra-angular foramen: The dentary is excluded from participation with the anterior supra-angular foramen as in "*Machaerosaurus*" *pawneensis* and Recent *Ophisaurus ventralis*. It differs from *Odaxosaurus piger*, in which the dentary participates with the anterior supra-angular foramen, and from *Pseudopus apodus*, in which the foramen emerges from the posterior part of the dentary below and anterior to the apex of the coronoid.

4) Dentary: The dentary is of moderate size. The posterior extremities exclude it from dorsal participation with the anterior supra-angular foramen. The posterior part of the tooth bearing region encroaches upon the anterior extremity of the coronoid.

5) Coronoid: The coronoid is moderately large, with a well developed lateral vertical mid-ridge as in glyptosaurines, "*Machaerosaurus*" *paw-*

neensis, and *Pseudopus apodus*. The ridge differs from the somewhat reduced coronoid ridge seen in *Ophisaurus ventralis*.

6) Surangular: The surangular extends anteriad of the anterior projection of the coronoid as in "*Machaerosaurus*" *pawneensis* and the Recent ophisaurs. It differs from the glyptosaurine condition in which both anterior extremities define an imaginary vertical line. It differs from "*Machaerosaurus*" *pawneensis* in that the anterior part of the surangular is not as broad and blunt; in this it agrees in morphology with Recent ophisaurs.

7) Jugal: Based on the anterior extension of the left jugal, this bone appears to be moderately well developed and similar to that of "*Machaerosaurus*" *pawneensis*.

8) Other skull bones: The fragmentary nature of the prefrontals, orbital osteoderm series, nasal, frontal, parietal, quadrate and epipterygoid makes interpretation of these bones impossible.

Discussion.—Gilmore (1928) established this species solely on the holotype (AMNH 5184, Pl. 1), which until recently was the only species referable to the genus *Machaerosaurus*. A recent study by Meszoely et al. (1978, p. 157) removed the species *Odaxosaurus (=Pancelosaurus) pawneensis* and transferred it to the genus *Machaerosaurus* on the basis of superficial and primitive similarities. The characters that they used for the inclusion of "*Machaerosaurus*" *pawneensis* within the genus *Machaerosaurus* included the simple conical unstriated tooth type and a similar morphology in the extension of the surangular on the labial surface of the jaw. The teeth in *Machaerosaurus torrejonensis* are laterally compressed and sharply recurved, whereas in "*Machaerosaurus*" *pawneensis* the teeth are of an inflated conical type and only slightly recurved. More importantly, the anterior extension of the surangular in the holotype of *Machaerosaurus torrejonensis* does not appear to be as broad or blunt as in "*Machaerosaurus*" *pawneensis* and glyptosaurines, but rather appears to be more like that of Recent anguines. However, the surangular extends in advance of the anterior projection of the coronoid on the labial surface, a primitive situation seen in all Recent anguines. It seems clear that the Oligocene species "*Machaerosaurus*" *pawneensis* probably belongs to a genus other than *Machaerosaurus* and accordingly this material is currently being redescribed (Sullivan, in prep.).

A few observations made by Gilmore (1928) regarding the morphology of the holotype of *Machaerosaurus torrejonensis* need to be reconsidered. Gilmore described the dermal armor of the frontal, parietal and the upper part of the maxilla as being a distinctive punctate-vermiculate sculpturing (Pl. 4.1, fig. 2). However, the previously mentioned sculpturing present on the upper part of the maxilla (Pl. 4.1, fig. 1 and 2) is a crenulation of the maxilla itself and is not part of the dermal armor. The distribution of dermal armor on the skull of anguines is similar to that of *Odaxosaurus*, being confined principally to the dorsal region of the skull, while in the glyptosaurines the entire skull is covered with dermal armor. The rugose sculpturing of *Machaerosaurus torrejonensis* agrees more closely with the kind seen in the Old World species *Pseudopus apodus*. The bone that Gilmore (1928) referred to as being the lachrymal appears to be the roughened sculptured part of the maxilla on the left side of the skull.

Machaerosaurus torrejonensis, until recently, has been known only from the holotype. I have referred a number of middle Paleocene (Torrejonian) dentaries and a single left frontal from Swain Quarry of Wyoming to this species (Sullivan, 1980). Comparison of the single frontal with the holotype indicates that based on dermal armor morphology, epidermal scale impression and size, they belong to the same species. The Swain Quarry frontal is important in that it reveals that the frontals of *Machaerosaurus torrejonensis* were paired. Referred dentaries have the identical unique tooth morphology that is characteristic of *Machaerosaurus torrejonensis*.

cf. MACHAEROSAURUS TORREJONENSIS

Referred material.—KU 7898, posterior part of left maxilla (associated with an indeterminate anguid dentary and urodele and unidentifiable bone fragments), from Kutz Canyon (KU New Mexico locality 13). The specimen is from Torrejonian strata of the Nacimiento Formation and was collected by Robert W. Wilson in 1948.

Discussion.—The maxilla is referred to the lizard *Machaerosaurus torrejonensis* based on the laterally compressed, unstriated recurved teeth and its similar size to the holotype. The lingual view of the teeth shows that the bases are somewhat swollen as in *Odaxosaurus* and the glyptosaurines.

Subfamily GLYPTOSAURINAE
Tribe GLYPTOSAURINI Sullivan, 1979
Genus PARAGLYPTOSAURUS Sullivan, 1979
PARAGLYPTOSAURUS YATKOLAI Sullivan, 1979
Text-fig. 4.4

Holotype.—AMNH 5181, nearly complete frontal.
Type locality.—Arroyo Blanco, north of Regina, New Mexico.
Horizon.—Regina Member, San Jose Formation.
Age.—Early Eocene (Wasatchian).

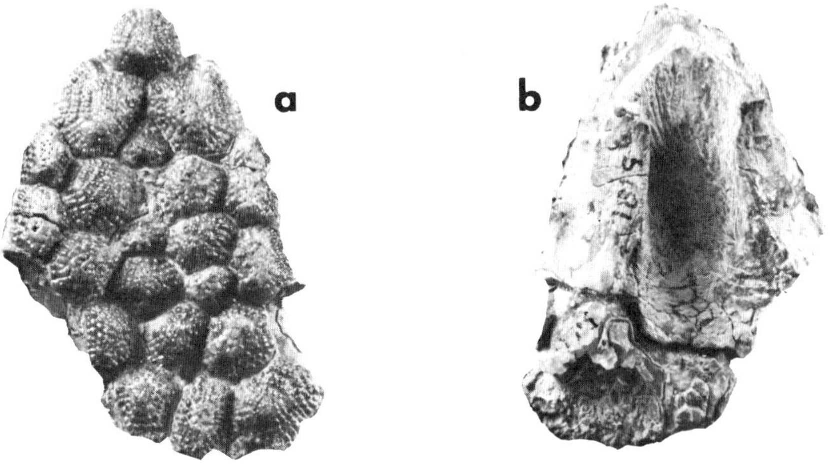

TEXT-FIG. *4.4—Paraglyptosaurus yatkolai* Sullivan. AMNH 5181 (holotype), nearly complete frontal. *a*, dorsal view; *b*, ventral view. Anterior direction is up. × 1.75

Collector.—Walter Granger and the expedition of the American Museum of Natural History, 12 June 1912.

Discussion.—*Paraglyptosaurus* is known by two species: *P. princeps* and *P. yatkolai* (Sullivan, 1979). The latter is known only from the type and is distinguished from *P. princeps* in having two prominent and large transversely elongated anterior frontal osteoderms and a poorly developed medial row of frontal osteoderms. In addition, these frontal osteoderms are somewhat inflated compared to the flat plate-like scutes of *P. princeps.*

GLYPTOSAURINAE Indeterminate

A few glyptosaurines from both the middle Paleocene and early Eocene of the San Juan Basin are also worth noting. AMNH 5187 is a very large glyptosaurine, represented by part of the left and right dentaries, part of the right maxilla and some unidentified bone fragments, from Torrejonian strata of the Nacimiento Formation at the head of Kimbeto Arroyo. The specimen was previously referred to *Peltosaurus* sp. by Gilmore (1928), and was designated cf. Glyptosaurinae (indet.) by Sullivan (1979, p. 45). The only other very large Paleocene glyptosaurine known to me is a specimen in the collections of the University of Michigan from Clarkforkian deposits of Wyoming. This specimen appears to belong to the tribe Melanosaurini, but more detailed study to determine its precise taxonomic position is needed. AMNH 5187 may also be a melanosaurinid, but its fragmentary nature prohibits any definitive taxonomic assignment.

Two additional Eocene glyptosaurine specimens have been recovered recently from the San Juan Basin, both of which are unnumbered osteoderms in the collections of the University of New Mexico; one from the Regina Member of the San Jose Formation, near Santos Peak, and the other from the Regina member of Arroyo Almagre, New Mexico. At present, osteoderm morphology is not diagnostic below the subfamily level (Sullivan, 1979). The latter specimen is from the same general area as Cope's "Placosauridae" osteoderms (Cope, 1877) as well as the same general area and stratigraphic horizon where the type of *Paraglyptosaurus yatkolai* was found. However, reference to *P. yatkolai* cannot be substantiated based on geologic and geographic occurrences alone.

SUMMARY

Table 4.1 summarizes the lizard taxa currently known from the San Juan Basin, New Mexico. As in a number of North American Paleogene localities, the anguid lizards represent a major part of the lizard fauna. Although rare, the fossil lizards known from this region are among some of the more important taxa from the Late Cretaceous and early Paleogene of North America. Two type specimens come from the deposits of the San Juan Basin: that of *Machaerosaurus torrejonensis* and *Paraglyptosaurus yatkolai.*

Further work in the Late Cretaceous and Paleogene deposits of the San Juan Basin promises to yield more lacertilian remains which undoubtedly will aid in the understanding of the early evolution and phylogenetic relationships of these lizard groups.

ACKNOWLEDGMENTS

I thank Dr. Eugene Gaffney and Ms. Charlotte Holton (American Museum of Natural History), Drs. Larry Martin and Hans Peter Schultz (University of Kansas) and Dr. Barry Kues (University of New Mexico) for the loan of specimens. Dr. Richard Estes (San Diego State University), Mr. Jacques Gauthier (University of California at Berkeley), and Dr. J. Alan Holman (Michigan State University) critically read the manuscript, and I thank them for their time and suggestions. Merald Clark made the text-figure drawings.

Locality data for the UNM specimens is on file at the Department of Geology, University of New Mexico, Albuquerque, New Mexico.

Table 4.1—List of the fossil lizards from the San Juan Basin.

TAXA	AGE	FORMATION
SAURIA		
Teiidae		
*Leptochamops denticulatus**	Campanian	Fruitland
cf. *Chamops segnis**	Campanian	Fruitland
Anguidae		
Incertae sedis		
Odaxosaurus cf. *piger*	Torrejonian	Nacimiento
?*Odaxosaurus* sp.	Puercan	Nacimiento
Glyptosaurinae		
Glyptosaurini		
Paraglyptosaurus yatkolai	Wasatchian	San Jose
Glyptosaurinae Indeterminate	Torrejonian–Wasatchian	Nacimiento and San Jose
Gerrhonotinae		
cf. *Gerrhonotus* sp.*	Campanian	Fruitland
Anguinae		
Machaerosaurus torrejonensis	Torrejonian	Nacimiento
cf. *Machaerosaurus torrejonensis*	Torrejonian	Nacimiento

*Listed by Armstrong-Ziegler (1978)

REFERENCES

Armstrong-Ziegler, J. G. 1978. An aniliid snake and associated vertebrates from the Campanian of New Mexico. J. Paleontol. 52:480–483.

Cope, E. D. 1877. Report on extinct vertebrates obtained in New Mexico by parties of the expedition of 1874. U.S. Geol. Surv. West of 100th Meridian, Wheeler Rept. 4:1–370.

Estes, R. 1964. Fossil vertebrates from the Late Cretaceous Lance Formation eastern Wyoming. Univ. Cal. Publ. Geol. Sci. 49:1–187.

Gilmore, C. W. 1928. Fossil lizards of North America. Mem. Nat. Acad. Sci. 22:ix + 201 p.

Matthew, W. D. 1937. Paleocene faunas of the San Juan Basin, New Mexico. Am. Philos. Soc. Trans. n.s.,30:510 p.

Meszoely, C. A. M. 1970. North American fossil anguid lizards. Mus. Comp. Zool. Bull., Harvard Univ. 139:87–149.

Meszoely, C. A. M., R. Estes and H. Haubold. 1978. Eocene anguid lizards from Europe and a revision of the genus *Xestops*. Herpetologica 34:156–166.

Sullivan, R. M. 1979. Revision of the Paleogene genus *Glyptosaurus* (Reptilia, Anguidae). Bull. Am. Mus. Nat. Hist. 163:1–72.

————. 1980. Lower vertebrates from Swain Quarry, Fort Union Formation, middle Paleocene (Torrejonian), Carbon County, Wyoming. Unpublished PhD dissertation, Mich. St. Univ. 56 p.

5

A skeleton of *Gillisonchus gillianus* (Mammalia; Condylarthra) from the Early Paleocene (Puercan) Ojo Alamo Sandstone, San Juan Basin, New Mexico, with comments on the local stratigraphy of Betonnie Tsosie Wash

J. Keith Rigby, Jr.
U.S. Bureau of Land Management, Albuquerque, New Mexico

INTRODUCTION

Significant fossil remains have been recently recovered from the Ojo Alamo Sandstone in Betonnie Tsosie Wash (Rigby & Lucas, 1977). These include a nearly complete skeleton of *Gillisonchus (=Anisonchus) gillianus*, a nearly complete juvenile skull of *Ectoconus*, and maxillae and dentary fragments of a variety of other Puercan fossil mammals. The *Gillisonchus* skeleton was eroding from between dark gray and light gray shale stringers encased within a white sandstone unit which forms the arroyo floor near the northern base of the classic Puercan locality Mammalon Hill in the SW¼ of Sec 35, T23N, R9W. Further exploratory quarry operations revealed additional material of *Gillisonchus gillianus* from the base of Mammalon Hill, as well as from other localities on the sandstone flat at the hill's base.

When discovered, most of the *Gillisonchus* skeleton had already disassociated and was scattered for some 1–2 m along a small gully. Sediment samples of approximately 100 kg were collected and screened to recover the disassociated remains. The skeleton was then pieced together under a microscope. Excavations initiated within shale lenses within an Ojo Alamo Sandstone tongue have produced additional specimens of *Gillisonchus gillianus* including a partial skull and jaw. Study of these new specimens suggests that *Anisonchus gillianus* should be removed from the genus *Anisonchus*. A new genus, *Gillisonchus*, is proposed for it.

ABBREVIATIONS

AMNH	American Museum of Natural History, New York
AW	Anterior tooth width (= trigonid width)
L	Tooth length
PW	Posterior tooth width (= talonid width)
UNM	Dept. of Geology, University of New Mexico, Albuquerque
W	Greatest transverse width

STRATIGRAPHY

During the summer of 1977, the U.S. Bureau of Land Management (BLM) commissioned a paleontological inventory of the central and southern parts of the San Juan Basin in preparation for an environmental impact statement dealing with the potential impacts of proposed strip mining of coal reserves in the Fruitland Formation upon paleontological and other resources (BLM, 1979). The field crew that completed this inventory under the direction of Drs. B. Kues, J. Froehlich, J. Schiebout, and Mr. S. Lucas located 1,157 fossil localities (Kues et al., 1977). Part of this work involved a preliminary survey of exposures of the lower part of the Nacimiento Formation, the Ojo Alamo Sandstone, and the upper part of the Kirtland Shale located along the northern margin of Betonnie Tsosie Wash in San Juan County, New Mexico (Text-figs. 5.1 & 5.2). Detailed stratigraphic descriptions appear in Appendix 5.1. These and subsequent investigations by the author and Spencer Lucas (Rigby & Lucas, 1977; Lucas & Rigby, 1979) verified the presence of two broadly defined fossiliferous horizons in the vicinity of Mammalon Hill (Text-fig. 5.3, Appendix 5.1). Sinclair and Granger (1914, Plate XXVI) figured this area and noted that fossils were recovered from the base of the hill located on the ridgecrest and from the stratigraphically lower white sandstone flat just north of the hill. Detailed stratigraphic studies (Rigby & Lucas, 1977; Lucas & Rigby, 1979) have shown that the lowest of these fossil localities occurs in shale stringers located within a sandstone tongue of the Ojo Alamo Sandstone (Text-fig. 5.3). The use of the rock-stratigraphic term "Ojo Alamo Sandstone" for the occurrence of these localities, however, has been contested (Lindsay et al., 1978, 1979, 1981).

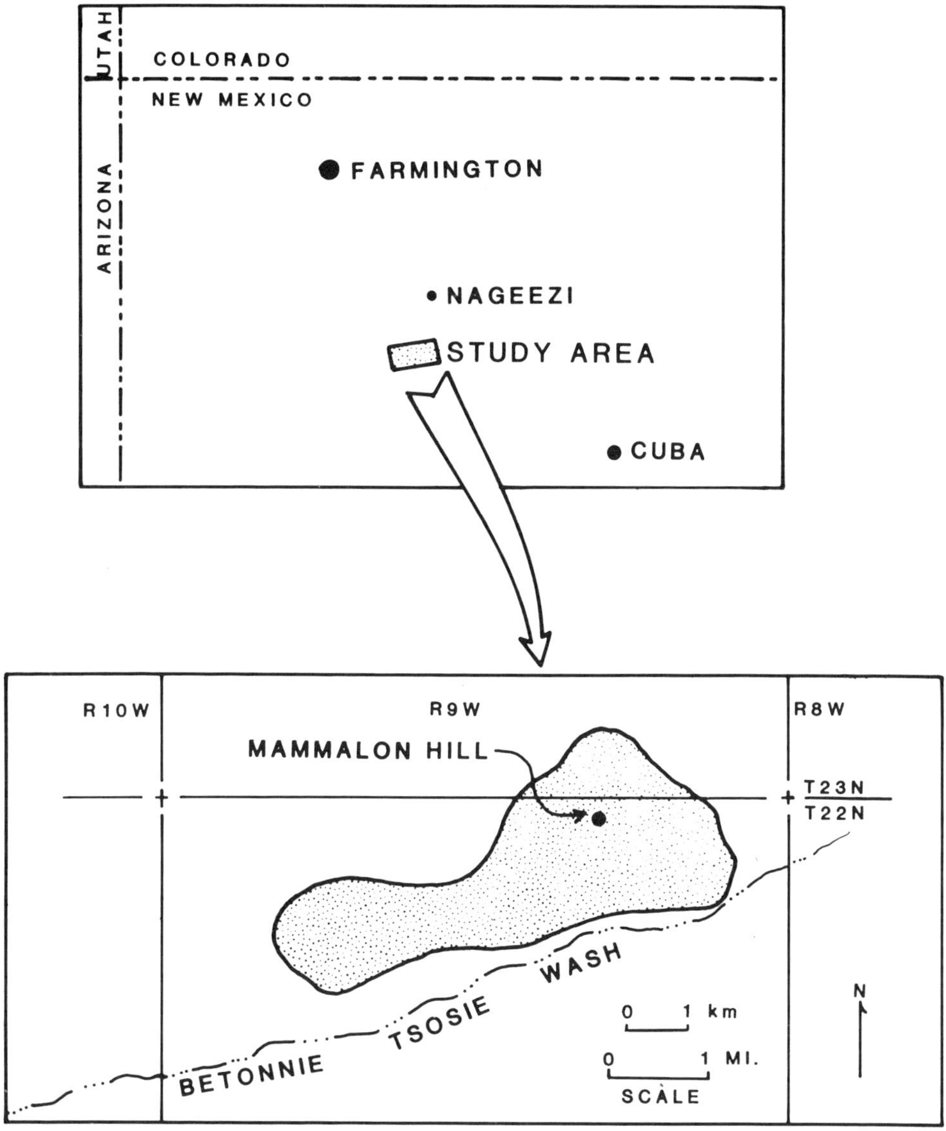

TEXT-FIG. 5.*1*—Index map showing the area of Upper Cretaceous and lower Tertiary exposures along the north side of Betonnie Tsosie Wash. The stippled area is shown as Text-fig. 5.2.

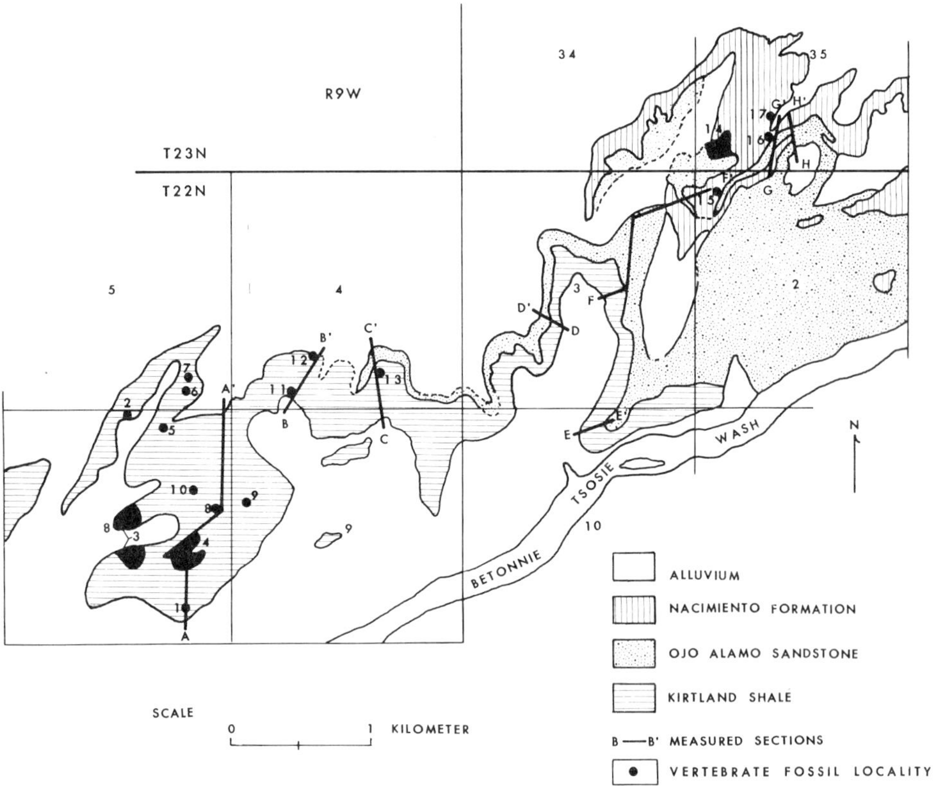

TEXT-FIG. 5.2—Geologic map of the exposures north of Betonnie Tsosie Wash, showing fossil localities and location of measured geologic sections. The *Gillisonchus gillianus* skeleton was recovered from locality 14 (UNM locality B77-7).

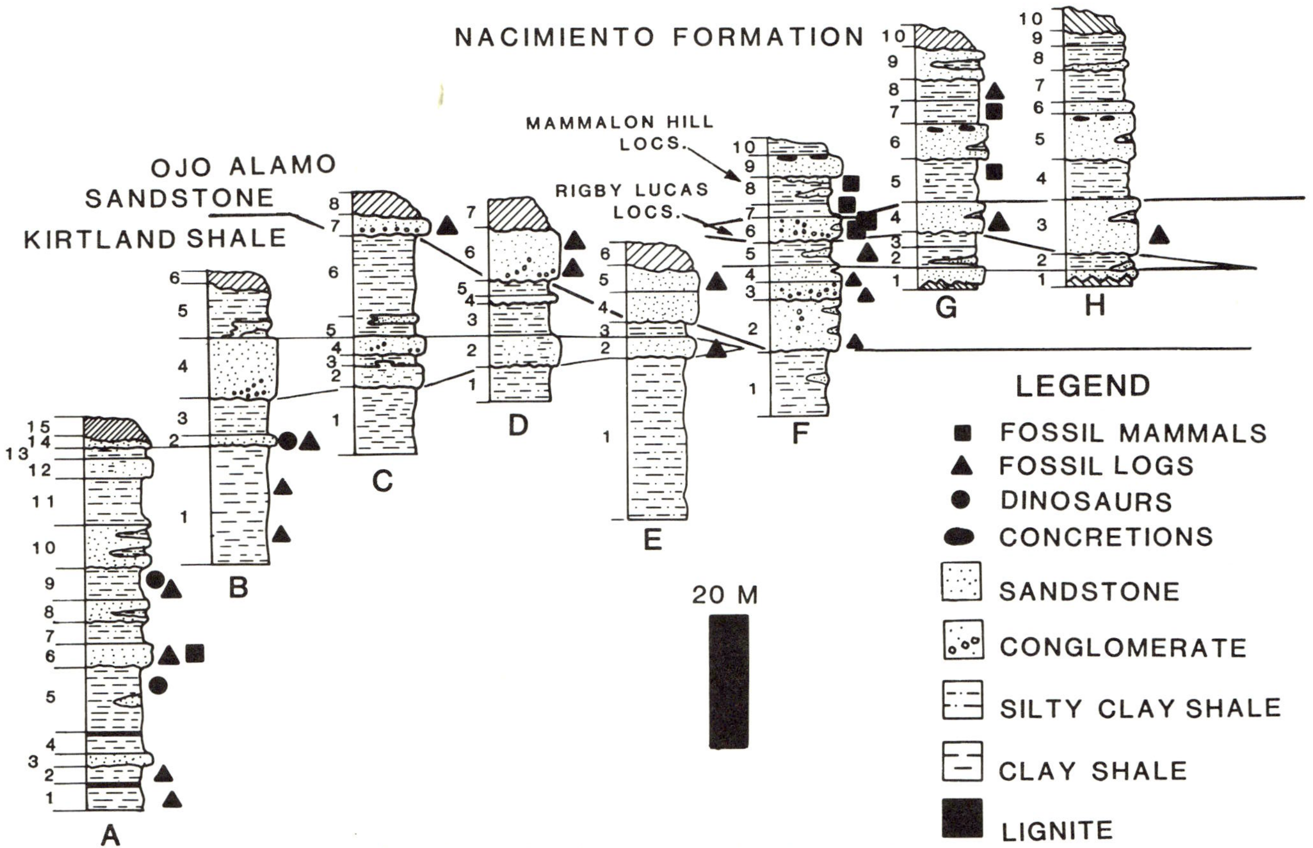

TEXT-FIG. 5.3—Measured geologic sections of rocks exposed north of Betonnie Tsosie Wash with connecting lines showing physical correlation. These sections are modified from Lucas and Rigby (1979). Detailed stratigraphic descriptions appear in Appendix 5.1.

SYSTEMATIC PALEONTOLOGY

Order CONDYLARTHRA Cope, 1881
Family PERIPTYCHIDAE Cope, 1882
Subfamily ANISONCHINAE Osborn and Earle, 1895

Remarks.—Matthew (1937, p. 152) revised Cope's original generic characters which separated *Anisonchus* from the other anisonchine genera *Hemithlaeus* and *Conacodon.* Recently, Rigby (1976, 1980) suggested, on the basis of lower molar morphology, that *Oxyacodon* is most closely related to anisonchines. D. Archibald (written communication, 1978) and Rigby (in BLM, 1979, Fig. II8b) independently identified the previously unknown upper dentition of *Oxyacodon.* These identifications confirmed the opinions of Van Valen (1978) and Hooker (1979), among others, that *Oxyacodon* is an anisonchine and not a hyopsodontid condylarth as it was previously considered. New remains of its upper dentition, currently under study by Archibald and Rigby, suggest a close relationship of *Oxyacodon* and *Conacodon.*

There is at least a bipartite division of Puercan and Torrejonian anisonchine genera: 1) *Oxyacodon*, *Conacodon* and perhaps *Haploconus*, and 2) *Anisonchus*, *Gillisonchus* and *Hemithlaeus. Oxyacodon*, *Conacodon* and *Haploconus* are most closely related to one another because they share the following derived characters: 1) The upper molar hypocone and anterior cingulum form a lingually exaggerated complete, or nearly complete, ring around the lingual base of the protocone, and 2) The protocone is lost on P^3. *Haploconus* is perhaps the most derived of all anisonchines because it has both lost the protocone on P^3 and has a substantially reduced lingual protocone slope on the upper molars and paraconid on the lower molars. *Haploconus* does not possess as complete a ring or as large an exaggeration of the hypocone base as in *Oxyacodon* and *Conacodon*, and the anterior cingulum in *Haploconus* is lingually terminated in a protostyle somewhat resembling that of *Hemithlaeus.* There is also a substantial reduction or loss of the lower molar and premolar paraconids.

Genus ANISONCHUS Cope, 1881

Remarks.—The relationships between *Anisonchus* and *Hemithlaeus* are unclear because of the inclusion of "A." *gillianus* within the genus *Anisonchus*; as Matthew (1937) suggested, the Puercan species "A." *gillianus* is more closely related to *Hemithlaeus*, whereas the Torrejonian species *A. sectorius* is substantially derived. Rigby (1976) and Van Valen (1978) suggested the strong possibility that at least two separate related groups are contained within the currently defined genus *Anison-*

chus. Species representing the two groups of *Anisonchus* (*A. sectorius* and *A. willeyi*) are found at Swain Quarry (Rigby, 1976), as well as at a number of other Paleocene localities in western North America (Russell, 1967).

Recently, Van Valen (1978) named two new species of *Anisonchus: A. athelas*, from the Gas Tank Hill locality in Emery County, Utah and *A. eowynae*, from the Wagonroad locality in Emery County, Utah. These two new species as well as *A. dracus* and *A. sectorius* were referred to Van Valen's new subgenus *Anisonchus*. A new subgenus, *Mithrandir*, was also named to include "A." *gillianus*, *A. oligistus*, *A. onostus* and *A. fortunatus*. Rigby (1980) subsequently described *A. willeyi* from the middle Paleocene (Torrejonian) Fort Union Formation at Swain Quarry, Wyoming.

A cladistic analysis of dental characters of *A. willeyi*, *A. sectorius* and *Gillisonchus gillianus* is shown in Text-figure 5.4. These species were chosen because they are representative of the various groups under consideration. Based upon this analysis, *A. fortunatus* is most closely related to *A. sectorius* because they share a number of derived features (see Table 5.1, part 5.a-e).

At present, the relationships of *A. eowynae* and *A. athelas* are unclear because their lower dentition is unknown. However, *A. eowynae* possesses the derived feature of preservation of the conules as crest confluences of the anterior and posterior conule wings and the protocone crests; its "paraconule" confluence is projected at or very near the apex of the protocone. The posterior metaconule wing interruption of the posterior cingulum of *A. eowynae* (Van Valen, 1978) appears to be a retained primitive feature because it is shared by many anisonchine species (some suggestion of which is seen in *A. willeyi*). This wing in *A. eowynae* meets the posterior cingulum but does not interrupt it as in *A. gillianus*. However, the position and separation of the hypocone, the lingual migration of the conules (or labial migration of the protocone apex), reduction of the parastyle and partial reduction of the anterior cingulum serve as derived features to unite *A. eowynae* with *A. willeyi*. *A. athelas* has a reduced lingual protocone slope, and has maintained rather prominent conules near the paracone and metacone bases. The conule wings remain high on the paracone and metacone slopes and do not approach the anterior or posterior cingula. These features serve to unite *A. athelas* and *A. sectorius*. The subgenus *Anisonchus* is here amended to include *A. sectorius*, *A. dracus*, *A. fortunatus* and *A. athelas*. The subgenus *Mithrandir* is consequently modified to include *A. oligistus*, *A. onostus*, *A. willeyi*, and *A. eowynae*.

The relationships within the genus *Anisonchus* are complicated by the inclusion of "A." *gillianus* (Text-fig. 5.4). Thus a new genus is proposed

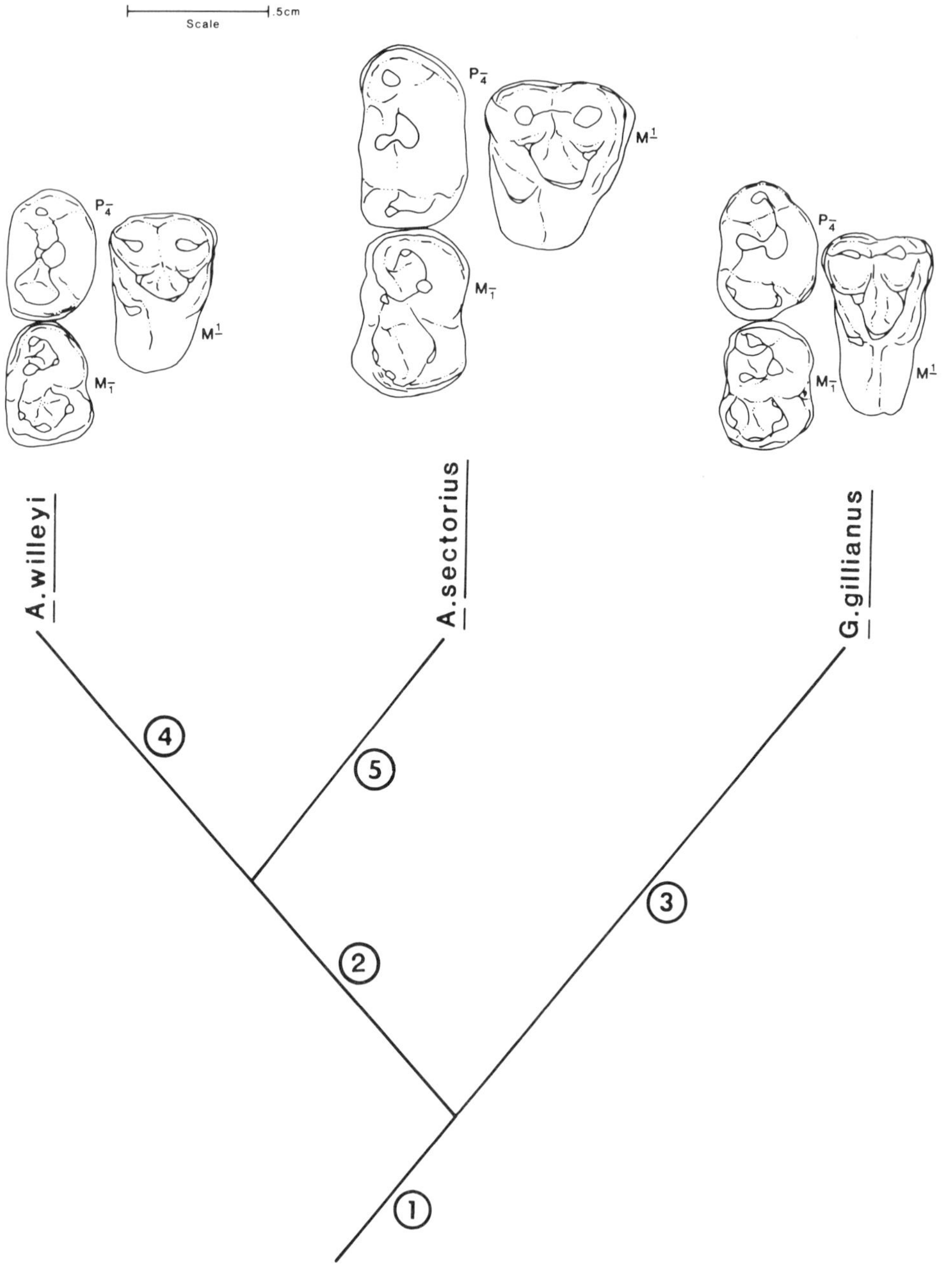

TEXT-FIG. 5.4—Cladistic relationships of *Anisonchus willeyi*, *A: sectorius* and *Gillisonchus gillianus* based on dental characters. These species are thought to be representative of at least three separate phylogenies previously contained within the genus *Anisonchus*. Line drawings of M^1 and P_4–M_1 (occlusal views) are shown for reference to the characters supporting the proposed relationships (*A. willeyi*, M^1, AMNH 87624; P_4–M_1, AMNH 100260; *A. sectorius*, M^1, AMNH 35717 (reversed), P_4–M_1, AMNH 100218 (reversed); *G. gillianus*, M^1, UNM-B029; P_4–M_1, UNM-NP1302). Numbers on the cladogram refer to groups of characters compiled in Table 5.1.

for *Anisonchus gillianus*. This reflects my judgement that "*A.*" *gillianus* is as different from other species of Anisonchus as the genera *Anisonchus*, *Conacodon*, *Haploconus*, *Hemithlaeus*, *Oxyacodon*, *Earendil*, and *Mimatuta* are from each other.

Genus GILLISONCHUS new genus

Type species.—Haploconus gillianus (Cope, 1882).

Diagnosis.—The hypocone and lingual part of the anterior cingulum closely approximate the apex of the protocone in the upper molars; the upper molar conules (crest confluences) are located near the bases of the paracone and metacone; the preparaconule wing and postmetaconule wing are confluent with, but do not interrupt, the anterior and posterior cingula; the lower premolars are inflated at the base and are compressed apically but not to the degree that the whole tooth is compressed in species of *Anisonchus*; the paraconid of P_{3-4} is low and prominent on the anterior median face.

Included species.—G. gillianus.

Etymology.—Named for the single referred species.

GILLISONCHUS GILLIANUS (Cope, 1882).
Pls. 5.1–5.8

Haploconus gillianus COPE, 1882, American Naturalist, v. XVI, p. 686.

Anisonchus gillianus COPE, 1884, Tertiary Vertebrata, pl. XXVe, plate description and figure 7; 1888, Trans. Am. Phil Soc. XVI, n.s., p. 352.

Type specimen.—AMNH 3543, fragments of upper and lower jaws and a few skeletal fragments.

Type locality.—Lower part of the Nacimiento Formation, San Juan Basin, New Mexico.

Referred material.—UNM-B029, a partial skeleton including skull fragments, upper and lower jaws (right maxilla with M^{1-3}, left edentulous maxilla, right dentary with P_3–M_3, left dentary with P_2–M_3, fragmentary axial skeleton (atlas, axis and several fragmented vertebrae), glenoid portion of scapulae, parts of the humerus, radius, pelvis, femur, tibia, fibula, and nearly complete astragalus, calcaneum, navicular, cuboid, and fragmented phalanges; UNM-NP1301, palate with left and right P^2–M^3; UNM-NP1302, right dentary fragment with P_3–M_1.

Description.—1) Skull: Most specimens referred to *G. gillianus* consist only of teeth and thus very little is known of the skull. Several specimens

(AMNH 3601, UNM–B029 and UNM–NP1301) are anterior portions of the skull. The palatal region contains an infraorbital foramen above P^3. The anterior border of the orbit is significantly excavated and terminates anteriorly above P^4 and M^1 posterior to the infraorbital foramen, as in *Haploconus angustus* (Matthew, 1937). Only the maxillary process of the zygomatic arch is preserved, and it is slender. The orbits are large and the snout above the palate appears to be highly arched as in *Haploconus.* However, the palate tapers strongly and evenly anteriorly and is not elongated with parallel dental rows, as in *Haploconus.*

2) Dentition: Most dental characters for *Anisonchus*, as well as for *G.* gillianus, are summarized in Text-figure 5.4 and Tables 5.1 and 5.2; illustrated in Plate 5.1, Plate 5.2, figure 1 and Plate 5.3, figures 1-3; and were discussed by Matthew (1937).

3) Axial Skeleton: The axial skeleton is badly weathered and the vertebrae are fragmented, with most of the arches broken beyond reconstruction. There are, however, approximately 20 centra that may be classified at least to type. These have been compared, where possible, to the only other known periptychid "condylarth" skeletons, those of *Ectoconus ditrigonus* (= *E. majusculus*) and *Periptychus carinidens* (= *P. rhabdodon*) (Matthew, 1937).

4) Cervical vertebrae: A lateral half of the atlas (Pl. 5.2, figs. 2-5) is preserved. It is similar in gross morphology to that of *Ectoconus.* However, the ventral medial portion is much more anteroposteriorly constricted, the wing-like transverse processes are thinner, posteroventrally directed, and they surround a proportionately larger transverse foramen. The atlantal foramen is large and located on the dorsal medial surface of the thin dorsal arch.

A large part of the anterior axis centrum, including odontoid process, is preserved (Pl. 5.2, figs. 6-9). It is very similar in proportions and features to that of *Ectoconus* but differs slightly in the increased dorsoventral attitude of the articular facets that contact with the atlas.

Two other questionably referred cervical centra are present. They differ from *Ectoconus* in having more broadly elliptical anterior and posterior acoelous facets. One of these centra has anteroposterior obliquely directed facets, but this is assumed to be due to distortion during preservation rather than a true morphological feature.

5) Thoracic vertebrae: Six thoracic centra are preserved in the skeleton; each preserves rib articular facets. Only one possesses part of the transverse process and associated articular facets (Pl. 5.3, figs. 4–7). The transverse process is similar to that of *Ectoconus* but differs in being relatively longer and proportionately thicker than those illustrated by Matthew (1937, Pl. XXVI, figs. 2–3).

6) Lumbar vertebrae: Two vertebral fragments are referred to the

Table 5.1—Derived characters used in the determination of cladistic relationships portrayed in Text-fig. 5.4. Polarity of characters was determined by out-group comparisons with *Oxyacodon*, *Hemithlaeus*, *Conacodon*, and *Haploconus*.

1. a. Molar pericone and lingual cingulum absent.
   b. Long relatively shallow protocone slope on upper molars.
   c. Upper molars have a relatively great transverse width.
   d. Molar conules present as confluences of crests and not as distinct cusps. (This may be a primitive anisonchine character.)
   e. Relatively large slightly protruding paraconid on P_{3-4}.
   f. Lingual and labial cingulids present on lower molars.
   g. This group may share a derived elongated neck on the astragalus, but, at present, only the astragali of *Gillisonchus gillianus* and *Anisonchus sectorius* (Matthew, 1937) are known.
   h. Lower premolars somewhat inflated but not to the extent in other periptychids.
2. a. Anterior cingulum, posterior cingulum and hypocone located low on the protocone slope.
   b. Hypocone enlarged and significantly separated from protocone apex.
   c. Hypocone base distinct from protocone base and forms a slight to noticable lobe on the postero-lingual molar profile.
   d. Upper molar transverse width reduced when protocone slope is steepened, resulting in the reduction of the lingual hypocone projection (reduced less in *A. eowynae*).
   e. "Conules" reduced due to reduced definition of conule crests.
   f. Protocone apex shifted anteriorly from medial anteroposterior position (very slightly in *A. eowynae* and *A. athelas*).
   g. Anterior cingulum does not extend to lingual margin of upper molars.
   h. Distance from protocone apex to paracone and metacone has been lessened, resulting in a subquadrate to subtriangular upper molar form.
   i. Lower premolars somewhat laterally compressed from condition seen in *G. gillianus* (unknown in *A. eowynae* and *A. athelas*).
   j. P_{3-4} talonid partially reduced (unknown in *A. eowynae* and *A. athelas*).
   k. Molar paraconid medially and not lingually placed as in *G. gillianus* (unknown in *A. eowynae* and *A. athelas*).
   l. Cusps of lower molar trigonid much more closely approximate each other, forming a more unified trigonid cone (unknown in *A. eowynae* and *A. athelas*).
3. a. Small hypocone and lingual part of anterior cingulum closely approximate protocone apex of upper molars.
   b. Upper molar conules (crest confluences) located near base of paracone and metacone.
   c. Lower premolars with a partially inflated base but base is not laterally compressed as in *A. sectorius* and *A. willeyi*.
   d. Paraconid of P_{3-4} low, medial, and prominent on anterior face.
4. a. Upper molar conules (crest confluences) located on crests of protocone and not at base of paracone and metacone.
   b. "Paraconule" confluence substantially higher on protocone slope than "metaconule." The projected confluence is at or very near the protocone apex in M^{2-3}, similar to the condition seen in *Oxyacodon*.
   c. Paraconid of P_{3-4} differentiated slightly from protoconid, yet positioned very high on protoconid slope.
   d. Lower premolar talonid posterior margin formed by a simple crest with little or no cusp development.

e. Lower molar trigonids somewhat restricted, forming a "closed" basin with a more posteriorly placed paraconid and better developed crests on the posterior slope of the paraconid and anterior slope of the metaconid.
f. Hypoconulid reduced in proportionate size compared to other talonid cusps, and medially placed.

5. a. Largest of *Anisonchus* species.
b. Upper molars show substantial "conule" reduction, yet retain them at or very near the bases of the paracone and metacone.
c. Hypocone further enlarged (see 2b and c above); its base forming a distinct lobe at the posterior lingual margin of the upper molar teeth.
d. Upper molar anterior cingulum uninterrupted by anterior paraconule wing.
e. Substantial shortening of distance between protocone and paracone and metacone.
f. P_{3-4} paraconid substantially enlarged and differentiated at protocone base similar to *G. gillianus*.
g. Lower molar talonids proportionately larger.
h. Lingual and labial cingulids generally lost (somewhat variable).
i. Lower molar hypoconulid reduced and shifted lingually.

Table 5.2—*Gillisonchus gillianus*. Measurements of dental elements from UNM locality B77-7 (all measurements in mm).

UNM-NP1301	Left		Right	
	L	W	L	W
P^2	4.2	3.9	4.1+ *	3.8
P^3	4.95	5.1	4.8	4.95
P^4	4.5	6.05	4.35	6.0
M^1	4.15	5.60	4.15	5.60
M^3	4.3*	7.0+ *	4.30	7.40
M^3	3.6	6.1	3.60	6.05
UNM-B029				
M^1			3.78	6.0
M^2			3.90	7.5
M^3			3.75	7.3

UNM-B029	L	AW	PW	L	AW	PW
P_2	4.18	2.88	——	——	——	——
P_3	4.32	3.15	——	4.28	3.1	——
P_4	4.55	3.62	——	4.50	3.58	——
M_1	4.20	3.06	3.12	4.26	3.10	3.12
M_2	4.23	3.80	3.85	4.22	3.78	3.78
M_3	5.62	3.70	——	5.70	3.76	——
UNM-NP1302						
P_3				4.66	3.45	——
P_4				4.76	3.72	——
M_1				4.48	3.05	3.22

*Denotes an approximate measurement because of broken or damaged specimens.

lumbar series from the skeleton. Only one preserves the pleurapophysis (Pl. 5.4, figs. 1–3), which is much more posteriorly directed than the presacral lumbar vertebra of *Ectoconus* illustrated by Matthew (1937, Pl. XXVI, fig. 1). This suggests that the *G. gillianus* specimen is an anterior lumbar vertebra, but the specimen is too poorly preserved to allow additional comparisons.

7) Sacral vertebrae: Two fragments (one is figured, Pl. 5.2, figs. 10–11) are referred to the sacrum, which is unfused (that of *Ectoconus* is fused). What may be the functional counterpart of the sacral intervertebral foramen is surrounded by the transverse process. Not enough of the vertebra is present to allow for additional comparison, but the presence of such a foramen is unusual.

8) Caudal vertebrae: Two proximal (one is figured, Pl. 5.3, figs. 8–11), two intermediate (one is figured, Pl. 5.3, figs. 12–14), and two fragments of distal caudal vertebrae are preserved. The caudal transverse processes of the proximal vertebrae are blade-like flanges that curl ventrally in their intermediate caudal portion. These flanges are substantially reduced or absent in the distal parts of the tail. The caudal vertebrae are very long in comparison to other vertebrae, suggesting a long tail. The most proximal of the vertebrae preserves either a groove or foramen in or near the base of the transverse process and near the midportion of the centrum, similar to the foramen present on fragments of the nonfused sacrum.

9) Scapula: Left and right glenoid fragments (Pl. 5.4, figs. 4–6) show a number of differences from the scapula of *Ectoconus*. The coracoid process in *Gillisonchus* is distally lengthened and parallels much of the glenoid curvature. A small facet is present on the lateral distal edge of the process for articulation with the clavicle. There is a scapular restriction posterior to the glenoid area in *Ectoconus*, which is not present to the same degree in *Gillisonchus*. Most other observable features are identical to those of *Ectoconus*, including the distal termination of the scapular spine.

10) Humerus: Although incomplete (Pl. 5.4, figs. 10–15), the humerus was probably 4 cm long and most closely resembles the humerus of *Periptychus*. The humerus of *Ectoconus* has an expansion in the distal third of the ectepicondylar crest, which is not seen in either *Gillisonchus* or *Periptychus*. The entepicondylar projection is proportionately larger in *Gillisonchus* and *Ectoconus* than in *Periptychus*. The radial trochlea is more rounded and ball-like in *G. gillianus*. The humeral head is substantially different in *Gillisonchus* in that all of the tuberosities are much more clearly defined than in *Periptychus* and *Ectoconus*.

11) Clavicle: A small clavicle (Pl. 5.4, figs. 7–9) is present in *G. gillianus*, as it is in *Ectoconus*. In the *Gillisonchus* skeleton, however,

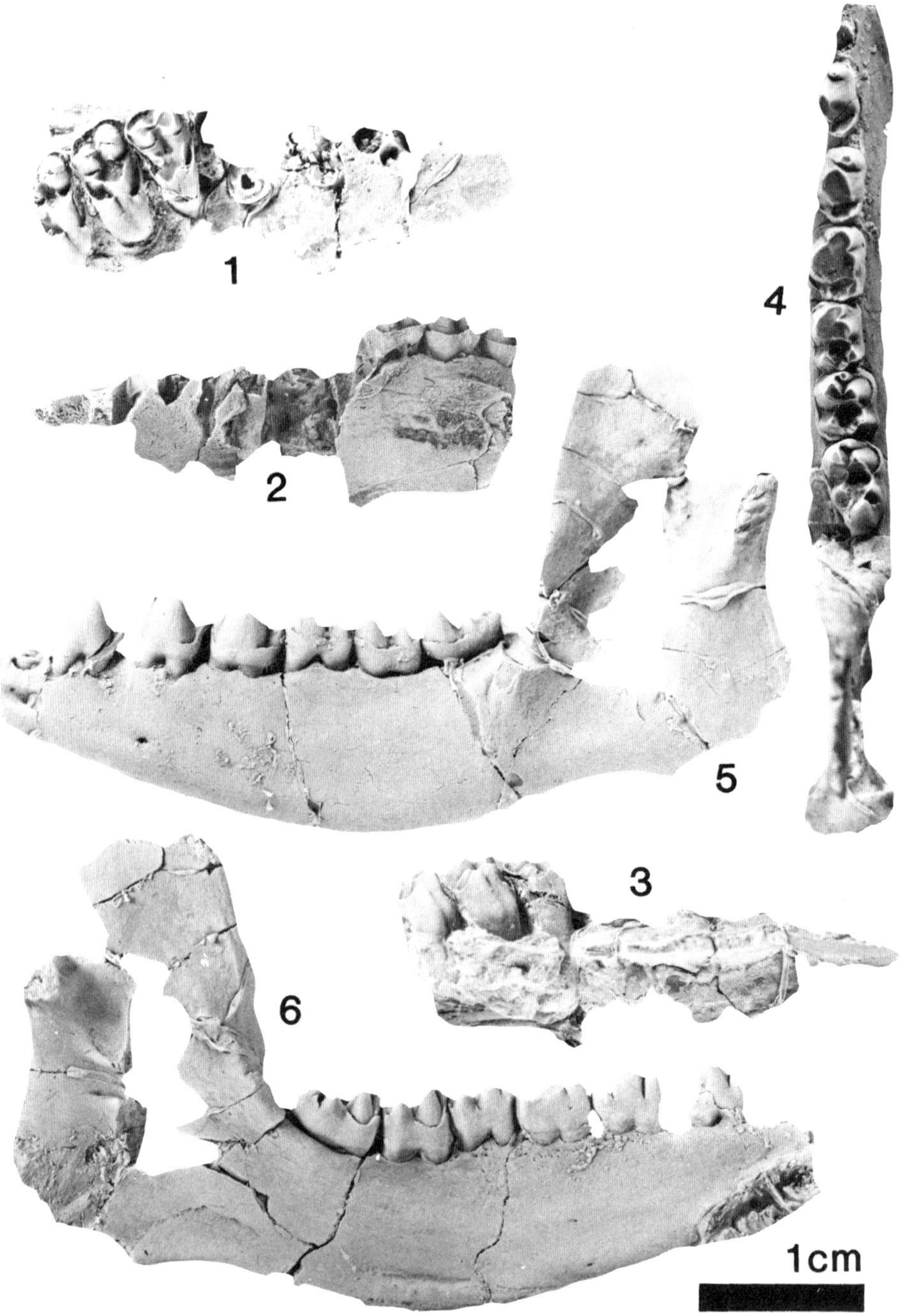

EXPLANATION OF PLATE 5.1

Gillisonchus gillianus (Cope) n. gen.

FIGS. *1–6*—UNM-B029, skeleton.

Maxilla. *1*, occlusal view; *2*, labial view; *3*, lingual view.

Dentary. *4*, occlusal view, *5*, labial view; *6*, lingual view.

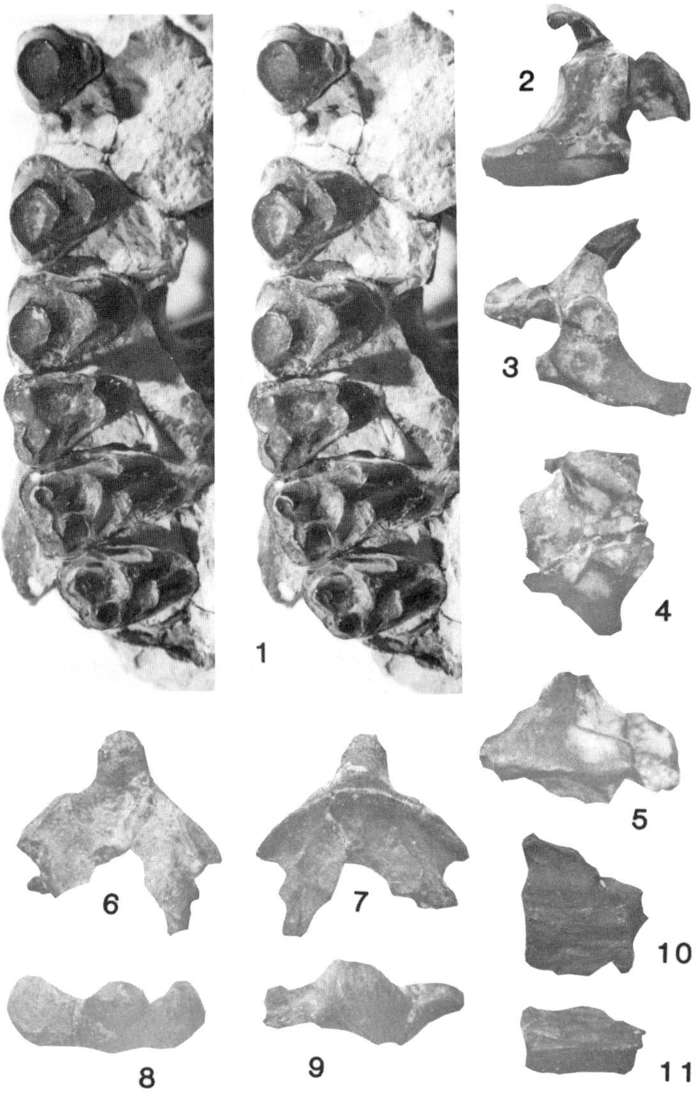

EXPLANATION OF PLATE 5.2

Gillisonchus gillianus (Cope) n. gen.

FIG. *1*—UNM-NP1301, Right side of palate illustrated (stereo pair) with P^2–M^3.

2–11—UNM-B029, skeleton.

Atlas. *2*, proximal view; *3*, distal view; *4*, lateral view; *5*, ventral view. All × 3.

Axis. *6*, dorsal view; *7*, ventral view; *8*, proximal view; *9*, lateral view. All × 3.

Sacral vertebra. *10*, dorsal view; *11*, lateral view. All × 3.

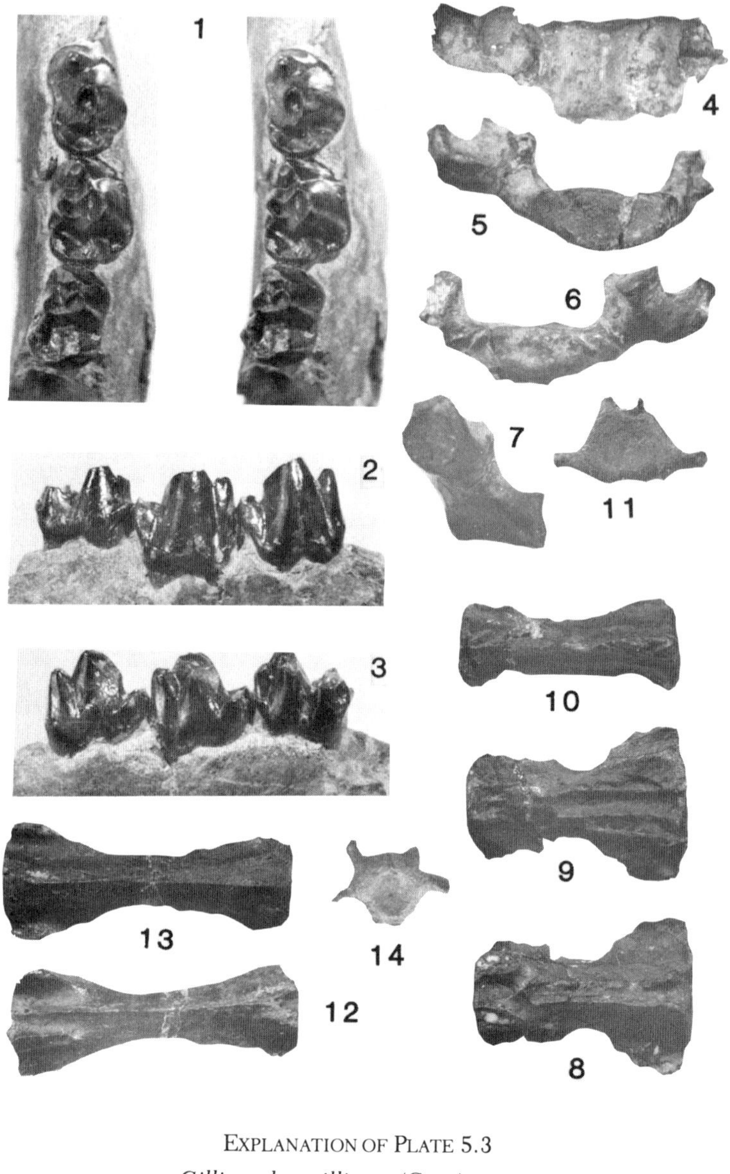

EXPLANATION OF PLATE 5.3

Gillisonchus gillianus (Cope) n. gen.

FIGS. *1–3*—UNM-NP1302, dentary fragment with P_3–M_1; *1*, occlusal view, stereo pair. *2*, labial view. *3*, lingual view. All × 3.

4–14—UNM-B029 skeleton.

Thoracic vertebra. *4*, dorsal view; *5*, proximal view; *6*, distal view; *7*, lateral view. All × 3.

Proximal caudal vertebra. *8*, dorsal view; *9*, ventral view; *10*, lateral view; *11*, proximal view. All × 3.

Intermediate caudal vertebrae. *12*, dorsal view; *13*, ventral view; *14*, proximal view. All × 3.

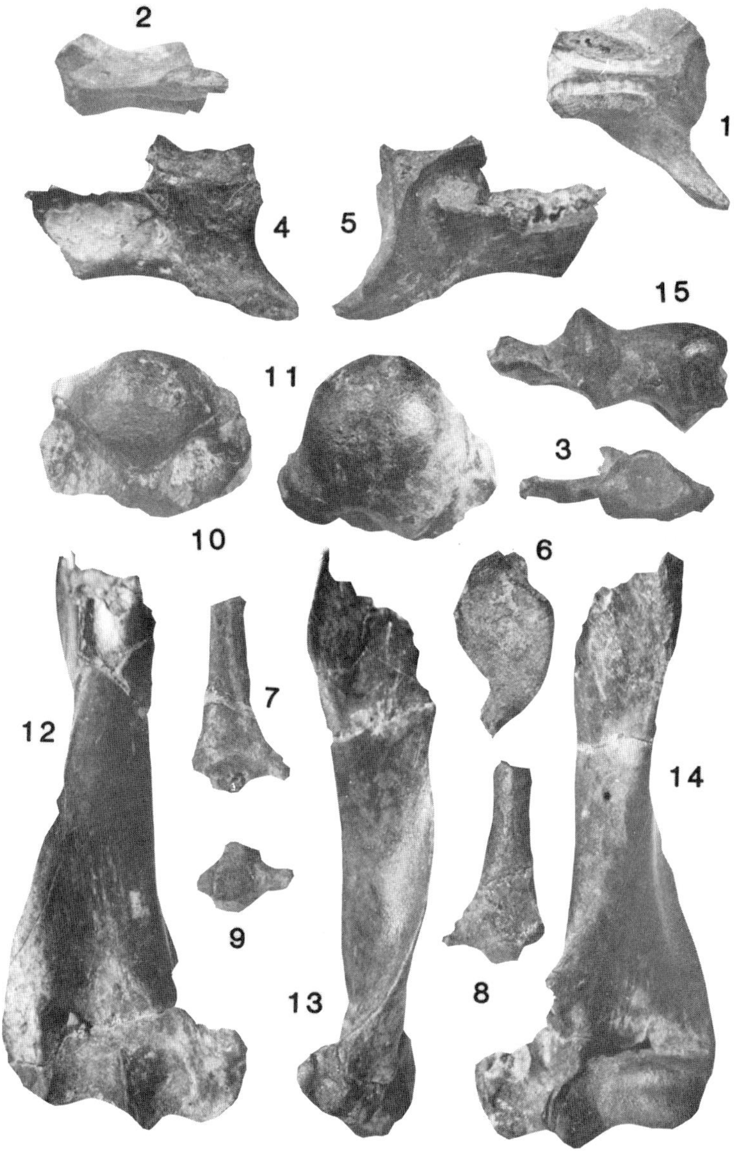

EXPLANATION OF PLATE 5.4

Gillisonchus gillianus (Cope) n. gen.

FIGS. *1–15*—UNM-B029, skeleton.

Lumbar vertebra. *1*, dorsal view; *2*, lateral view; *3*, distal view. All × 3.

Scapula. *4*, internal view; *5*, external view; *6*, glenoid cavity; distal view. All × 3.

Clavicle. *7*, internal view. *8*, external view; *9*, proximal view. All × 3.

Humerus. *10*, humeral head internal view; *11*, humeral head dorsal view; *12*, posterior view; *13*, external view; *14*, anterior view; *15*, distal view. All × 3.

only the proximal half is preserved. The clavicle differs from that of *Ectoconus* in having the articular facets better defined and in having one thin process and another thicker tuberosity near the articular facet, both of which are lacking in *Ectoconus*.

12) Radius: The radius of *G. gillianus* (Pl. 5.5, figs. 6–9) is proportionately more slender than that of *Ectoconus* or *Periptychus*. All three forms have an elliptical humeral facet and massive distal trochlear head. Although only fragments are present, the length is estimated to have been 3 cm or longer. The distal radial head shows two well developed facets for the scaphoid and lunar, like those in *Ectoconus*.

13) Pelvis: Fragments of right and left inominate bones are preserved. They appear to differ substantially from those of *Pantolambda* (Matthew, 1937). Although the evidence is somewhat fragmentary, the large projections anterior to the acetabular fossa in *Pantolambda* do not occur in *G. gillianus*. There is, however, a thin posterior bladelike projection which appears to be much more slender than that of *Pantolambda*.

14) Femur: Fragments of both left and right femora are present (Pl. 5.5., figs. 1–5; Pl. 5.6, figs. 1–6). If complete, they would have been at least 5.5–6.0 cm long. There is an enormous digital fossa located proximally between the greater trochanter and the femoral head, similar to that in *Ectoconus* but developed to a much greater degree. The excavation of the digital fossa extends to the base of the lesser trochanter, which projects posteriorly more than in either *Ectoconus* or *Periptychus*. The greater trochanter is substantially separated from the femoral head and projects above it. The third trochanter is located similarly to that of *Ectoconus*. The patellar trochlea is incised as in *Periptychus* and is not flattened as it is in *Ectoconus*. A groove separates the tibial trochlear facets and is deeply incised, as in *Ectoconus*, but differs from the relatively flat laterally continuous unseparated trochlea of *Periptychus*.

15) Tibia and Fibula: The proximal head of the tibia and cnemial crest are expanded very similar to the condition in *Ectoconus*. Very few differences exist between the tibia of *G. gillianus* (Pl. 5.7, figs. 1–6; Pl. 5.8, figs 1–2) and that of *Ectoconus*, even in the morphology of the distal articular facets. The proximal part of the fibula is unknown in *G. gillianus*, but it would probably be identical to that of *Ectoconus*. The distal fibula (Pl. 5.6, figs. 7–10) of *G. gillianus* has a slightly larger and more posteriorly projecting anteroexternal crest than in *Ectoconus*.

16) Pes: Significantly, study of the skeleton of *Gillisonchus gillianus* provides additional information regarding the tarsus of early ungulates. For tarsal comparisons I have relied upon the works of Matthew (1937), Schaeffer (1947), Szalay & Decker (1974), and Szalay (1977). Postcranial remains of primitive mammals are rarely found in association with dental material which, to a large extent, has formed the basis of

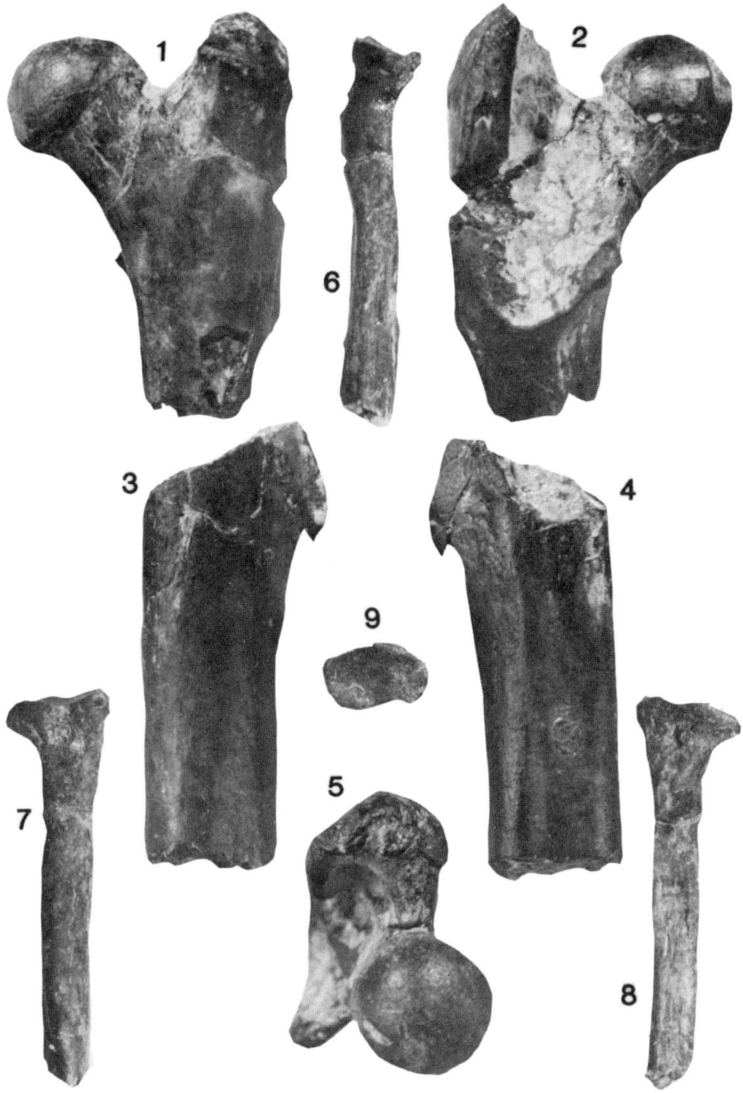

EXPLANATION OF PLATE 5.5

Gillisonchus gillianus (Cope) n. gen.

FIGS. *1–9*—UNM-B029, skeleton.

Femur. *1* and *3*, anterior view; *2* and *4*, posterior view; *5*, proximal view. All × 3.

Proximal radius. *6*, external view; *7*, anterior view; *8*, posterior view; *9*, proximal view. All × 3.

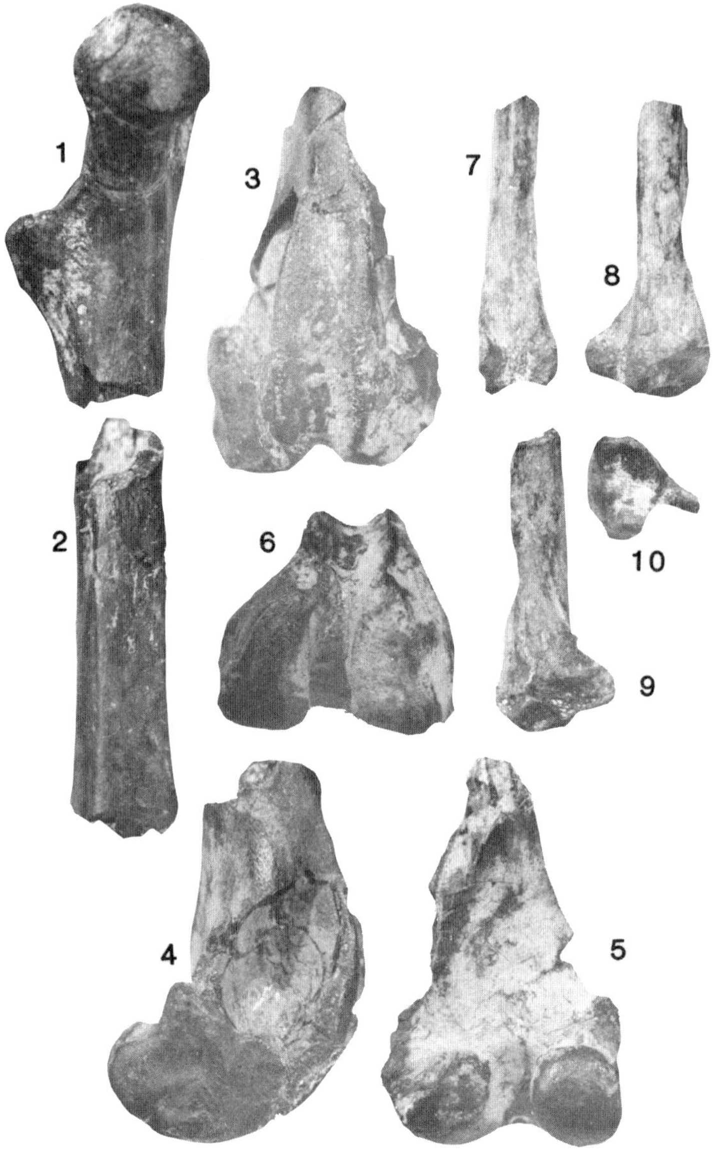

EXPLANATION OF PLATE 5.6

Gillisonchus gillianus (Cope) n. gen.

FIGS. *1–10*—UNM-B029, skeleton.

Femur. *1*, *2*, and *4*, internal view; *3*, anterior view; *5*, posterior view; *6*, distal view. All × 3.

Fibula. *7*, anterior view; *8*, external view; *9*, internal view; *10*, distal view. All × 3.

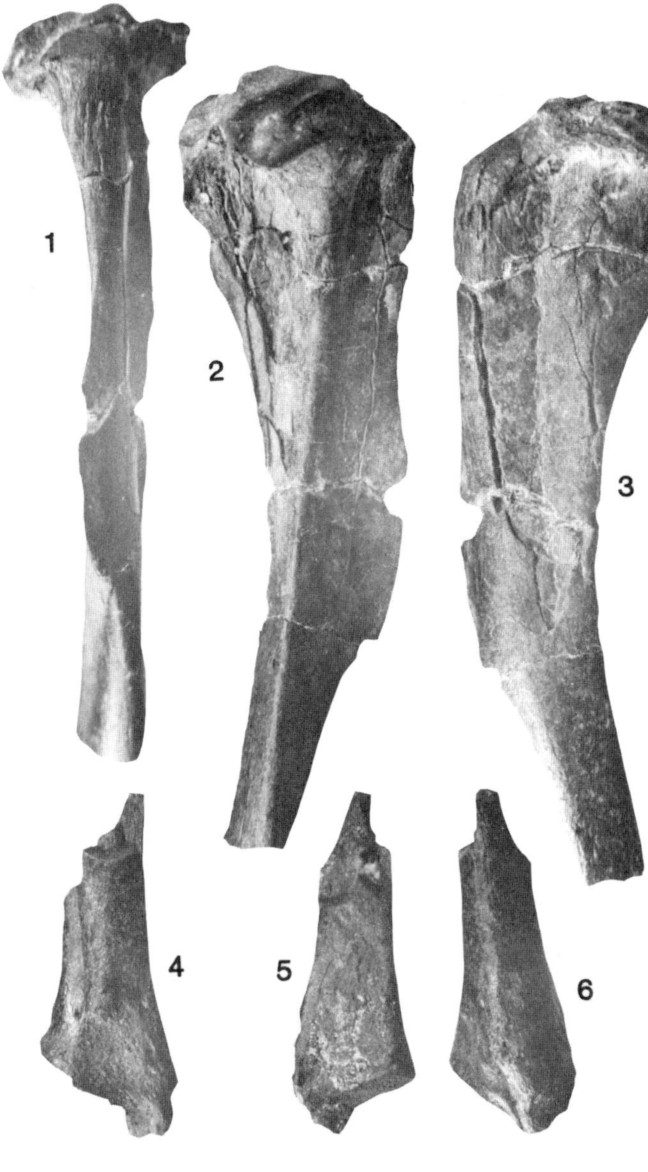

EXPLANATION OF PLATE 5.7

Gillisonchus gillianus (Cope) n. gen.

FIGS. *1–6*—UNM-B029, skeleton.

Tibia. *1* and *4*, anterior view; *2* and *5*, internal view; *3* and *6*, external view. All × 3.

fossil mammalian systematics. A potential problem exists in the identification of tarsal elements, particularly in *Protungulatum* and *Procerberus* for which, to my knowledge, no articulated remains have been found (Szalay & Decker, 1974). Comparisons using these postcranial remains are typological in nature because of the general lack of adequate samples with which to gauge variability. Thus, at present, little is known about the morphologic variability of tarsal elements of primitive mammals.

In general proportions, the *Gillisonchus* astragalus (Text-fig. 5.5; Pl. 5.8, figs. 3–8) is similar to that of *Protungulatum*. Most of this similarity is probably due to shared primitive characters. The neck and head of the astragalus appear to be elongated compared with other known primitive mammals. The ratio of tibial trochlear length to astragalar length is 0.43 in *Gillisonchus* compared to 0.54 for *Protungulatum*, 0.49 for *Procerberus*, and 0.66 for *Plesiadapis*, as reported by Szalay and Decker (1974). The *Gillisonchus* tibial trochlear facet is very broad and is approximately as wide as the astragalar head. This feature is considered a primitive character because other assumed derived forms, such as plesiadapid primates, have a proportionately reduced (25%) astragalar head width. The astragalar tibial facet crests of *Gillisonchus*, particularly the lateral crest, are low and somewhat rounded, similar to *Protungulatum*. *Procerberus* has an expanded and sharpened lateral crest border as a derived feature, and plesiadapid primates have elevated and sharpened crest borders (Szalay, 1977). Such trochlear definition is considered to be derived with respect to the primitive eutherian state.

The angle that the median groove of the astragalar tibial trochlea makes with the main axis of the astragalus and the angle that this same groove makes with the long axis of the calcaneum seems to be an important character: in most mammals the medial trochlear groove is oriented

EXPLANATION OF PLATE 5.8

Gillisonchus gillianus (Cope) n. gen.

FIGS. *1–25*—UNM-B029, skeleton.

Tibia. *1*, distal view; *2*, proximal view. All ×4.

Astragalus. *3*, dorsal view; *4*, external view; *5*, internal view; *6*, ventral view; *7*, distal view; *8*, proximal view. All ×4.

Calcaneum. *9*, dorsal view; *10*, internal view; *11*, external view; *12*, ventral view; *13*, posterior view; *14*, anterior view. All ×4.

Navicular. *15*, proximal view; *16*, distal view; *17*, external view; *18*, internal view; *19*, ventral view; *20*, dorsal view. All ×4.

Cuboid. *21*, proximal view; *22*, distal view; *23*, dorsal view; *24*, internal view; *25*, ventral view. All ×4.

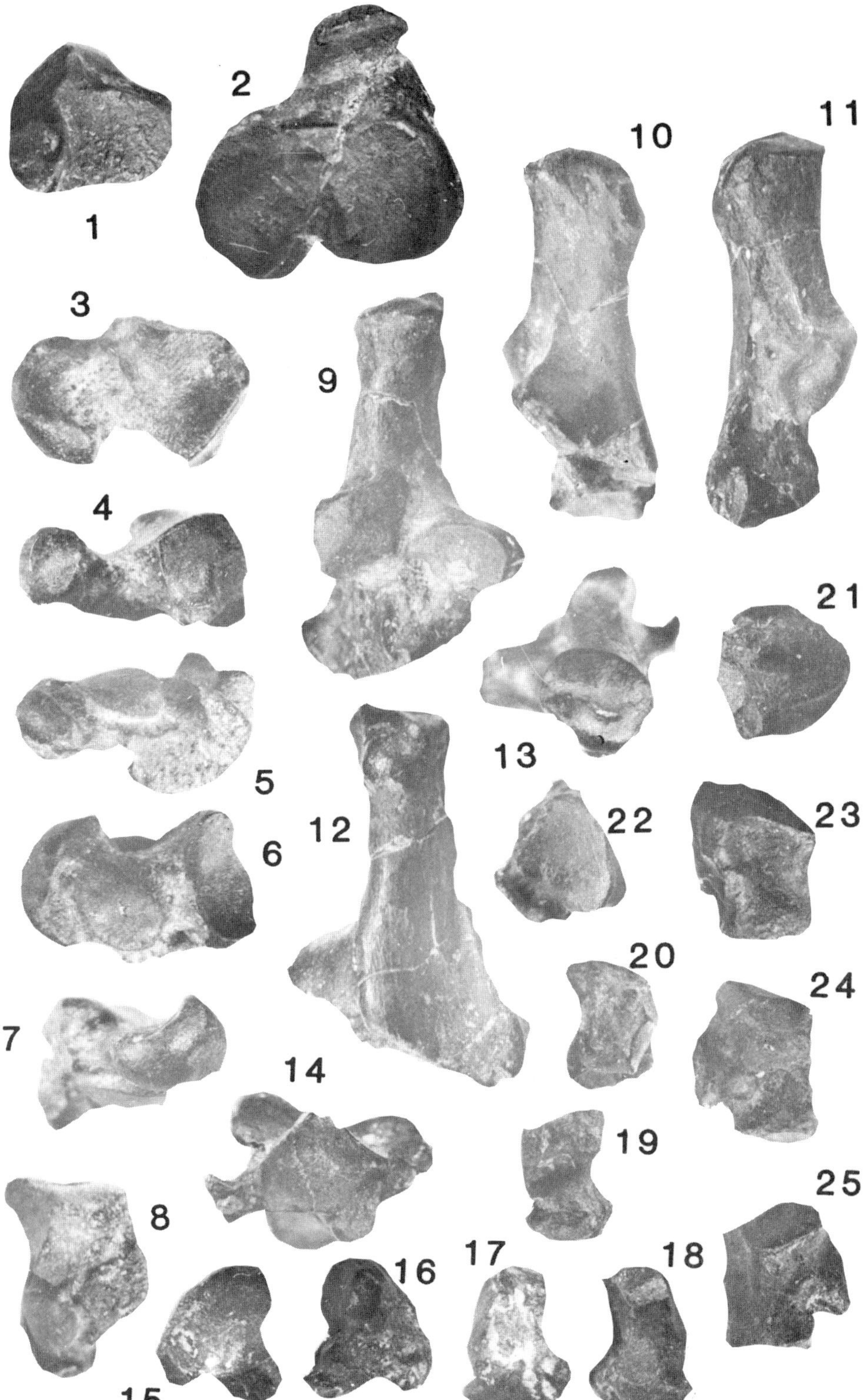
1
2
3
4
5
6
7
8
9
10
11
12
13
14
15
16
17
18
19
20
21
22
23
24
25

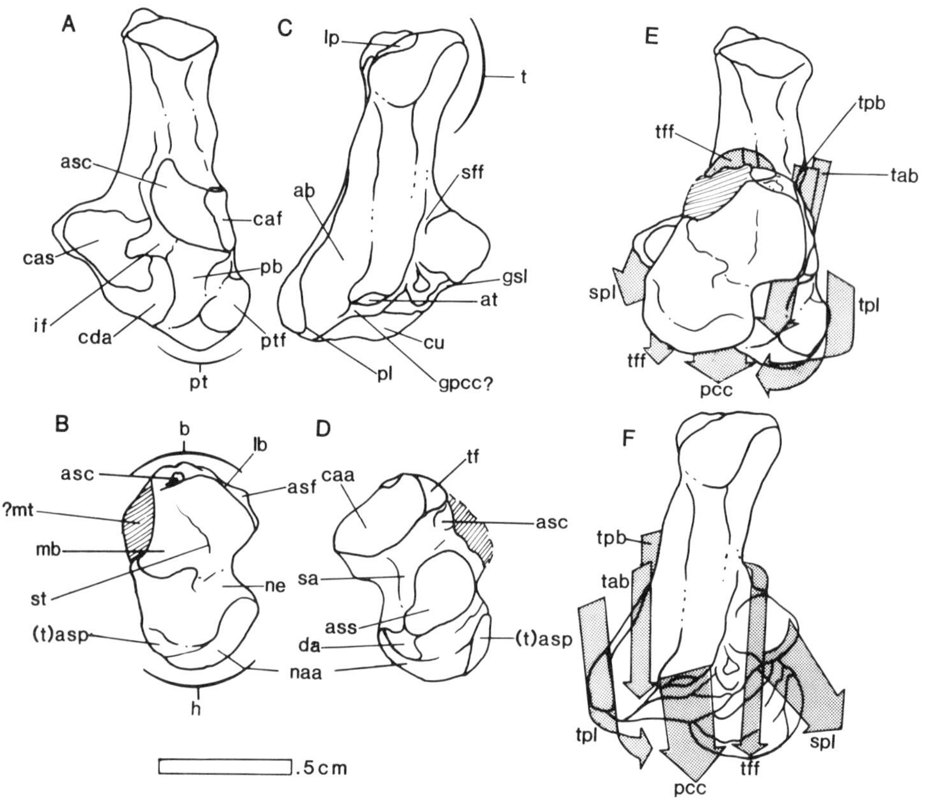

A
asc
caf
cas
pb
if
cda
ptf
pt
C
lp
t
ab
sff
gsl
at
cu
pl
gpcc?
E
tpb
tff
tab
spl
tpl
tff
pcc
B
b
lb
asc
asf
?mt
mb
st
ne
(t)asp
naa
h
D
tf
caa
asc
sa
ass
da
(t)asp
naa
F
tpb
tab
tpl
spl
tff
pcc
.5cm

TEXT-FIG. 5.5—Morphology and some muscle-tendon relationships of the astragalocalcaneal complex of *Gillisonchus gillianus*. Dorsal (*A*) and ventral (*C*) view of the calcaneum; dorsal (*B*) and ventral (*D*) views of the astragalus; and dorsal (*E*) and ventral (*F*) views of the astragalocalcaneal complex. The nomenclature used is that of Schaeffer (1947) and Szalay & Decker (1974).

Abbreviations seen on the calcaneum: *ab*, groove for m. abductor digiti quinti?; *asc*, astragalocalcaneal facet; *at*, anterior plantar tubercle; *caf*, calcaneal fibular facet; *cas*, calcaneal sustentacular facet; *cda*, distal astragalocalcaneal facet; *cu*, cuboid facet; *gpcc*?, groove for plantar calcaneocuboid ligament; *gsl*, sustentacular groove for attachment of "spring" ligament; *if*, interosseous fossa; *lp*, lateral process of tuber calcanei; *pb*, groove for tendon of m. peroneus brevis; *pl*, groove for tendon of m. peroneus longus; *pt*, peroneal tubercle; *ptf*, flange of the peroneal tubercule; *sff*, sustentacular groove for tendon of m. flexor digitorum fibularis; *t*, tuber of the calcaneum.

Abbreviations seen on the astragalus: *asc*, astragalar canal; *asf*, astragalar fibular facet; (*t*) *asp*, astragalar "spring" ligament facet or tibiale facet; *ass*, astragalar sustentacular facet; *b*, body; *caa*, calcaneoastragalar facet; *h*, head; *lb*, lateral border (crest) of trochlea; *mb*, medial border (crest) of trochlea; *mt*?, medial tibial facet of trochlea; *naa*, naviculoastragalar facet; *ne*, neck of astragalus; *sa*, sulcus astragali; *st*, superior tibial facet of trochlea; *tf*, trochlear groove for tendon of m. flexor digitorum fibularis.

Abbreviations seen on the astragalocalcaneal complex: *pcc* (long) plantar calcaneocuboid ligament; *spl*, "spring" ligament (calcaneonavicular ligament); *tab*, tendon or fleshy fibers of abductor digiti quinti or other muscles; *tff*, tendon m. flexor (digitorum) fibularis; *tpb*, tendon m. peroneus brevis; *tpl*, tendon m. peroneus longus.

parallel to the long axis of the calcaneum; thus the angle that it makes with the long axis of the astragalus is a measurement of the lateral displacement of the astragalar head. *Protungulatum* has a displacement of approximately 20–25°, *Gillisonchus* 20–25°, *Plesiadapis* approximately 20°, *Diacodexis* 2–5°, *Claenodon* 15°, and *Choeroclaenus* 30° (Schaeffer, 1947; Szalay & Decker, 1974).

Those animals that have a trochlear-astragalar axis angle beyond 25° appear to be derived with respect to primitive ungulates. Most arctocyonid "condylarths" have reduced the angle. This lends some support to the argument that some of the arctocyonids should be considered the sister group of the artiodactyls.

Gillisonchus possesses a large astragalar foramen similar in proportion, size and orientation to that of *Protungulatum*. Schaeffer (1947) concluded that the presence of tissues in the foramen did not substantially reduce the movement arc of 75° for *Claenodon*. He noted that the trochlear surfaces continued proximally from the foramen for some distance. A similar relationship for *Protungulatum* was noted by Szalay & Decker (1974). However, such trochlear extensions do not appear to continue past the foramen in *Gillisonchus* and *Ectoconus*. This feature, in fact, may be derived for some periptychid condylarths with respect to primitive ungulates.

The astragalus of *G. gillianus* has an elongated neck and an enlarged and more clearly differentiated head when compared to *Protungulatum* but is not as restricted as in *Procerberus* or *Claenodon*. The *Gillisonchus* astragalus possesses a narrow, deeply incised, interarticular sulcus that is confluent with the astragular foramen, similar to that of *Protungulatum*. *Choeroclaenus* retains the narrow dimension of the sulcus while the depth of the incision is reduced. *Claenodon* and *Procerberus* have a broadened sulcus, and the depth of the incision is reduced (Szalay & Decker, 1974).

The sustentacular facet of *Gillisonchus* is elongated and confluent with a distal astragalar facet which proximally borders, although it is not confluent with, the naviculoastragalar facet. The facet in *Protungulatum* is similar, although less distinct as a separate facet. The presence of a distal astragalar facet is probably a shared primitive character in ungulates and may even be a shared primitive character for eutherians. *Procerberus* has lost the distal facet and has a differentiated distal border of the sustentacular facet. *Claenodon* and *Choeroclaenus* both have the distal facet, which is enlarged in *Choeroclaenus*. The sustentacular facet is elongated in *Gillisonchus*, as it is in *Protungulatum* and other primitive ungulates; however, the angle formed by the long axis of the facet elongation and the main axis of the astragalus through the neck and head is substantially different. That angle in *Protungulatum*, *Choeroclaenus*, and *Claenodon* is approximately 30°, whereas in *Gillisonchus* it is approximately 45°. *Procerberus* does not have a substantially elongated sustentacular facet, but it more closely approximates the naviculoastragalar facet than any primitive ungulate.

The naviculoastragalar facet is prominent and large in *Gillisonchus* and is dorsoventrally and mediolaterally wrapped around the astragalar head. It is largest and most bulbous on the medioventral and distal surface and tapers laterally and ventrally. This facet is much more concentrated on the medial surface of the head in early primates and is not as drawn out as in the ungulates. *Choeroclaenus* displays a similar condition. However, *Claenodon* has a substantially reduced facet, in all parameters, and the facet curvature is also substantially reduced, representing a derived feature when compared to the primitive condition seen in *Protungulatum* and *Gillisonchus*. A second facet, designated the astragalar ligament facet by Szalay & Decker (1974), is also present in *Gillisonchus*.

Matthew (1937, Pl. V, fig. 2) illustrated for *Claenodon*, and Schaeffer (1947) suggested that an additional ossified element may be present in *Claenodon* as the primitive ungulate condition. A tibiale, possibly present as a reptilian holdover, is present in *Claenodon* (Schaeffer, 1947), and is probably present in *Gillisonchus*, *Ectoconus*, *Protungulatum*, and *Procerberus*. All have an appropriate facet on the astragalus, but the tibiale is not known. Even in forms where the "complete" foot is known (*Ectoconus*), such a bone is not present; if present, it would have had to form at least a singly attached ossification, as in *Claenodon*, because the navicular and cuboid of both *Ectoconus* and *Gillisonchus* show no additional facets. Szalay & Decker (1974) identified the facet as associated with the calcaneonavicular ligament ("spring" ligament) in plesiadapid primates. This facet is large in plesiadapid primates, is somewhat smaller in *Procerberus*, and is smaller and much more poorly defined in *Protungulatum*. Whether the tibiale or the "spring" ligament was the structure associated with the observed facet in *G. gillianus* is not known. In *Gillisonchus* the facet is large and only slightly smaller in total surface area than the naviculoastragular facet. The large size of this facet may be another feature unique to the periptychid "condylarths."

The calcaneoastragalar facet is longer and more circular in *Gillisonchus* than in *Protungulatum* and *Procerberus*, although all possess about the same amount of curvature. The facet is slightly broader in *Gillisonchus* than in *Protungulatum*, but about the same as in *Procerberus*. It is narrower in *Claenodon* and is not as long in *Choeroclaenus*. In *Gillisonchus* the facet forms a 55-60° angle with the long axis of the neck and head of the astragalus. This feature is shared by *Protungulatum*, *Claenodon*, and *Choeroclaenus*. This angle is significantly less (30–35°) in plesiadapid primates and somewhat greater in *Procerberus* (60-65°).

The astragalar facet in *Gillisonchus* is similar to that in *Protungulatum* and *Procerberus* in that it retains a substantial continuation onto the calcaneum. However, in *G. gillianus* there is a sharp differentiation between the calcaneal astragalar facet and the calcaneal portion of the astragalar fibular facet. This differentiation is not seen in *Protungulatum*.

The calcaneum of *G. gillianus* (Text-fig. 5.5; Pl. 5.8, figs. 9–14) shows a striking resemblance to that of *Protungulatum*. This gross morphological similarity is undoubtedly due to a large number of shared primitive ungulate and eutherian characters.

The astragalocalcaneal facet forms a lateral angle of 45° with the medial long axis of the calcaneum in *Gillisonchus* but only 30° in *Protungulatum*, *Procerberus*, and *Claenodon*. Pleasiadapid primates have a substantially reduced angle of 10–15°. In *G. gillianus* the dorsoventral angle of the facet, formed by a line from the distal and proximal end points of the facet and the medial long axis of the calcaneum, is 25–30° and is virtually the same in *Protungulatum*.

Gillisonchus possesses a well-defined small calcaneal fibular facet, which is continuous with the same functional facet on the astragalus, as in *Protungulatum*. The facet is small and sharply differentiated, in contrast to the more rounded margin of the astragalocalcaneal facet in *Protungulatum*. The facet is not present in *Procerberus*. The sustentacular facet is moderately to strongly concave, in contrast to the shallow and progressively less extensive facets in *Protungulatum*, *Choeroclaenus* and *Claenodon*.

The peroneal tubercule in *Gillisonchus* is as large as or larger than that of *Protungulatum*. Most other primitive ungulates preserve some part of the tubercule. Frequently the only part preserved is the bladelike flange located at or near the distal lateral calcaneal border. The groove for the plantar calcaneocuboid ligament, present in *Protungulatum*, is almost lost in *Gillisonchus*; however, the small pit for the attachment of the ligament, present in *Protungulatum*, is substantially larger in *Gillisonchus* and somewhat laterally displaced. Many ungulates, such as *Hyopsodus*, *Claenodon*, *Periptychus*, and to a lesser extent, *Ectoconus*, possess only the bladelike flange while the tubercule is not retained. *Gillisonchus* has an enlarged and better defined crest from the posterior margin of the calcaneal tuber to the tubercule, which is only partially developed in *Protungulatum*. The cuboid facet forms the whole distal surface of the calcaneum in much the same manner as in *Protungulatum*. It is similarly symmetrically rhomboidal, slightly concave and only slightly shifted laterally from the condition seen in *Protungulatum.* The lateral shift of the facet is a derived feature with respect to primitive ungulates because *Claenodon*, *Periptychus*, *Hyopsodus*, and others shift the facet medially as the peroneal tubercule is reduced. The facet is obliquely oriented in the mediolateral and dorsoventral aspects, as it is in *Protungulatum.* Plesiadapid primates and other primitive ungulates tend to reduce the dual oblique nature and tend to have the facet mutually perpendicular to the long axis of the calcaneum.

The *Gillisonchus* calcaneum also has a small, but distinct distal articular facet that is located at the distal and ventral margin but is not confluent with the astragalocalcaneal facet. The feature is much the same as in *Claenodon* but is smaller, yet more distinct, than in *Protungulatum.*

The calcaneal tuber possesses a plantar tuberosity on the posterior margin of the shaft. It forms the posterior termination of a rounded longitudinal swelling confluent with the anterior tubercule. It is larger in *Protungulatum* but is similarly positioned. The assumed homologue in plesiadapid primates is laterally placed (Szalay & Decker, 1974). The anterior tubercule of *Gillisonchus* is nearly identical to that of *Protungulatum.*

The navicular (Pl. 5.8, figs. 15–20) and cuboid (Pl. 5.8, figs. 21–25) of early mammals are rarely recovered and are unknown for *Protungulatum* and *Procerberus.* Left and right naviculars and a right cuboid are present in the *G. gillianus* skeleton. In *Gillisonchus* a cuboid and astragalar partial articulation is present forming approximately 25% of the total surface that articulates with the naviculoastragalar facet. This nonserial arrangement is similar to that of *Ectoconus.*

Schaeffer (1947) observed the S-shaped articulation between the navicular and cuboid in *Claenodon* and concluded that no movement was possible. This is probably true for anteroposterior motion. However, in *Gillisonchus* the common facets of the two bones are of different sizes, those of the cuboid being slightly larger and slightly convex. Such a morphology suggests that at least limited rotation of the cuboid and navicular bones about the median axis of the foot was possible.

The navicular and cuboid form a relatively strong pit with high ridges for articulation with the astragalus. Such articular control is weaker in other animals where the relationships are known, such as *Hyopsodus*, where the ridges are reduced. *Ectoconus* has the nonserial articulation, reduced crests, and the navicular is curled around the astragular head with little cuboid-astragalar contact.

Szalay & Decker (1974) provided a list of what they consider to be primitive eutherian characters. I agree with their assessment and suggest, as did Matthew (1937), that the nonserial or alternating tarsus is the primitive eutherian condition. Based upon additional evidence from

Gillisonchus, this is the case. Such a relationship suggests that Cope's (1881) serial tarsus, presented as the primitive condition, represents a derived state and may, in fact, serve to unite *Phenacodus* with the perissodactyls. This may be complicated, however, by the possibility that a serial tarsus could have been derived several times within separate lineages. The cladistic relationships of some primitive mammals based on tarsal characters are shown in Text-figure 5.6 (See also Table 5.3).

Fragments of what are assumed to be fore and hind foot phalanges are present in the collection. However, they cannot be identified with confidence as belonging to either a fore or hind foot. Comparisons with other such skeletal elements (i.e., of *Ectoconus*) show no conspicuous differences.

TEXT-FIG. 5.6—Cladistic relationships of the tarsus of several early mammals. The numbers on the diagram refer to groups of characters compiled in Table 5.3.

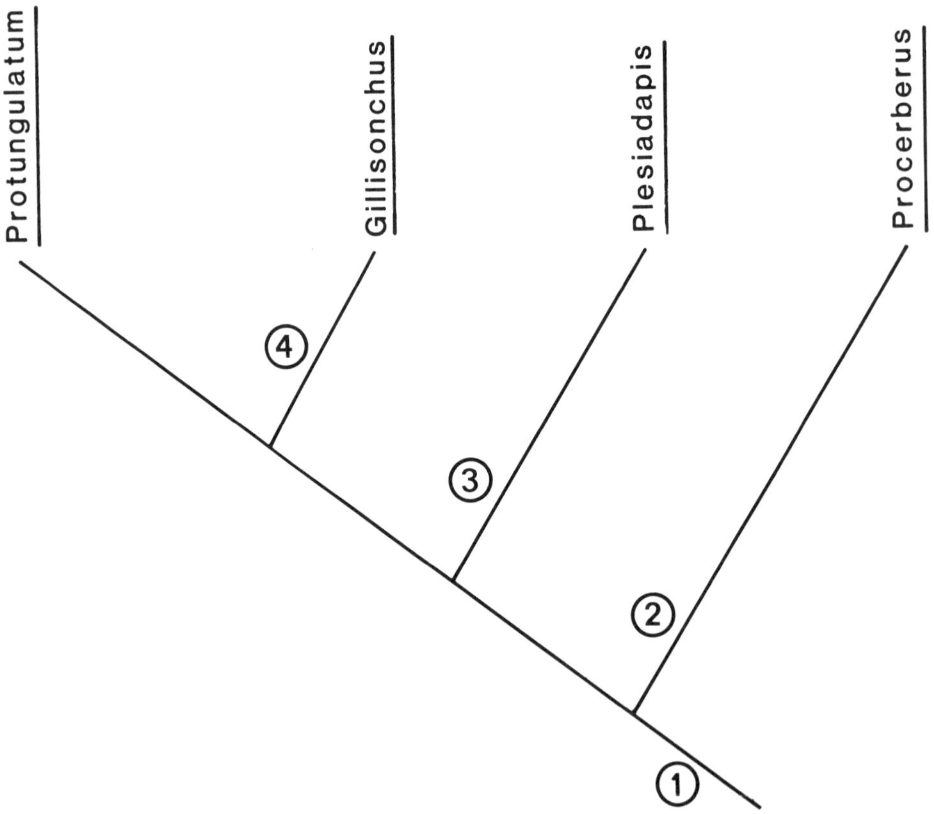

Table 5.3—Tarsal characters used in constructing the cladistic relationships in Text-figure 5.6.

1. Primitive Eutherian Characters
   a. Distal peroneal tubercule present.
   b. Cuboid facet oblique to long axis of calcaneum.
   c. Posterior astragalocalcaneal articulation forming a relatively large angle (35 °) with long axis of calcaneum.
   d. Astragalocalcaneal articulation having oblique axis of rotation.
   e. Short calcaneal body present anterior to astragalocalcaneal facet.
   f. Plantar anterior tubercule present on calcaneum.
   g. Low, slightly grooved, and short tibial trochlea present on astragalus.
   h. Large astragalar canal pierces body.
   i. Cuboid contacts astragalus (nonserial or alternating type of tarsus).
   j. Astragalar sustentacular facet continuous with a distal facet.
   k. Astragalar head wide, laterally thickened, and rotated slightly dorsolaterally.
   l. "Tibiale" facet present on dorsolateral surface of astragalar head.

2. *Procerberus* Derived Characters
   a. Loss of astragalar canal.
   b. Enlarged and sharply defined lateral astragalar tibial facet.
   c. Isolated astragalar sustentacular facet. (See text discussion and 1j.)
   d. Loss of functional calcaneal fibular facet.
   e. Much reduced "tibiale" facet retained on astragalar head.
   f. Angle of 50–60 ° between calcaneoastragalar facet and long astragalar axis.

3. *Plesiadapis* Derived Characters
   a. Reduced astragalar head width and reduced lateral head displacement.
   b. Elevated and sharpened astragalar crests on tibial trochlear facets.
   c. Nearly circular cuboid facet that is nearly perpendicular to the long and transverse calcaneal axis.
   d. Well defined sustentacular tendinous groove.
   e. Loss of "tibiale" facet on astragalar head.
   f. Strongly reduced distal astragalar facet.
   g. Angle of 30–35 ° between calcaneoastragalar facet and long astragalar axis.
   h. Lateral angle of 10–15 ° between astragalocalcaneal facet and medial axis of calcaneum.
   i. Strongly reduced peroneal tubercle.
   j. Presence of a tuberosity laterally placed on the calcaneal tuber anterior to a similar feature in *Protungulatum*.

4. *Gillisonchus* Derived Characters
   a. Shortened astragalar trochlear length not extending posteriorly past astragalar canal.
   b. Elongated and slightly restricted astragalar neck.
   c. Angle of 45 ° formed between astragalar sustentacular facet and astragalar axis.
   d. Strong differentiation of astragalar fibular facet.
   e. Angle of 45 ° formed between astragalocalcaneal facet and long axis of calcaneum.
   f. Enlarged and more sharply defined peroneal tubercle crest.
   g. Slight lateral shift of cuboid facet, tendonous pit and enlarged peroneal tubercle.
   h. Loss of groove at base of anterior tubercle for tendon of the plantar calcaneocuboid ligament. It is preserved only as a notch at the lateral base of the anterior tubercle.

CONCLUSIONS

A new genus, *Gillisonchus*, is proposed for "*Anisonchus*" *gillianus* because "*A.*" *gillianus* possesses dental characters which remove it from the two *Anisonchus* subgenera of Van Valen (1978). The postcranial skeleton of *G. gillianus* is similar to *Ectoconus ditrigonus* in most respects, but differs from it in many details. A mixture of primitive and derived features of the hind limb (femur, tibia, and fibula) are shared between *G. gillianus* and *E. ditrigonus*.

The tarsus of *G. gillianus* is remarkably similar to that of *Protungulatum* as described by Szalay & Decker (1974). This overall resemblance is attributed to a large number of shared primitive ungulate, and perhaps even eutherian, characters. The tarsal character complex has derived features which probably remove *G. gillianus*, at least, from a close relationship to artiodactyls or perissodactyls.

ACKNOWLEDGMENTS

Barry Kues has loaned specimens under his care at the University of New Mexico to aid in this study. J. W. Froehlich and several students in the UNM Anthropology Department aided in the collection of additional fossil material. Valuable discussions on anisonchine relationships were had with David Archibald. David Archibald, Coleman Robison, Leigh Van Valen, Donald Wolberg, and Spencer Lucas have reviewed the manuscript but are not responsible for its shortcomings. The Albuquerque District staff of the Bureau of Land Management aided in the preparation of Text-figures and the manuscript.

REFERENCES

Bureau of Land Management. 1979. Star Lake–Bisti Regional Coal. FES 79-11. Washington, D.C., 749 p.

Cope, E. D. 1881. On the origin of foot structures of the ungulates. Am. Nat. 15:269–273.

Hooker, J. J. 1979. Two new condylarths (Mammalia) from the early Eocene of southern England. Bull. Brit. Mus. Nat. Hist. 32(1):43–56.

Kues, B. S., J. W. Froehlich, J. A. Scheibout, and S. G. Lucas. 1977. Paleontological survey, resource assessment, and mitigation plan for the Bisti–Star Lake area, northwestern New Mexico. Unpublished Report to the U.S. Bureau of Land Management, Albuquerque District Office, 1,525 p.

Lindsay, E. H., L. L. Jacobs, and R. F. Butler. 1978. Biostratigraphy and magnetostratigraphy of Paleocene terrestrial deposits, San Juan Basin, New Mexico. Geology 6:425–429.

Lindsay, E. H., R. F. Butler, and N. M. Johnson. 1979. Reply to "Comment on 'Biostratigraphy and magnetostratigraphy of Paleocene terrestrial deposits, San Juan Basin, New Mexico.' " Geology 7:326.
———. 1981. Magnetic polarity zonation and biostratigraphy of Late Cretaceous and Paleocene continental deposits, San Juan Basin, New Mexico. Am. J. Sci. 281:390–435.
Lucas, S. G. and J. K. Rigby, Jr. 1979. Comment on "Biostratigraphy and magnetostratigraphy of Paleocene terrestrial deposits, San Juan Basin, New Mexico." Geology 7:323–325.
Matthew, W. D. 1937. Paleocene faunas of the San Juan Basin, New Mexico. Trans. Am. Philos. Soc., n. s., 30:1–510.
Rigby, J. K., Jr. 1976. Swain Quarry of the Fort Union Formation, middle Paleocene (Torrejonian), Carbon County, Wyoming: Geologic setting and mammalian fauna. Unpublished Ph.D. dissertation, Columbia Univ., N.Y., 296 p.
———. 1980. Swain Quarry of the Fort Union Formation, middle Paleocene (Torrejonian), Carbon County, Wyoming: Geologic setting and mammalian fauna. Evol. Monogr. 3, 179 p.
Rigby, J. K., Jr., and S. G. Lucas. 1977. Fossil mammals from the Ojo Alamo Sandstone. New Mexico Geol. Soc. Guidebook, 28th Field Conf., Supplement, p. 55–56.
Russell, D. E. 1967. Le Paleocene continental d'Amerique du Nord: Mem. Mus. Nat. Hist. Nat. (Paris) n. s., ser. C., 16:39–99.
Schaeffer, B. 1947. Notes on the origin and function of the artiodactyl tarsus. Am. Mus. Novitates 1356, 24 p.
Sinclair, W. J. and W. Granger. 1914. Paleocene deposits of the San Juan Basin, New Mexico. Bull. Am. Mus. Nat. Hist. 33:297–316.
Szalay, F. S. 1977. Phylogenetic relationships and a classification of the eutherian Mammalia. p. 315–374 *in:* M. K. Hecht, P. C. Goody and B. M. Hecht (eds.), Major patterns in vertebrate evolution, NATO Advanced Study Institute, Plenum Publ. Co., N.Y.
Szalay, F. S. and R. L. Decker. 1974. Origins, evolution, and function of the tarsus in Late Cretaceous Eutheria and Paleocene primates. p. 223–250 *in:* F. A. Jenkins, Jr. (ed.), Primate Locomotion, Acad. Press, N.Y.
Van Valen, L. 1978. The beginning of the age of mammals. Evol. Theory 4:45–80.

APPENDIX 5.1: MEASURED STRATIGRAPHIC SECTIONS

Localities and descriptions of stratigraphic secitons are given below.

Section A

Measured in the eastern half of Sec. 8, T22N, R9W. The base of the section is at an east-west road. The section continues northward through low badlands and climbs to the plateau on the north flank of Betonnie Tsosie Wash.

Thickness (Meters)		Top
± 3.0-6.1	15.	Stabilized, unconformable dune sand and soil, probably of Quaternary age, covers the top of the east to west trending plateau separating Kimbeto and Betonnie Tsosie Washes.

Kirtland Shale:

0.6 14. Sandstone, brown, coarse-grained, platy. This unit disappears westward into the overlying soil. The unit thickens slightly eastward and is a prominent bench-former.

1.5 13. Clay-shale, olive green, silty.

3.0 12. Sandstone, gray white, medium-grained, clayey. Some black carbonaceous mottling is present on bedding planes.

6.1 11. Clay-shale, buff brown mottled, silty. Forms a slope.

5.8 10. Sandstone, multistoried, with limited interbeds of clay-shale. Weathers brown and forms a distinct ridge. Medium-grained with some darker brown spheroidal concretions weathering out. Upper .3m is white with orange brown ferric staining.

4.6 9. Clay-shale, olive green, silty. Upper half weathers mottled brown and gray. Petrified wood and unidentified dinosaur bones (Localities 8, 9 and 10) common.

3.0 8. Sandstone, gray white, medium-grained; and clay-shale, buff brown, silty, interbedded. Sandstone has orange brown ferric staining.

3.0 7. Clay-shale, buff and gray, silty.

3.0 6. Sandstone, gray brown, medium-grained. Some clayey lenses and platy black brown ledges. Much petrified wood and dinosaur bones (Localities 3, 4, 5, 6 and 7). Also fossil mammals and other microvertebrates. (Kansas University locality #16).

9.1 5. Clay-shale, mottled buff, gray, olive green. Occasional stringers of friable, gray white, medium-grained sandstone. Some orange brown ferric staining. Limonitic nodules common on surface. Locality 3 (dinosaur bones) begins in the upper part of this unit and continues into unit 6. Locality 2 (dinosaur bones) also correlated to the upper part of this unit.

3.0 4. Lignite, brown black, and clay-shale, buff and gray, petrified wood.

1.5 3. Sandstone, gray white, medium-grained. Clayey lenses and orange brown ferric staining. Caps ridge. Locality 1.

2.4 2. Clay-shale, buff and gray, petrified wood.

3.9 1. Lignite (black) and clay-shale (buff and gray), petrified wood.

Section B

Measured in the SW¼, Sec. 4, T22N, R9W. The base of the section is in the bed of a north to south arroyo, and the top is the plateau on the north flank of Betonnie Tsosie Wash.

Thickness (Meters) Top

± 3.0-6.1 6. Stabilized unconformable dune sand and soil (see Section A, unit 15).

Kirtland Shale:

6.0 5. Clay-shale, olive green and gray, silty. Basal thin, platy, poorly sorted sandstone and brown "cannonball" sandstone concretions and much petrified wood.

8.8 4. Sandstone, buff brown but weathers gray, coarse-grained with some local conglomeratic lenses. Large-scale crossbedding and some spheroidally weathering concretions. Forms a prominent cliff and has a locally unconformable base due to channeling. Locality 12.

4.6	3.	Clay-shale, gray and buff, silty.
1.5	2.	Sandstone, brown, coarse-grained, platy. Thickens and thins laterally and forms prominent benches. Petrified wood (Locality 11).
16.2	1.	Clay-shale, buff brown, olive green and gray. Petrified wood.

Section C

Measured in the NE¼, Sec. 9 and SE¼, Sec. 4, T22N, R9W. The base of the section is in the bed of a northeast to southwest trending arroyo, and the top is on the plateau on the north flank of Betonnie Tsosie Wash.

Thickness (Meters)		Top
± 3.0-6.1	8.	Stabilized unconformable dune sand and soil (see Section A, unit 15).
Ojo Alamo Sandstone (restricted):		
2.1	7.	Sandstone, gray but weathers copper brown, coarse-grained. Conglomeratic near base, well-indurated, caps ridges. Much petrified wood. Thickness is estimated due to weathering of top of unit. Unconformity, erosional.
Kirtland Shale:		
11.3	6.	Clay-shale, gray, olive green and buff, silty.
2.7	5.	Clay-shale, olive green and gray, silty and brown coarse-grained sandstone.
2.4	4.	Sandstone, buff brown but weathers gray, coarse-grained with some conglomeratic lenses.
1.5	3.	Clay-shale, gray and buff, very silty with white, medium-grained sandstone stringers. Locality 13 (turtle carapace fragments) and petrified wood.
3.0	2.	Sandstone, gray white, clayey, medium-grained. Some orange brown ferric staining.
9.1	1.	Clay-shale, buff, olive green and gray. Forms a slope.

Section D

Measured in the southern half of Sec. 3, T22N, R9W beginning in the bed of a north to south trending arroyo and ending southeast of a cluster of Navajo dwellings on the plateau separating Kimbeto and Betonnie Tsosie Washes.

Thickness (Meters)		Top
± 3.0-6.1	7.	Stabilized unconformable dune sand and soil (see Section A, unit 15).
Ojo Alamo Sandstone (restricted):		
7.9	6.	Sandstone, cross-bedded, gray but weathers yellow and copper brown, coarse-grained with conglomeratic lenses of chert and jasper nodules. Many petrified logs, some with diameters greater than 1.5 meters. Locally unconformable base due to channeling. The unit forms a prominent ridge. Unconformity, erosional.
Kirtland shale:		
2.1	5.	Clay-shale, olive green and bluish gray, silty.
0.9	4.	Sandstone, gray white and buff, clayey, medium-grained, friable.
4.0	3.	Clay-shale, olive green, purple and bluish green, silty.
4.3	2.	Sandstone, gray white, medium to coarse-grained, clayey. Some resistant ledges but generally friable.
± 5.0	1.	Clay-shale, gray and olive green, silty. Mostly covered with soil.

Section E

Measured in the NE¼, Sec. 10, T22N, R9W. The bottom of the section is in the bed of a north to south trending tributary of Betonnie Tsosie Wash; the top is on a north to south trending ridge that has several Navajo dwellings on it.

Thickness (Meters) — Top

±3.0-6.1 6. Stabilized unconformable dune sand and soil (see Section A, unit 15).

Ojo Alamo Sandstone (restricted):

3.0 5. Sandstone, gray and buff, coarse-grained, much weathered, with petrified wood. Forms a ridge but top is truncated by overlying unconformable dune sand, and the basal contact is not precisely defined due to weathering.

4.0 4. Sandstone, gray white and yellow, medium-grained. Some orange brown ferric staining.
Unconformity, erosional.

Kirtland shale:

1.5 3. Clay-shale, buff and gray, silty.

2.7 2. Sandstone, buff, medium to coarse-grained. Petrified wood. Forms a cliff.

22.0 1. Clay-shale, buff brown, gray and olive green. Silty and covered with sandstone debris and soil.

Section F

Measured mostly in the eastern half of Sec. 3, T22N, R9W, except for the upper eight meters, which were measured in the NW¼, NW¼, Sec. 2, T22N, R9W. The bottom of the section is in a north to south trending tributary of Betonnie Tsosie Wash; the top is in the Nacimiento Formation at the tip of "Mammalon Hill."

Thickness (Meters) — Top

Nacimiento Formation (in part):

1.8 10. Clay-shale, buff brown, silty. Forms tips of "Mammalon Hill."

3.0 9. Sandstone, gray white, medium-grained. Friable with black stained nodules and resistant sandstone "tubes."

4.0 8. Clay-shale (buff brown and olive green, silty) and interbedded gray white medium-grained sandstone lenses. Base of this unit is base of "Mammalon Hill." Locality 15 (Puercan mammals).

1.5 7. Clay-shale, dark gray, light gray and black, carbonaceous, silty. Caps a long ridge just NE of "Mammalon Hill."

Ojo Alamo Sandstone (restricted), upper part:

3.7 6. Sandstone, gray white, medium-grained. Friable with clayey lenses and some orange brown ferric staining. Large petrified logs, some over 1.5 m long. Underlies the flat north side of "Mammalon Hill." Locality 14 at top of this unit. The following is a revised faunal list (cf. Rigby & Lucas, 1977): *Oxyclaenus simplex*, *Gillisonchus gillianus*, *Conacodon entoconus*, *Hemithlaeus kowalevskianus*, *Ectoconus ditrigonus*, phenacodont cf. *Tetraclaenodon* and *Periptychus sp.*

Nacimiento Formation, tongue:

3.0 5. Clay-shale (gray, dark gray, black and green, silty) and sandstone (gray to dark green, weathers buff brown, medium-grained). The sandstone forms an irregular prominent ledge about 0.5 m thick within the clay-shale.

Ojo Alamo Sandstone (restricted), lower part:

2.4	4.	Sandstone, gray and buff, medium to coarse-grained. Friable with clayey lenses and some petrified wood and ironstone concretions.
2.4	3.	Sandstone, dark copper brown, coarse-grained with lenses of chert and jasper pebbles. Cross-bedded, well indurated, caps a prominent pillar. Some petrified wood.
7.0	2.	Sandstone, gray white, medium-grained. Orange brown ferric staining and petrified wood present. Some clayey lenses. Locally unconformable base due to channeling. Unit forms a cliff. Unconformity, erosional.

Kirtland shale:

9.1	1.	Clay-shale, olive green, purple, blue and red, some silty lenses.

Section G

Measured on the north-central edge of Sec. 2, T22N, R9W, starting in an arroyo bed and continuing to the top of a northeast to southwest trending ridge in Sec. 35, T23N, R9W.

Thickness (Meters) Top

± 3.0-6.1	10.	Stabilized unconformable dune sand and soil (see Section A, unit 15).

Nacimiento Formation (in part):

4.2	9.	Sandstone (mottled, gray, purple and buff clayey), and clay-shale (mottled, buff, and purple).
2.7	8.	Clay-shale, gray and black. Lighter colored near top with some silty lenses. Some petrified wood (locality 17) and gypsum.
3.0	7.	Clay-shale, buff brown, silty.
4.6	6.	Sandstone, black brown to gray white, with some clayey lenses, medium-grained. Weathers into ledges and knobs.
6.1	5.	Clay-shale, gray and black, buff. Forms shoulder of ridge. Locality 16 (= Mammalon Hill fossiliferous horizon) near top of unit.

Ojo Alamo Sandstone, upper part:

4.0	4.	Sandstone, gray white, medium-grained, some clayey lenses and spheroidal weathering concretions, with petrified wood.

Nacimiento Formation, tongue:

2.7	3.	Clay-shale, gray black, carbonaceous.
1.8	2.	Clay-shale, purple with a sandstone lens, dark green, some ferric staining, medium-grained.

Ojo Alamo Sandstone, lower part:

2.4	1.	Sandstone, gray white, buff brown and gray medium- to coarse-grained.

Section H

Measured in the SW¼, Sec. 35, T23N, R9W. Starting in an arroyo bed and climbing the same ridge as in Section G.

Thickness (Meters) Top

± 3.0-6.1	10.	Stabilized unconformable dune sand and soil (see Section A, unit 15).

Nacimiento Formation (in part):

1.5	9.	Clay-shale, purple and red, silty.
3.0	8.	Clay-shale, gray, brown, silty with some petrified wood. A thin lenticular sandstone, gray white, medium-grained, clayey, is at the unit's base.

4.0 7. Clay-shale, gray and black, some silty lenses.

1.5 6. Sandstone, gray, medium-grained. Some brown ferric staining and clayey lenses.

6.1 5. Sandstone, gray white, medium-grained. Some brown ferric staining and olive green and gray clayey lenses. Weathers to form knots and "tubes" of black resistant sandstone.

5.5 4. Clay-shale, gray and black, some silty lenses and brown and light green mottling.

Ojo Alamo Sandstone, upper part:

7.6 3. Sandstone, gray white, medium-grained. Petrified wood and some gray and olive green clayey lenses.

Nacimiento Formation, tongue:

2.4 2. Clay-shale, gray with orange brown ferric staining, silty. Some carbonaceous lenses and a thin bed of dark green sandstone. This unit pinches out several meters to the east.

Ojo Alamo Sandstone, lower part:

2.0+ 1. Sandstone, buff brown and gray, medium- to coarse-grained. Forms resistant ledges in arroyo cuts.

6 Systematics of Paleocene Multituberculates from the San Juan Basin, New Mexico[1]

Robert E. Sloan
University of Minnesota
Minneapolis, Minnesota

INTRODUCTION

This paper reports the multituberculates from a series of collections made in the Paleocene Nacimiento Formation, San Juan Basin, New Mexico, before 1965. From 1948 to 1955, while at the University of Kansas, Robert W. Wilson made important collections in the middle Torrejonian *Deltatherium* "zone." These were at the "Little Pocket" or Kansas University Quarry 9 locality in Sec. 2, T23N, R9W, and at the Quarry 13 locality in Sec. 4, T27N, R11W, in Kutz Canyon. These specimens were briefly reported by Wilson (1956). During the early 1960's, William A. Clemens, Jr. also collected from these two localities for the University of Kansas. In 1958, a party from the American Museum of Natural History made a significant collection by screen washing sediment from "Mammalon Hill" in Tsosie rincon, Betonnie Tsosie Wash, Sec. 2, T22N, R9W, in the basal Puercan *Hemithlaeus* "zone."

TERMINOLOGY

Intact lower jaws and partial skulls of multituberculates occur sufficiently rarely so that most new taxa must be defined on the basis of

[1]Publication No. 1016 from the School of Earth Sciences, University of Minnesota.

isolated teeth. Fortunately, the most distinctive teeth are also the most common. Isolated first and second molars, from the various species of the common neoplagiaulacids, overlap in cusp formulae and dimensions to such a degree that it is an almost hopeless task to assign them to species. With additional collections, this will surely be remedied. For the present, neoplagiaulacid first and second molars will be neglected unless they occur in jaws. Far more useful are the upper and lower shearing blades, P^4 and "P_4" of various authors. On the basis of several jaws of species from areas outside of the San Juan Basin, the upper blade does have a deciduous precursor and thus is a premolar. It is conventionally called P^4, although the blade is the 6th post-canine tooth in the oldest known multituberculate skull. The lower blade is not a premolar but is instead a molar. In the very large collections of loose teeth from Bug Creek Anthills, Swain Quarry, and Shotgun Quarry, the only lower blades are those also found in adult jaws. In specimens of *Ptilodus montanus*, *Neoplagiaulax hunteri*, *Mesodma formosa*, *M. thompsoni*, *Parectypodus childei*, and *Meniscoessus robustus* the permanent and only lower blade erupts at the same time as the deciduous incisor and is already worn when the permanent incisor erupts. To avoid confusion with the older literature resulting from renumbering the lower molars, the lower blade (P_4 of previous authors) is here redesignated M_b. This terminology was used previously by Schiebout (1974) and Sloan (1979). Dental terminology and measurements follow Krause (1977) and are diagrammed in Text-figure 6.1.

INSTITUTIONAL ABBREVIATIONS

The names of institutions referred to in this paper are abbreviated as follows:

AMNH American Museum of Natural History, New York.
BM British Museum of Natural History, London.

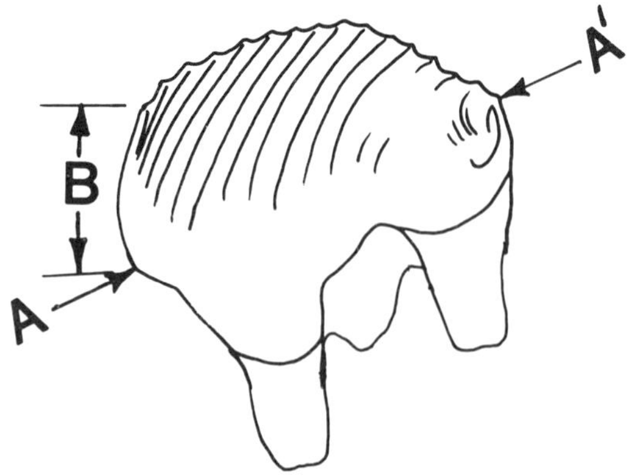

TEXT-FIG. *6.1*—Standard measurements of M_b's: A–A′ is standard length, B is the height of the first serration, and B/A–A′ is relative height of the first serration. Figured specimen is an unworn M_b, *Neoplagiaulax hunteri* AMNH 33993, from Scarritt Quarry, Crazy Mountain Field, Montana.

KU	University of Kansas, Lawrence.
MNHN	Museum National d'Histoire Naturelle, Paris.
PU	Geological Museum, Princeton University, New Jersey.
UCM	University of Colorado Museum, Boulder.
UCMP	University of California Museum of Paleontology, Berkeley.
UMVP	University of Minnesota Vertebrate Paleontology Collection, Minneapolis.
USNM	United States National Museum, Washington.

SYSTEMATIC PALEONTOLOGY

Class MAMMALIA
Subclass PROTOTHERIA Gill, 1872
Order MULTITUBERCULATA Cope, 1884
Suborder PTILODONTOIDEA Sloan & Van Valen, 1965
Family NEOPLAGIAULACIDAE Ameghino, 1902

Remarks.—The definition of genera in the Neoplagiaulacidae is not aided by their general occurrence as isolated teeth. A principal problem of neoplagiaulacid taxonomy is the high incidence of familial parallelism so that lower blades I refer to *Neoplagiaulax* and to *Ectypodus* are very similar. All three major genera, *Ectypodus*, *Neoplagiaulax*, and *Parectypodus*, have been placed in synonymy at one time or another. However, as a result of the intensive screen washing operations at numerous Paleocene quarry sites during the past two decades, enough new species have been discovered to show that the type species of each described genus pertains to a distinctive set of species, and thus each described genus can be validated.

Genus MESODMA Jepsen, 1940

Remarks.—Clemens (1963) adequately redefined this genus except for the following additions. The M_b of *Mesodma* is relatively lower than that of *Neoplagiaulax*, *Ectypodus*, *Mimetodon*, and *Parectypodus*, and the relative height of the first serration is less than one-third of the standard length. The upper blade agrees with that of *Ectypodus* in having the principal cusp at the junction of the anterior and posterior slopes in profile view and in having a triangular rather than arcuate profile. *Mesodma* upper blades commonly have from three to five cusps in the anterolabial external row rather than the one or two seen in *Ectypodus*. *Mesodma formosa* seems surely to be the ancestor of *Ectypodus*. *Mesodma formosa* or *M. thompsoni* may be the ancestor of *Neoplagiaulax* and *Parectypodus*. *M. thompsoni* is surely the ancestor of *Mimetodon*.

MESODMA FORMOSA (Marsh, 1889)
Text-fig. 6.2, Table 6.1

Halodon formosus MARSH, 1889, p. 179.
Mesodma formosa: CLEMENS, 1963, p. 31.

Referred specimens.—AMNH 58466, 58504, 59911, right M_b's; 58374, 58287, 59844, left M_b's; 58291, left P^4; all from the lower part of the Nacimiento Formation at Mammalon Hill, Betonnie Tsosie Wash, San Juan Basin, New Mexico.

Discussion.—The present specimens cannot be distinguished from the common small Lancian to Puercan neoplagiaulacid *Mesodma formosa* as characterized by Clemens. They represent an important, but not surprising, range extension for this species.

MESODMA THOMPSONI Clemens, 1963
Text-fig. 6.3, Table 6.1

Mesodma thompsoni CLEMENS, 1963, p. 39; SLOAN & VAN VALEN, 1965, p. 222.

Referred specimens.—AMNH 58286, right M_b; 58668, 58288, left M_b's; 58760, right P^4; all from the lower part of the Nacimiento Formation at Mammalon Hill, Betonnie Tsosie Wash, San Juan Basin, New Mexico.

Discussion.—The present specimens cannot be distinguished from the Lance and Hell Creek Formation specimens of *Mesodma thompsoni* I have seen. As in the case of *Mesodma formosa*, they represent an important range extension of this species.

Genus NEOPLAGIAULAX Lemoine, 1882

Remarks.—*Plagiaulax eocaenus* (Lemoine, 1880), from the late Paleocene of France, is the type species of the genus *Neoplagiaulax* Lemoine (1882). These were the first described genus and species of the Neoplagiaulacidae. All discussion of generic synonymy must, therefore, begin with this species. Lemoine (1885) later described *Neoplagiaulax copei* from the same locality as *Plagiaulax eocaenus*. Russell (1964) refigured the type specimens and additional material, retaining *N. copei* in this genus with a query. North American species presently referred to this genus include: *N. macintyrei*, a new species from the early Paleocene of New Mexico; *N. macrotomeus* (Wilson, 1956) from the Torrejonian *Deltatherium* "zone" of New Mexico; an undescribed species from the Dragon fauna; another undescribed species from the Puercan *Taeniolabis* "zone" of Montana and the middle Paleocene of Montana,

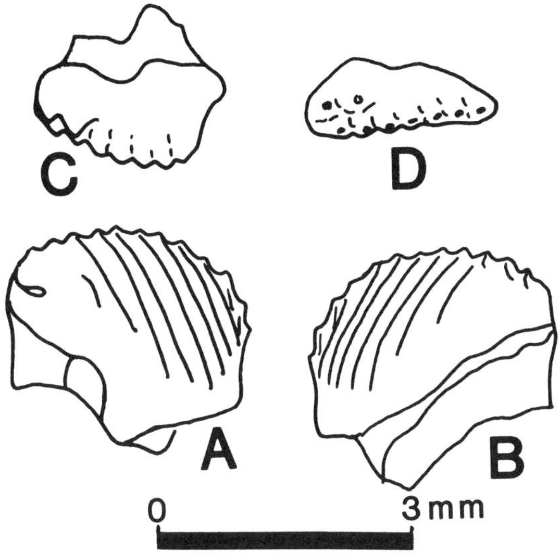

TEXT-FIG. 6.2—*Mesodma formosa*, AMNH 59911, right M_b, labial view (*A*) and lingual view (*B*); AMNH 58291, left P^4, labial view (*C*) and lingual view (*D*).

Wyoming and Alberta; *N. hunteri* (Simpson, 1936b) and *N. nanophus* (Holtzman, 1978) from the middle Tiffanian; *N. grangeri* (Simpson, 1935) from the middle Paleocene; and *N. hazeni* (Jepsen, 1940), *N. fractus* (Dorr, 1952), and two undescribed species from the late Tiffanian.

In this genus, as it is presently defined, the upper blade is the most diagnostic tooth because the penultimate or antepenultimate cusp of the principal cusp row is farthest from the base of the enamel rather than the final cusp, as in *Mesodma* and *Ectypodus*. The posterior slope of the blade ascends in a straight line from the final cusp to the base of the enamel in contrast to the concave slope characteristic of *Parectypodus*.

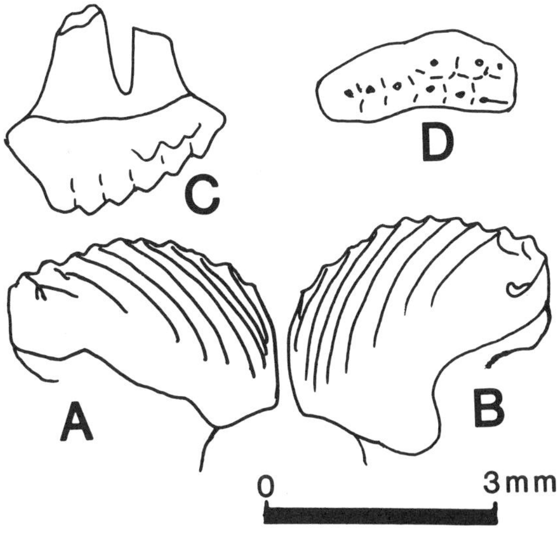

TEXT-FIG. 6.3—*Mesodma thompsoni*, AMNH 58668, left M_b, lingual view (*A*) and labial view (*B*); AMNH 58670, right P^4, labial view (*C*) and occlusal view (*D*).

Table 6.1—Blade dimensions of Neoplagiaulacidae from the Puercan, Tsosie rincon, New Mexico (length M_b and P^4 in mm).

Specimen		M_b Length	Number of Serrations	P^4 Length	Cusp Formula
Mesodma formosa					
AMNH	58466	2.8	10	——	——
AMNH	58504	3.2	10	——	——
AMNH	59911	3.1	12	——	——
AMNH	58374	3.1	12	——	——
AMNH	58291	——	——	2.4	2:7
Mesodma thompsoni					
AMNH	58286	3.7	13	——	——
AMNH	58668	3.7	11	——	——
AMNH	58670	——	——	2.7	3:6
Neoplagiaulax macintyrei					
AMNH	58283	3.6	10	——	——
AMNH	58373	3.6	10	——	——
AMNH	58487	3.6	10	——	——
AMNH	58642	3.6	10	——	——
AMNH	58642	3.6	10	——	——
AMNH	58667	3.7	11	——	——
AMNH	58228a	3.5	10	——	——
AMNH	58228c	3.3	10	——	——
AMNH	58394	——	——	3.2	2:7
Parectypodus vanvaleni					
AMNH	58465	4.0	11	——	——
AMNH	58506	3.9	11	——	——
AMNH	58507	3.8	12	——	——
AMNH	58641	3.9	11	——	——
AMNH	58669	4.0	11	——	——
AMNH	59847	4.0	10	——	——
AMNH	59848	3.9	10	——	——
AMNH	58666	——	11	——	——
AMNH	58598	4.0	11	——	——
AMNH	58372	——	——	3.5	2:7
AMNH	58673	——	——	3.5	2:7
AMNH	59880	——	——	3.9	2:7
AMNH	58371	——	——	3.5	2:7
AMNH	58489	——	——	3.6	3:7

The ratio of first serration height to lower blade (M_b) standard length (relative height) is from 0.30 to 0.45; generally a higher ratio than in *Mesodma*, lower than in *Parectypodus*, but similar to that of *Ectypodus*. In *Neoplagiaulax*, the highest serration above the line along which the length is measured is generally the fifth, sixth, or seventh, whereas, in *Ectypodus* the highest serration is generally the fourth. The anterior profile of *Ectypodus* M_b's is generally more convex than in *Neoplagiaulax*. In *Neoplagiaulax*, the ratio of M_b to M_1 length ranges from 1.55 (*N. grangeri*) to 2.05 (*N. macrotomeus*). In most species of *Neoplagiaulax*, this ratio is larger than in *Ectypodus*. P_3 is not lost in any species of this genus other than *N. eocaenus*.

Neoplagiaulax may be redefined as follows: Neoplagiaulacidae of medium to very large size; standard length of M_b 3.2–8.0 mm; 10 to 15 serrations; relative height of first serration from ⅓-½ of standard length; posterior angle between plane of molar occlusion and anterior face of M_b greater or equal to a right angle; labial height of M_b at exodaenodont lobe less than the standard length; fifth to seventh serration from front of M_b generally highest above baseline for measurement of standard length; ratios of M_b:M_1 length medium to large for neoplagiaulacids (1.55–2.05, mode 1.75); P^4 about ¾ the length of M_b; P^4 cusp formulae 0–4:7–11; modal value 2:8–9 (later species generally show a reduction in the number of cusps in the external row); posterior slope of P^4 profile generally short and straight; anterior slope long and convex with penultimate or antepenultimate cusp highest above base of enamel; posterior basal cuspule generally present.

NEOPLAGIAULAX MACINTYREI n. sp.
Text-fig. 6.4, Table 6.1

Type material.—Holotype, AMNH 58667, left M_b; paratypes, 58642, 58373, 58283, 58476, 58487, 58375 left M_b's; 58228a, 58228c, right M_b's; 58394, right P^4; all are from the *Hemithlaeus* "zone" in the lower part of the Nacimiento Formation at Mammalon Hill, Betonnie Tsosie Wash, San Juan Basin, New Mexico.

Etymology.—This species is named for my friend and colleague Giles T. MacIntyre in honor of his yeoman work in collecting these and other specimens.

Diagnosis.—A species of *Neoplagiaulax* with mean length of M_b 3.56 mm; modal number of serrations 11; relative height of first serration about 0.39; posterior external cuspule a simple ridge descending from the last serration and angling forward; labial striation descending from the penultimate serration, short and curved sharply posteriorly; length of P^4 3.2 mm; cusp formula 2:7.

Discussion.—This is the oldest species of the genus. The sole known upper blade is clearly referable to this species on the basis of profile, cusp formula, size, and by process of elimination. It seems likely that this species was derived from *Mesodma formosa* or *M. thompsoni* and that it gave rise to the later species of *Neoplagiaulax*.

Genus PARECTYPODUS Jepsen, 1930a

Remarks.—This genus, the third neoplagiaulacid genus to be described, has *Parectypodus simpsoni* as its type species. This species lacks

TEXT-FIG. 6.4—*Neoplagiaulax macintyrei*, AMNH 58667, holotype, left M_b, labial view (*A*) and lingual view (*B*); AMNH 58394, right P^4, labial view (*C*) and occlusal view (*D*).

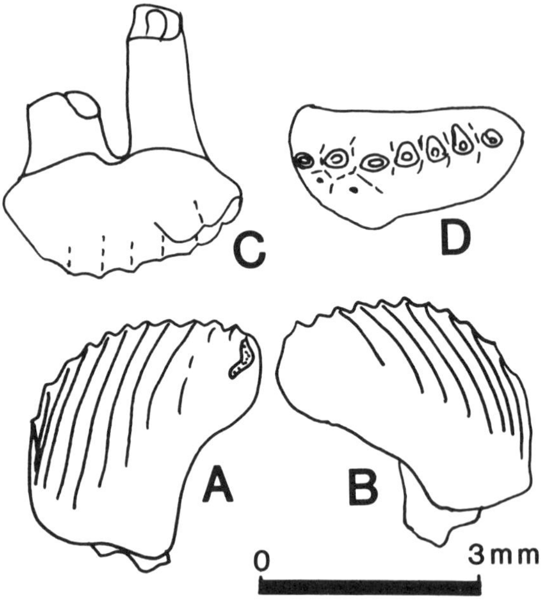

P_3, as do later species of *Ectypodus*, *Neoplagiaulax*, and another undescribed genus and species. Jepsen (1930a) emphasized this missing tooth in his description and Simpson (1937) regarded this lack as the principal defining feature of the genus. Jepsen (1940) temporarily reduced the genus to synonymy with *Ectypodus* in its broad sense. Sloan (1966) resurrected and redefined the genus to include *Parectypodus lovei* from the late Eocene of Wyoming, ?*Ptilodus sinclairi* from the middle Paleocene of Montana, and *Ectypodus laytoni* from the late Paleocene of Wyoming, in addition to the type species.

The generic diagnosis of *Parectypodus*, slightly revised from Sloan (1966), is as follows: Neoplagiaulacidae with relative height of first serration more than 0.45 (in most late Paleocene species the relative height is greater than 0.50); labial height of enamel at exodaenodont lobe approximately equal to or greater than standard length; posterior angle between plane of occlusion of molars and anterior face of M_b a right or usually acute angle; third or fourth serration of M_b highest above base line for standard length; known values for ratio M_b:M_1 length high for neoplagiaulacids (1.8–2.0); length of P^4 80–85 percent of M_b length; P^4 cusp formulae 2–5:7–9 (additional external cusps occur in advanced species); posterior slope of P^4 short, steep, and slightly concave; anterior slope convex, antepenultimate cusp usually highest; posterior basal cuspule usually absent.

Parectypodus was probably derived from the immediate ancestor of *Neoplagiaulax macintyrei* since the earliest species, *P. vanvaleni*, shows a closer approximation to that species than to any earlier or contemporary species.

PARECTYPODUS VANVALENI n. sp.

Text-fig. 6.5, Table 6.1

Type material.—Holotype, AMNH 58641, left M_b; paratypes, 59847, 58666, 58639, 58507, 58465, right M_b's; 59848, 58669, 58506, 58598, left M_b's; 58372, 58673, 58371, right P^4's; 59880, 58489, left P^4's; all from the *Hemithlaeus* "zone" in the lower part of the Nacimiento Formation at Mammalon Hill, Betonnie Tsosie Wash, San Juan Basin, New Mexico.

Etymology.—This species is named for my esteemed friend and co-worker Leigh Van Valen in honor of his yeoman work in collecting these and other specimens.

Diagnosis.—A species of *Parectypodus* with M_b mean length 3.95 mm; 10 or 11 serrations, mode of 11; labial striations more convex forward than in other species; posterolabial cuspule J-shaped; penultimate labial striation short and curved sharply posteriorly; third serration highest above base line for measurement of standard length; relative height of anterior serration about 0.55; mean length of P^4 3.60 mm; P^4 cusp formulae 2–3:7 (2–7 modal); posterobasal cuspule present in line with principal row; three posterior cusps of main row equally high.

Discussion.—This is the oldest and, in some respects, the most primitive species of the genus. Despite the absence of dentulous jaws referable to this species, it seems likely, from the lower blade profile and roots, that the anterior face of the tooth was at a right angle to the molar plane of occlusion instead of angling forward, as in later species. In this

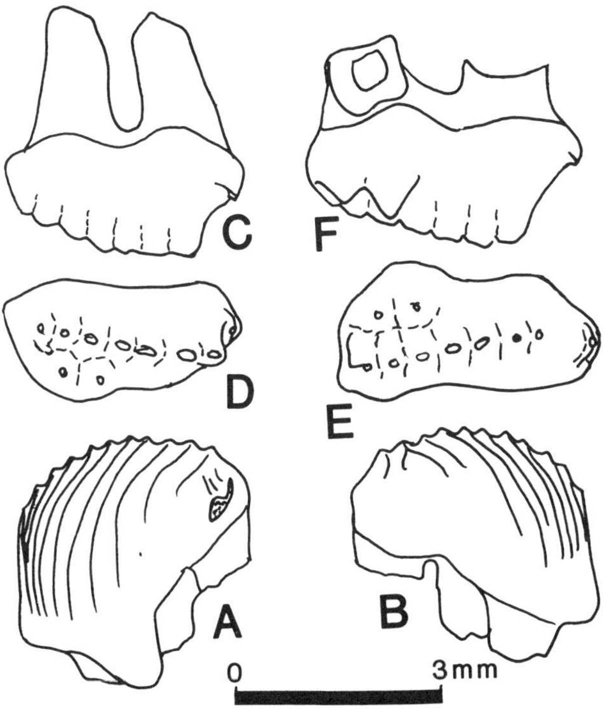

TEXT-FIG. 6.5—*Parectypodus vanvaleni*, AMNH 58641, holotype, left M_b, labial view (*A*) and lingual view (*B*); AMNH 58372, slightly worn right P^4, lingual view (*C*) and occlusal view (*D*); AMNH 59880, left P^4, occlusal view (*E*) and labial view (*F*).

feature, and the cusp formula of P^4, this species is closer to *Neoplagiaulax* than any other known species of *Parectypodus*. Nothing in the known morphology of this species prevents it from being ancestral to all of the later species of the genus. It was probably derived from *Neoplagiaulax macintyrei* or its ancestor.

PARECTYPODUS CLEMENSI n. sp.
Text-figs. 6.6, 6.7

Ectypodus c., RIGBY, 1980, p. 43

Holotype.—KU 16001, crushed rostrum and lower jaws from KU locality 13, Torrejonian *Deltatherium* "zone" of the Nacimiento Formation, Kutz Canyon, San Juan Basin, New Mexico.

Referred specimens.—KU 14032, left M_b from KU locality 9 ("Little Pocket"), Torrejonian *Deltatherium* "zone" of the Nacimiento Formation, San Juan Basin, New Mexico; AMNH 87855, right lower jaw with M_b-M_1; UMVP 6034, 6035, 6036, 6037, right M_b's; 6038, right jaw with M_1-M_2; 6039, right P^4; all from the Torrejonian *Pantolambda* "zone" at Swain Quarry, Sec. 3, T15N, R92W, Carbon County, Wyoming.

Etymology.—Named for my esteemed colleague William A. Clemens, Jr., who collected the type specimen.

Diagnosis.—A species of *Parectypodus* medium in size; M_b–M_2 length about 6.4 mm; M_b length about 3.4 mm; 10 to 12 serrations, modal value 12; M_1 length about 2.2 mm. width 1.0 mm, cusp formulae 6–7:4–5; M_2 length about 1.4 mm, width 1.1 mm, cusp formula 3:2; P^4 length about 2.3 mm, cusp formula 4:8; M^1 length 2.8 mm, cusp formula 7:?:?.

Description.—The type specimen has a crushed rostrum, broken posteriorly at the frontoparietal suture and anteriorly just in front of the maxillopremaxillary suture. The right mandible is nearly complete with the tip of the incisor broken and the coronoid and articular processes broken away. Both mandibles are in fully occluded and retracted positions. The left mandible is broken away horizontally at the level of the exodaenodont lobe of the M_b; the inner side of M_1 and M_2 can be seen. Cusp formulae in the type specimen cannot be determined due to the tightness of occlusion. P^1 and P^2 are tricuspid, P^3 is quadricuspid, P^4 has a cusp formula of 4:7 (left side) and 4:8 (right side), is 2.4 mm long, and is 1.1 mm wide. The P^1–P^4 length is 6.1 mm. The root of the I^2 can be seen in cross section in the fragment of remaining right premaxillary with its centerline 1.3 mm medial to the line of the premolars and molars. The dental ramus of the maxillary is strongly arched, as is the M^1 profile. The zygomatic ramus is broken away at the level of the infraorbital foramen. Length of the skull from the front of the maxillary to the frontoparietal suture is 13.7 mm. Estimated skull length is 25 mm, assuming proportions similar to *Ectypodus tardus* (Sloan, 1979). The estimated orbital diameter is about 2.5 mm; interorbital width is 7.8 mm.

Overall length of left M_b to M_2 is 6.4 mm; left M_2 length is 1.35 mm. The estimated length of the right M_b is 3.5 mm; therefore, the relative height of the first serration is about 0.48. The dorsal surface of the frontals is flat; the interorbital wall of the frontals is vertical. The posteroventral flange of the mandible for the attachment of the masseter and pterygoid muscles is normal to the surface of the ramus; the pterygoid flange is about 2 mm wide. The referred M_b, KU 14032, is 3.2 mm long, with 10 serrations, and the relative height of the anterior serration is 0.41.

The specimens from Swain Quarry appear to belong to the same species. The mean length of six M_b's is 3.47 mm with an observed range of 3.2 to 3.8 mm; 11 and 12 serrations were observed. The P^4, UMVP 6039, is 2.2 mm long, 1.0 mm wide, with cusp formula 4:8. The M_1 and M_2 of UMVP 6038 are 2.4 and 0.9 mm long respectively, with cusp formulae of 7:4 and 3:2. In AMNH 87855, the M_b is 3.2 mm long with 11 serrations and the M_1 is 2.3 mm long, 0.9 wide, with cusp formula 7:5.

PARECTYPODUS CHILDEI (Kuehne, 1969), n. comb.
Pl. 6.1, figs. 1–4; Text-fig. 6.8

Ectypodus tardus: MCKENNA, 1960 a, p. 33–38 (in part), figures 12 a, b, c, e, 13b, c.

Parectypodus sp. A VAN VALEN & SLOAN, 1966, p. 261–278; BOWN, 1979, p. 52–53.

Parectypodus sp. SLOAN, 1966, p. 313; DELSON, 1971, p. 315; WEST, 1973, p. 749–750.

Charlesmooria childei KUEHNE, 1969 (in part), p. 199–202.

Type material.—Lectotype (designated herein), BM M26617, right M_b, from early Eocene Sparnacian Blackheath beds, London Clay, Abbey Wood, Kent, England; topotype, BM M15131, right M_b, from the same locality.

Discussion.—There are two small neoplagiaulacids in the early Eocene, early Graybullian of North America and Sparnacian beds of England and France that have lower blades of the same size and serration count—*Ectypodus tardus* (Jepsen, 1930a) and *Parectypodus childei* (Kuehne, 1969). They frequently occur together and are often confused because of their similar size. *Ectypodus tardus* is known from a well-preserved skull and from lower jaws (Sloan, 1979); *Parectypodus childei* has never been well described. Since it might be expected to occur in the

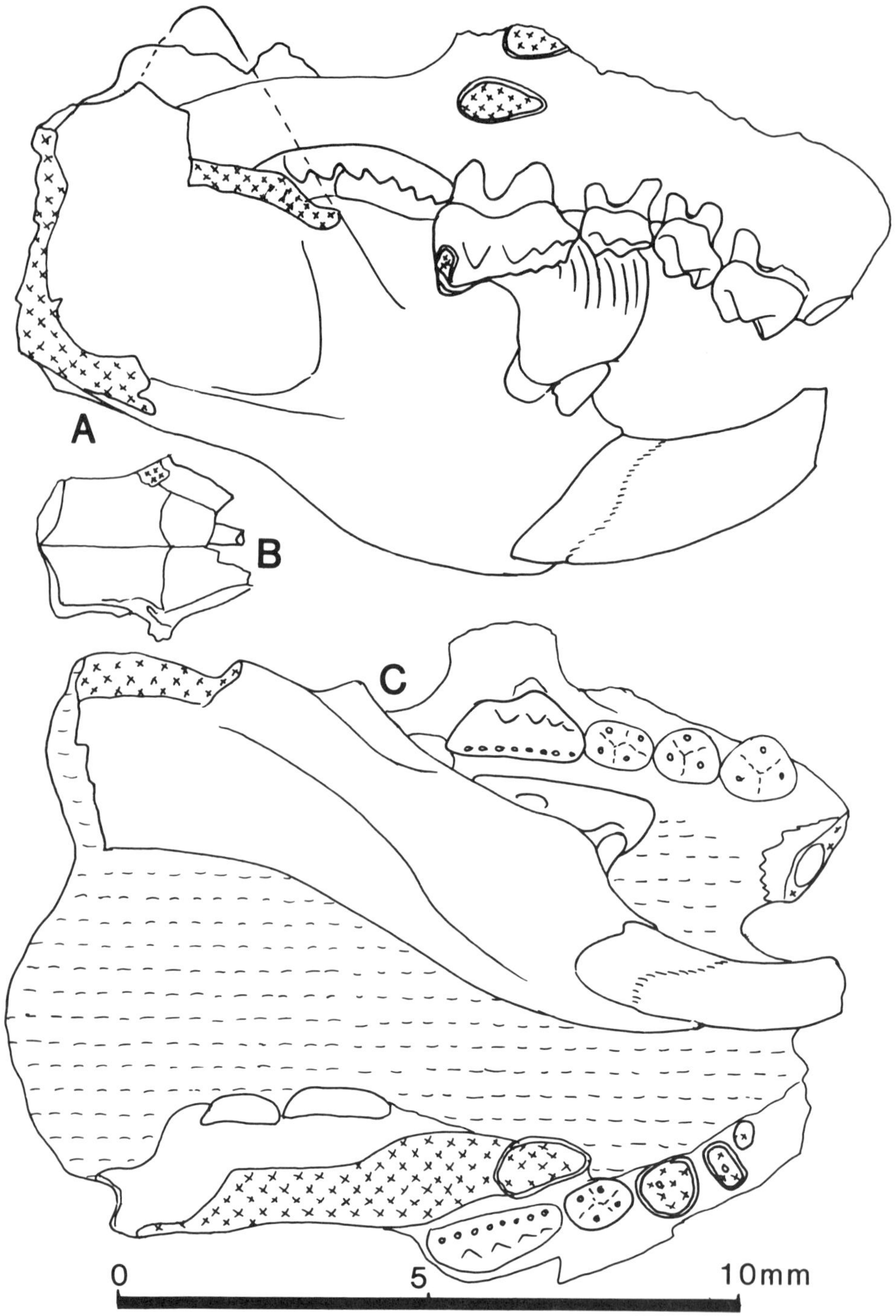

TEXT-FIG. 6.6—*Parectypodus clemensi*, holotype, KU 16001, from the *Deltatherium* "zone," lateral view (*A*), frontal view (*B*), and palatal view (*C*). The "x" symbol represents broken surfaces, dashes represent matrix, and dashed lines represent valleys in the teeth.

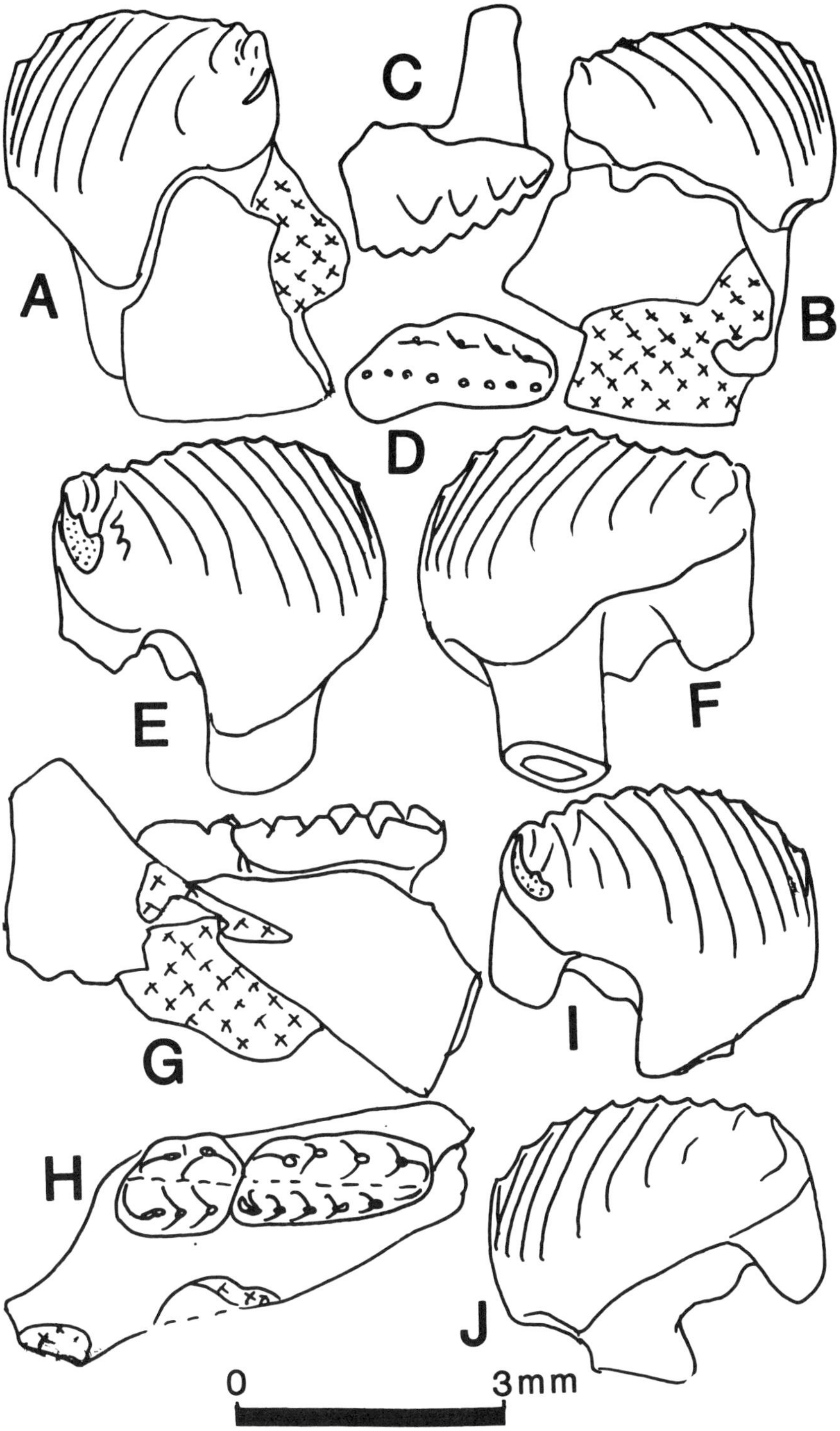

TEXT-FIG. 6.7—*Parectypodus clemensi*, referred specimens, KU 14032 from the *Deltatherium* "zone," labial view (*A*) and lingual view (*B*). Other specimens are from Swain Quarry; UMVP 6039, right P^4, labial view (*C*) and occlusal view (*D*); UMVP 6034, right M_b, labial view (*E*) and lingual view (*F*); UMVP 6038, jaw with M_1–M_2, labial view (*G*) and occlusal view (*H*); UMVP 6035, right M_b, labial view (*I*) and lingual view (*J*).

early Eocene San Jose Formation of New Mexico, and because there is more material of this species than of any other in the genus, it is described here.

This most common species of earliest Eocene neoplagiaulacids first became known from the Four-Mile local fauna of earliest Eocene age from Moffat County, Colorado (McKenna, 1960a). It is now known from the contemporary basal "Wasatch" Formation of the Powder River Basin in Johnson County, Wyoming (Delson, 1971); the basal portion of the Willwood Formation, Bighorn Basin, Washakie County, Wyoming (Bown, 1979), and the contemporary Golden Valley Formation, Stark County, North Dakota (West, 1973). These specimens were recognized as belonging to *Parectypodus* (Sloan, 1966; Van Valen & Sloan, 1966).

Kuehne (1969) gave the name *Charlesmooria childei* to a pair of multituberculate teeth from the London Clay without designating a type specimen. The two specimens do not belong to the same species; in fact, they are from different suborders. The incisor (BM M26618, Pl. 6.1, figs. 5–7) is a specimen of *Microcosmodon* (Jepsen 1930b), and probably belongs to a new species of that genus being described by Krause. The lower blade cannot be distinguished from the common North American species *P. childei*. This tooth, BM M26617 (Kuehne, 1969, fig. 1), is here designated the lectotype of *Charlesmooria childei*. Another M_b of this species, BM M15131, had been collected from the same locality earlier. *Charlesmooria* (Kuehne, 1969) thus becomes a junior synonym of *Parectypodus* (Jepsen, 1930a). The proper name for the most common species of early Eocene multituberculates in North America therefore becomes *Parectypodus childei* (Kuehne).

The two British specimens (Pl. 6.1, figs 1–4) have standard lengths of 2.9 and 3.0 mm and serration counts of 11 each. The largest sample of M_b's available to me is 72 blades from the East Alheit pocket of the Four-Mile local fauna, housed at the UCMP and AMNH. The observed range of the standard length is 2.8–3.8 mm, the mean is 3.24 mm, and the standard deviation is 0.2 mm. The number of serrations is 9–12; 10 is the modal number. Thirteen M_b specimens from nearby Sand Quarry and Timberlake Quarry have a mean standard length of 3.4 mm. The average relative height of the first serration of East Alheit specimens of *P. childei* is 0.56; the species lacks P_3. The mean length and width of 12 East Alheit specimens of M_1 is 1.92 mm and 0.98 mm respectively; the cusp formulae are 6–8:4 with 7:4 modal. The mean length of 12 specimens of P^4 from East Alheit is 2.34 mm with cusp formulae 2–4:5–6; the modal cusp formula is 2:6. The mean length and width of six East Alheit specimens of M^1 are 2.47 mm and 1.30 mm, respectively; the cusp formulae are 7–9:8–19:4–7; the modal formula is 7:10:5.

The size distribution of isolated neoplagiaulacid M_b's from East Alheit is bimodal with a small area of overlap between those of *P. childei* and those of *Ectypodus tardus*. P_3 is generally present in *E. tardus*, although missing in the type. The mean length of 26 East Alheit M_b's is 3.06 mm, the observed range is from 2.5 to 3.5 mm. Nine to twelve serrations are present with 10 the modal number. The mean relative height of the first serration is 0.40. Three lower jaws with M_b to M_2 (PU 13265, 16522, 17675), a skull (PU 14724; Sloan, 1979), and two maxillaries with P^4 to M^2 (PU 16522, 17674) are known from the early Graybullian Willwood Formation of the Bighorn Basin. One maxillary is associated with a lower jaw. In these specimens of *E. tardus*, the M_b's average 2.7 mm long, ranging from 2.3 mm to 3.0 mm; all have 10 serrations. The M_1's average 1.80 mm long; the cusp formulae are 6–8:4. The M_2's average 1.03 mm long with cusp formulae 3–4:2. The P^4's average 1.8 mm long with cusp formula 3:6. The M^1's average 2.23 mm long with cusp formulae 8–9:9–10:4. The M^2's average 0.93 mm long with cusp formula 1:3:3. *Ectypodus tardus* is slightly smaller than *Parectypodus childei* and can best be distinguished by the shapes of upper and lower blades.

P. childei is also known from the Sparnacian of Mutigny, France (29 specimens in the MNHN) and a slightly larger species closer to the type of the genus, *P. simpsoni* (Jepsen, 1930a), is known from the younger locality at Avenay, France (20 specimens in the MNHN).

Russell (1964) found no potential ancestors for this species in the late Paleocene of France, despite a large collection rich in multituberculates. The described North American species closest to the ancestry of this form are the late Tiffanian *Parectypodus laytoni* (Jepsen, 1940) and the early Tiffanian *P. sloani* (Schiebout, 1974). This species is close to the ancestry of *P. lovei* (Sloan, 1966).

Genus MIMETODON Jepsen, 1940
MIMETODON KRAUSEI n. sp.
Text-fig. 6.9

Type specimen.—KU 16002, crushed left maxillary containing P^1–M^2, KU locality 9 ("Little Pocket"), Torrejonian *Deltatherium* "zone" of the Nacimiento Formation, San Juan Basin, New Mexico.

Etymology.—This species is named for my esteemed colleague David W. Krause, who was the first to recognize and define the upper dentition of *Mimetodon.*

Diagnosis.—Medium sized *Mimetodon*, intermediate in size between *M. silberlingi* and *M. churchilli*; length P^1–M^2 13.5 mm; P^3 tricuspid; P^4 length 3.2 mm, width 1.25 mm, cusp formula 1:6; M^1 length 3.9 mm, width 1.5 mm, cusp formula 9:13:7; M^2 length 1.6 mm, width 1.4 mm, cusp formula 1:4:4.

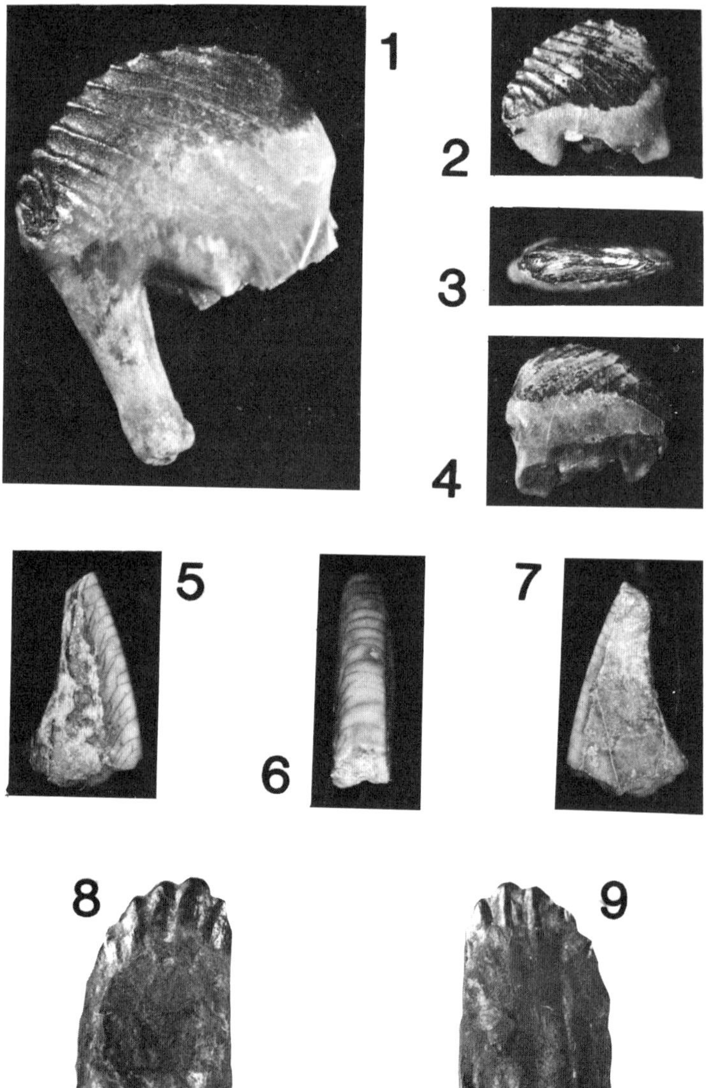

EXPLANATION OF PLATE 6.1

FIGS. *1–4*—*Parectypodus childei*. *1*, topotype (BM M15131), lower blade, × 14.4; lectotype (BM M26617), lower blade, *2*, labial view; *3*, occlusal view; *4*, lingual view. All × 7.2.

5–7—*Microcosmodon* sp., BM M26618, incisor from the Blackheath beds of Kent. *5*, lateral view; *6*, anterior view; *7*, medial view. All ×8.

8–9—*Taeniolabis taoensis*, AMNH 729a, unworn tip of incisor. *8*, lingual view; *9*, labial view. All ×4.

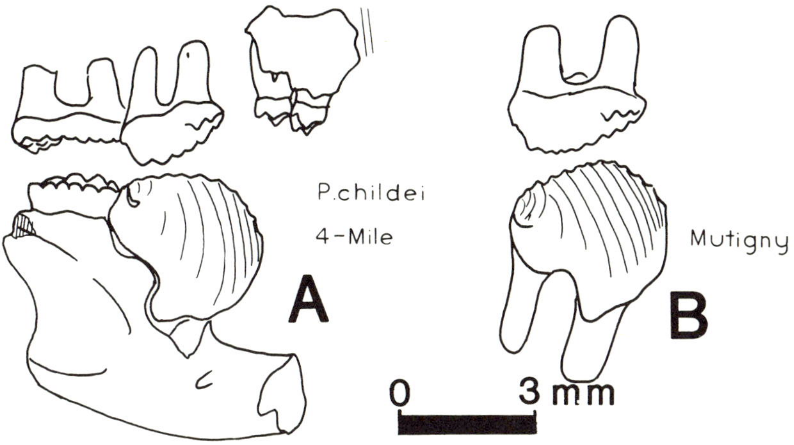

TEXT-FIG. 6.8—*Parectypodus childei*, referred specimens from East Alheit pocket, Four Mile fauna, Colorado and Mutigny, France. A composite restoration of the dentition includes: AMNH 80094, lower jaw and blade; 80420, M_1 in jaw; 59613, M^1 in maxillary; 80400, P^4; 80057, P^1, P^2 (A). MNHN Mu5589, M_b, and Mu6575, P^4 are from Mutigny (B).

Discussion.—This maxilla has the characteristic neoplagiaulacid vertically curved row of premolars with anteroventrally facing upper blade. Tooth proportions and cusp formulae agree with those of *M. silberlingi* (Krause, 1977), although the size is about 30% larger than that species and an estimated 10–20% smaller than the very much later *M. churchilli* (Jepsen, 1940). While no teeth are known in common with this species and *M. churchilli*, the estimated proportions of this species differ significantly from *M. churchilli*. This is the earliest species yet known of the genus.

Genus ECTYPODUS Matthew & Granger, 1921

Remarks.—*Ectypodus musculus* (Matthew and Granger, 1921) was the fourth neoplagiaulacid species and second genus described. The species was defined from bones in a presumed single carnivore den, the Mason Pocket, near Tiffany, Colorado. It is the only species of multituberculate from that locality; there is thus no doubt of the association of upper and lower jaws. This genus is clearly valid on the basis of the unique shape of the upper blade in the type species which has anterior and posterior slopes of equal length and a much higher crown than usual. No other known neoplagiaulacid genus has a similar blade, and, in particular, *Neoplagiaulax* and *Parectypodus* do not. There is a lineage of species that has lower blades comparable to that of *E. musculus* and upper blades intermediate in shape between that species and *Mesodma*

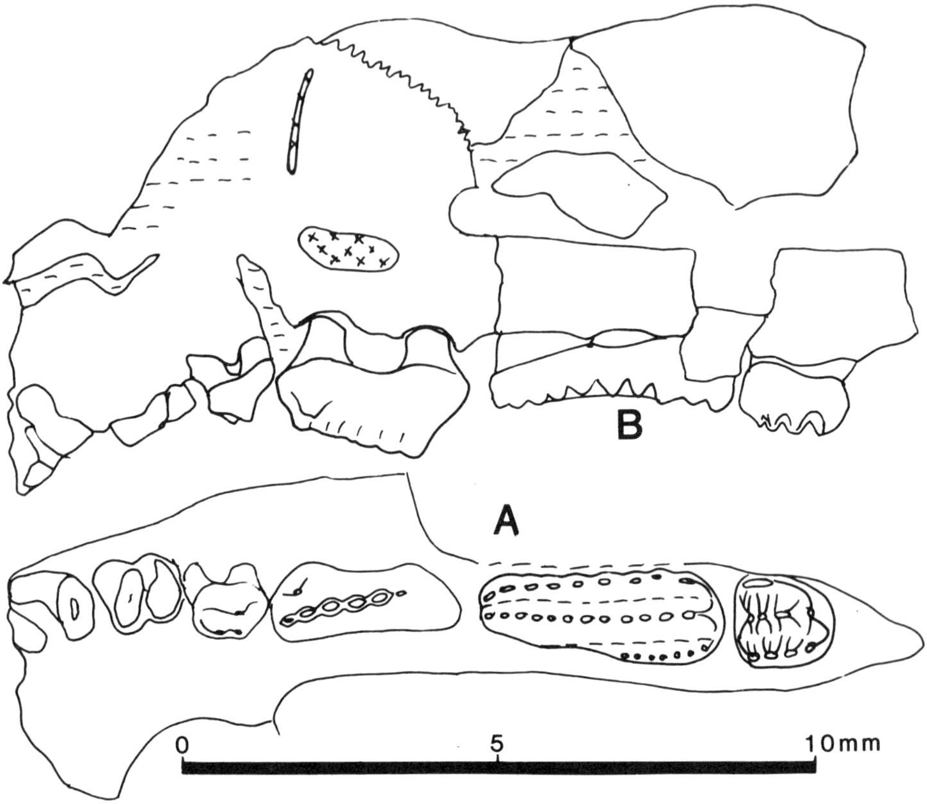

TEXT-FIG. 6.9—*Mimetodon krausei*, holotype, KU 16002 from "Little Pocket," *Deltatherium* "zone" in palatal (*A*) and lateral view (*B*).

formosa. This lineage is composed of *E. szalayi* n. sp. from the middle Torrejonian *Deltatherium* "zone" of New Mexico, the late Torrejonian *Pantolambda* "zone" of Gidley and Swain Quarries of Montana and Wyoming, and the early Tiffanian Shotgun Quarry of Wyoming; *Ectypodus powelli* (Jepsen, 1940) from many middle and late Tiffanian localities; and *Ectypodus tardus* (Jepsen, 1930a) from the early Eocene. It appears likely that this lineage was derived from *Mesodma formosa*. I would derive *E. musculus* from an undescribed late Torrejonian species descended from *E. szalayi*.

Ectypodus may be redefined as follows: Neoplagiaulacidae of small size, standard length of M_b 2.6–4.0 mm, 10 to 14 serrations; relative height of first serration between ⅓ and ½ the standard length of M_b; labial height of enamel at exodaenodont lobe less than standard length; posterior angle between plane of occlusion of molars and anterior face of the M_b approximately a right angle; fourth serration from front generally highest above base line for standard length; known length ratio

$M_b:M_1$ low to moderate for neoplagiaulacids (1.37–1.80, mode 1.60); P^4 about ⅔ length of M_b; P^4 cusp formulae 0–3:6–7; last cusp in principal row highest; P^4 very trenchant with both anterior and posterior slopes straight in profile; posterior basal cuspule usually weak or absent; P_3 missing in advanced species as in *Neoplagiaulax* and *Parectypodus*.

ECTYPODUS SZALAYI n. sp.
Text-fig. 6.10, Table 6.2

?*Ptilodus sinclairi*: SIMPSON, 1937, p. 27–98, (in part).
Ectypodus? sp. WILSON, 1956, p. 82.

Type material.—Holotype, AMNH 35536, left and right associated lower jaws, each with I_1 to M_2, from the Torrejonian *Pantolambda* "zone" of Gidley Quarry, Lebo Shale, Sweetwater County, Montana; paratypes, AMNH 35533, USNM 9792, 9796, 9797, jaws with P_3–M_2; AMNH 35518, 35524, 35525, USNM 9725, isolated M_b's all from the above type locality.

Referred specimen.—KU 9535, palate with left M_b, left I_1, right P^1–P^2 and roots of P^3–P^4, from KU locality 13, Torrejonian *Deltatherium* "zone" of the Nacimiento Formation, Kutz Canyon, San Juan Basin, New Mexico.

Etymology.—This species is named for my esteemed colleague Frederick S. Szalay.

Diagnosis.—P_3 present; M_b length 2.69 mm, range 2.6–2.8 mm, 10–13 serrations (11 modal); M_1 length 1.91 mm, range 1.8–2.0 mm, width 1.0, range 0.9–1.1 mm, cusp formulae 8–9:4–6, M_2 length 1.1 mm, range 1.0–1.1 mm, width 1.1 mm, range 1.0–1.1 mm, cusp formulae 4:2–3; M_b:M_1 length ratio 1.33–1.51, modal ratio 1.40.

Discussion.—This is presently the earliest known species of *Ectypodus*. Specimens of this species were included in the highly composite aggregation of species called ?*Ptilodus sinclairi* by Simpson (1937). In addition to the holotype, now included in *Parectypodus*, this aggregation also included specimens belonging to *Mimetodon silberlingi*, an undescribed species of *Mesodma* and another species of *Ectypodus*, as yet undescribed. All are closely similar in size, serration count, and cusp formulae, yet differ in shapes and ratios. The present species is clearly related to *Mesodma formosa*, *Ectypodus powelli*, and *E. tardus* and forms a link in this lineage of four species. *E. szalayi* is also known from Swain Quarry, from which an uncatalogued AMNH coprolite collected by Szalay shows the association of the upper and lower dentitions.

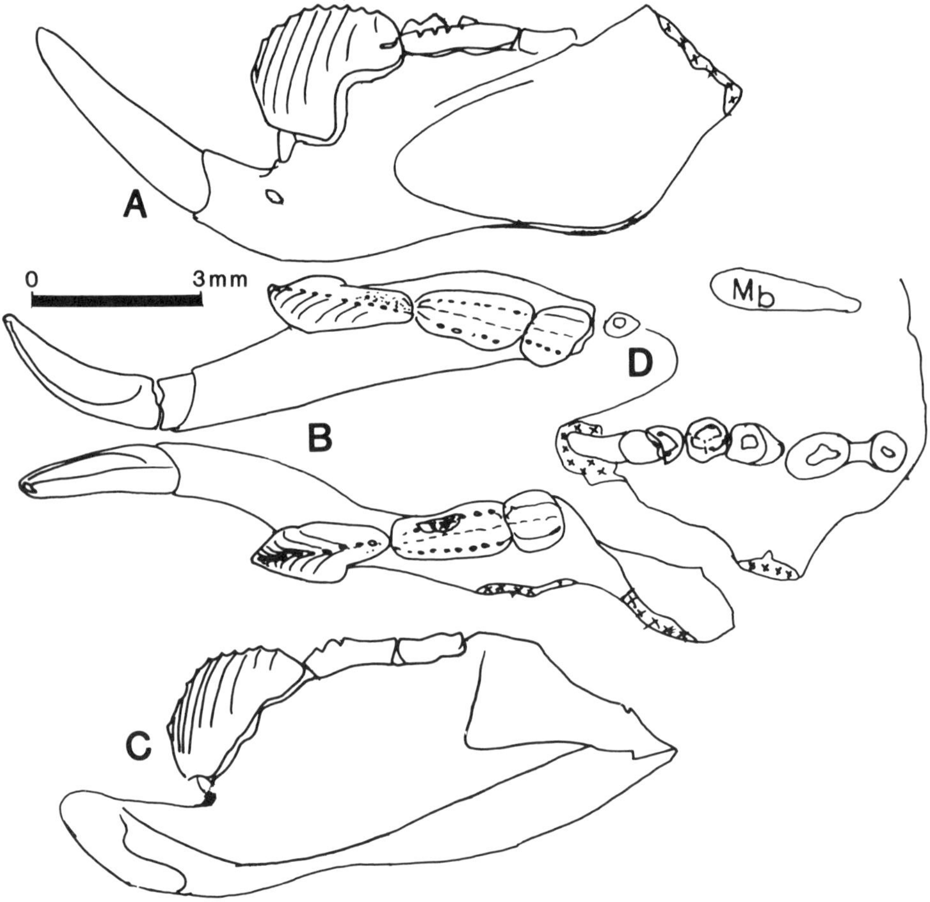

TEXT-FIG. 6.10—*Ectypodus szalayi*, holotype, AMNH 35536 from Gidley Quarry, labial (*A*) and occlusal (*B*) views; AMNH 35533, associated lower jaw with all teeth (*C*); and KU 9535, palate and M_b from Kutz Canyon, *Deltatherium* "zone" (*D*).

Table 6.2—Dental characters of *Ectypodus szalayi* (all dimensions in mm).

	M_b		M_1			M_2		
	Length	Number of Serrations	Length	Width	Cusp Formula	Length	Width	Cusp Formula
AMNH 35533	2.8	11	1.9	0.95	--	1.05	1.1	--
AMNH 35536 right	2.8	11	2.0	0.90	9:5	1.00	--	-:2
AMNH 35536 left	2.8	11	2.0	--	9:?	1.00	--	-:2
AMNH 35525	2.8	10						
AMNH 35518	2.8	12						
AMNH 35524	2.8	12						
USNM 9792	2.8	13	1.85	?	?:4	1.00	--	--
USNM 9796	2.6	11	1.85	--	--	1.10	--	--
USNM 9797	2.6	10	1.9	0.9	8:6	1.00	0.95	4:3
USNM 9725	2.7	10						
	Length	Width						
KU 9535								
P^1	0.9	0.6						
P^2	0.6	0.6						
P^3	0.8	0.6						
P^4	2.1	0.75						

Family CIMOLODONTIDAE Marsh, 1889
Genus ANCONODON Jepsen, 1940
ANCONODON GIDLEYI (Simpson, 1935)
Text-fig. 6.11

?*Ptilodus gidleyi* SIMPSON, 1935, p. 225; 1937, p. 95–97
Anconodon gidleyi: JEPSEN, 1940, p. 250

Referred material.—KU 12673, left M_b from KU locality 13, Torrejonian *Deltatherium* "zone" of the Nacimiento Formation, Kutz Canyon, San Juan Basin, New Mexico.

Discussion.—This rather badly worn specimen (length 6.0 mm with 14 serrations) extends the geographic and stratigraphic range of the species. At present this is the oldest known specimen of *Anconodon*.

TEXT-FIG. *6.11—Anconodon gidleyi*, KU 12673 from the *Deltatherium* "zone," Kutz Canyon.

Family PTILODONTIDAE Simpson, 1927

Remarks.—The resurrection of Neoplagiaulacidae and of Cimolodontidae requires that the definition of Ptilodontidae be revised. P^4 is the most characteristic tooth of this family. Its lateral profile is essentially straight with all cusps equally high and continuous with the line of cusps on molars and anterior premolars. The external row of cusps is always well developed on P^4 and there is frequently an additional external partial row of cusps. The ratio between the lengths of M_b and M_1 is greater than 1.6. Other characteristics are as in other families of the suborder Ptilodontoidea.

Genus KIMBETOHIA Simpson, 1936

Revised generic diagnosis.—Medium sized ptilodontids having four upper premolars with dental rows curving medially and ventrally, the latter curvature as in neoplagiaulacids rather than as in *Ptilodus*; P^2 tricuspid; P^3 quadricuspid; P^4 low and wider transversely than in other ptilodontids, with modal cusp formula of 1:4:7; P_3 inferred to be present; M_b profile lower than in *Ptilodus*; relative height of first serration about 0.33; one or two median labial striations often ventrally bifurcate; last two serrations, which descend in a posteriorly directed arc, having more prominent striations than others.

KIMBETOHIA CAMPI Simpson, 1936
Text-fig. 6.12, Table 6.3

Kimbetohia campi SIMPSON, 1936a, p. 2.
?*Mesodma* sp. CLEMENS, 1963, p. 43.

Holotype.—UCMP 31305, skull fragment with right P^2–P^4 and broken right and left P^1's, Puercan of Nacimiento Formation, Tsosie Arroyo, New Mexico.

Referred specimens.—AMNH 58672, 58540, 58468, right P^4's; 58789, left P^4; 58489, left P^4 (questionably referred); 58659, 58499, 59940, 27636, right M_b's; 58285, 58282, 58339, 58398, 58538, 58537 partial right M_b's; 58486, 58392, UCMP 36496, 89693, left M_b's; AMNH 58538, 58601, 58228b, partial left M_b's; all from the Puercan *Hemithlaeus* "zone" of the lower part of the Nacimiento Formation of Mammalon Hill, Betonnie Tsosie Wash, San Juan Basin, New Mexico. Also known from UCMP locality V5620, Late Cretaceous Lance Formation, Niobrara County, Wyoming.

Discussion—Simpson's (1936a) illustration of the type dentition of *Kimbetohia campi* shows the occlusal view of each tooth drawn separately. In fact, the upper premolars do not have their wear surfaces in a single plane. These teeth are aligned in a vertical curve roughly parallel to the profile of M_b as in neoplagiaulacids, quite in contrast to the straight row of *Ptilodus*. The type specimen of *K. campi* includes the root of the right zygoma, which originates opposite P^3 and P^4 as in *Ptilodus*. The palatal vacuity resembles that of *Ptilodus*. In palatal view, the line of upper premolars is bowed labially with P^2 and P^3 implanted more labially than P^1 and P^4, again a similarity to *Ptilodus*.

A fair number of M_b's of appropriate size for occlusion with the type and referred P^4's of *K. campi* have now been collected. The M_b's agree with P^4's of *Kimbetohia* in being intermediate in morphology between *Mesodma thompsoni* and those of the various species of *Ptilodus*. Clemens (1963, p. 43) referred teeth of this type, from the Lance Formation of Wyoming, to ?*Mesodma* sp. These lower blades are larger than those of any known species of *Mesodma* and differ from *Mesodma* blades in having a series of diagonal labial wrinkles slanting posteroventrally toward the posteroexternal cusp. The lateral profile is somewhat higher than in any species of *Mesodma* but lower than in any species of *Ptilodus*. The relative height of the first serration is about 0.33, again intermediate between *Mesodma* and *Ptilodus*. In addition, the medial striations on the labial face of the blade are frequently ventrally bifurcate. The break in arc of the lateral profile of the lower blade at the first serration is more pronounced in *Kimbetohia* than in *Ptilodus* and about as in some species of *Mesodma*. An additional species of *Kimbetohia* is known from the late Puercan Purgatory Hill local fauna of Montana. *K. campi* is at present the most likely ancestor for the species of *Ptilodus* and most probably was derived from a large species of *Mesodma*, such as the Campanian *M. primaevus*.

Genus PTILODUS Cope, 1881
PTILODUS TSOSIENSIS n. sp.
Text-fig. 6.13, Table 6.3

Type material.—Holotype, AMNH 59821, left P^4; paratypes, 58464, 58658, 59872, 59874, left M_b's; 59800 right M_b; all from Puercan *Hemithlaeus* "zone" of the lower part of the Nacimiento Formation in Betonnie Tsosie Wash, San Juan Basin, New Mexico.

Etymology.—This species is named for Betonnie Tsosie Wash.

Diagnosis.—Smallest species of *Ptilodus*; mean length of M_b 6.27 mm, range 6.0–6.7 mm, 10–12 serrations; differs from *P. douglassi* and *P. kummae* in profile; P^4 4.5 mm long, mean width 2.75 mm, width range 2.4–3.1 mm, cusp formulae 2-4:8:8.

Discussion.—This species is the smallest and earliest species of *Ptilodus*. It is the closest to the late Torrejonian *P. douglassi* and the late Tiffanian *P. kummae*, in size and proportions. It most likely is ancestral

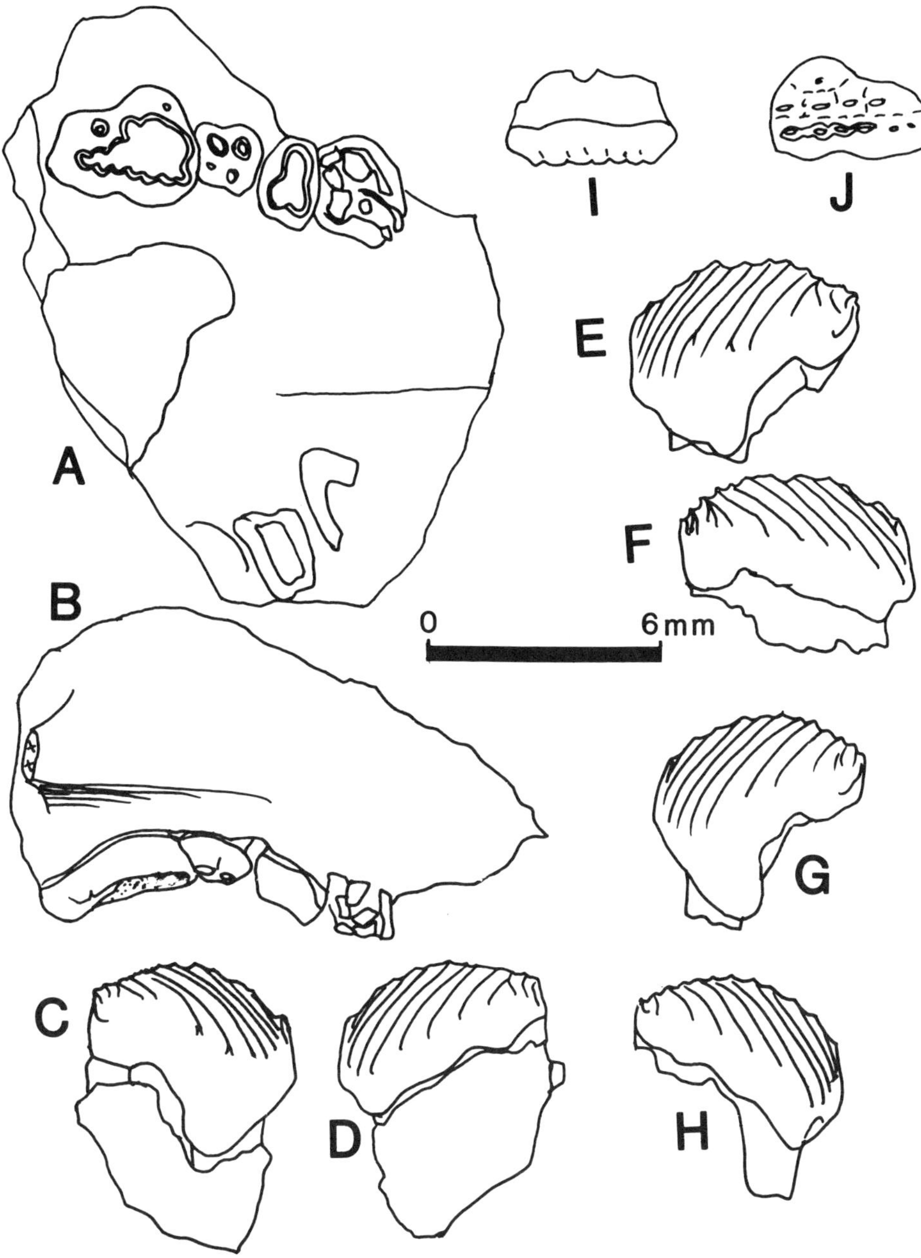

TEXT-FIG. *6.12*—*Kimbetohia campi*, holotype, UCMP 31305, palate with right P^{1-4}, broken roots of left P^{1-2}, and partial alveolus for left P^{3}, palatal view (*A*) and right lateral view (*B*) showing shape of maxilla and profile of P^{1-4}; AMNH 58499, right M_b in mandible fragment, labial view (*C*) and lingual view(*D*); UCMP 36496, left M_b, labial view (*E*) and lingual view (*F*); AMNH 58392, left M_b, labial view (*G*) and lingual view (*H*); AMNH 58789, left P^4, lingual view (*I*) and occlusal view(*J*).

Table 6.3—Blade dimensions of Ptilodontidae from the Puercan of Tsosie rincon, New Mexico (length and width of M_b and P^4 in mm).

Specimen	M_b Length	Number of Serrations	Bifurcation on Labial Striation	P^4 Length	P^4 Width	Cusp Formula
Kimbetohia campi						
AMNH 58392	5.8	12	(absent)	--	--	--
AMNH 27636	5.8	14	8,9	--	--	--
AMNH 58499	5.9	12	7,8	--	--	--
AMNH 58659	6.0	12	7	--	--	--
AMNH 58486	6.1	12	?	--	--	--
UCMP 36496	6.3	13	7,9	--	--	--
AMNH 59940	6.5	10	(absent)	--	--	--
AMNH 58228b	--	--	7,8	--	--	--
AMNH 58672	--	--	--	3.8	2.7	1:4:7
UCMP 31305	--	--	--	4.1	2.8	1:4:7
AMNH 58540	--	--	--	4.2	2.7	2:4:8
AMNH 58789	--	--	--	4.3	2.6	1:4:7
UCMP 89693	6.6	11	7,8	--	--	--
AMNH 58468	--	--	--	--	3.1	4:4+:5+
Ptilodus tsosiensis						
AMNH 58464	6.7	12	--	--	--	--
AMNH 58658	6.1	10	--	--	--	--
AMNH 59800	6.0	11	--	--	--	--
AMNH 59821	--	--	--	4.5	2.4	2:8:8

to both that lineage and the other better known lineage of *Ptilodus ferronensis*, *P. mediaevus*, *P. montanus*, *P. wyomingensis*, and *Prochetodon cavus*.

Suborder TAENIOLABIDOIDEA Sloan & Van Valen, 1965
Family EUCOSMODONTIDAE Jepsen, 1940
Genus EUCOSMODON Matthew and Granger, 1921
EUCOSMODON PRIMUS Granger & Simpson, 1929, new rank
Pl. 6.2, figs. 1–4

Eucosmodon americanus primus GRANGER & SIMPSON, 1929, p. 649.

Type material.—Holotype, AMNH 16327, right M_b; paratype, AMNH 16328, broken left I_2; both from the Puercan *Hemithlaeus* "zone" in the lower part of the Nacimiento Formation "3 miles east of Kimbetoh," Betonnie Tsosie Wash, San Juan Basin, New Mexico.

Referred specimens.—AMNH 16325, partial skeleton from the above locality; 58005, right jaw with broken M_b; 58085, 59999, right I_1's;

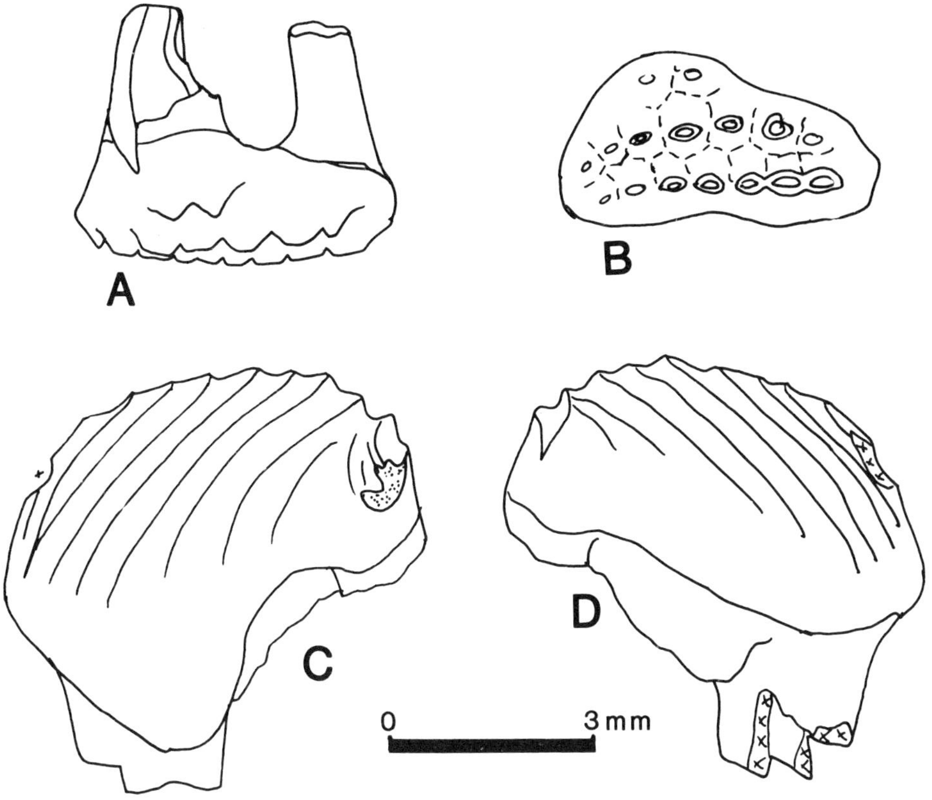

TEXT-FIG. 6.13—*Ptilodus tsosiensis*, holotype, AMNH 59821, left P^4, labial view (*A*) and occlusal view (*B*); AMNH 58658, right M_b, labial view (*C*), (X where first serration is broken) and lingual view (*D*).

59934, right upper incisor; all from the Puercan *Hemithlaeus* "zone" in the lower part of the Nacimiento Formation at Mammalon Hill, Betonnie Tsosie Wash, San Juan Basin, New Mexico.

Diagnosis.—Smallest species of *Eucosmodon*, about 12% smaller than *E. americanus* from the *Taeniolabis* "zone"; mean length of M_b 9.2 mm, observed range 9.0–9.4 mm, 10 serrations; mean height of lower incisor 4.17 mm, observed range 4.0–4.3 mm; mean width of lower incisor 2.03 mm, observed range 1.9–2.2 mm; maximum width of upper incisor 3.6 mm.

Discussion.—The only known species of *Eucosmodon* are *E. primus* from the Puercan *Hemithlaeus* "zone," *E. americanus* from the later Puercan *Taeniolabis* "zone," and *E. molestus* from the Torrejonian. They are known only from the San Juan Basin. No specimens have ever been found elsewhere, published references to the contrary. In *Eucosmodon*, the M_b:M_1 length ratio is changing in the opposite direction from that in *Stygimys* and *Microcosmodon*. In addition, the M_b and P^4 have lost their shear function but function as additional grinding teeth in a most unusual fashion. The major change in the *Eucosmodon* lineage is increase in size. The two known M_b's of *E. primus* are 9.1 mm (AMNH 16327) and 9.0 mm (AMNH 58005) long. Those of *E. americanus* are 10.6 and 10.5 mm (AMNH 3028) and 10.0 mm (AMNH 16326) long, whereas those of *E. molestus* are 15.0 mm and 15.1 mm long in the two known specimens.

Family TAENIOLABIDIDAE Granger and Simpson, 1929
Genus CATOPSALIS Cope, 1882
CATOPSALIS FOLIATUS Cope, 1882

Type specimen.—AMNH 3035, fragment of right lower jaw with M_b–M_2, probably from the Puercan *Hemithlaeus* "zone," of the Nacimiento Formation, San Juan Basin, New Mexico.

Referred specimen.—AMNH 3030, edentulous left ramus from Puercan deposits of New Mexico.

Discussion.—*Catopsalis* is never a very common animal throughout its range from the Late Cretaceous to the late Tiffanian (Kielan-Jaworowska & Sloan, 1979). The type of *C. foliatus* was described at a time when Puercan and Torrejonian levels in the San Juan Basin were not recognized. Granger and Simpson (1929) concluded that the specimen was from the Torrejonian, apparently on the basis that *C. fissidens* occurs in the Torrejonian, and that at that time few genera were thought to occur in both Puercan and Torrejonian sediments. I have not been able to find a valid record of *C. foliatus* from the Torrejonian, Granger and Simpson (1929, p. 662) to the contrary. AMNH 3030 is a left ramus collected by Baldwin from the Puercan deposits. Cope (1888, pp. 307, 308, 361) referred this specimen to *Eucosmodon molestus*. Granger and Simpson (1929, p. 650) showed clearly that this identification was wrong, but suggested no other. McKenna (1960b) pointed out that the stump of the lower incisor was present and resembled that of *Taeniolabis* in cross section but indicated that the more posterior of the two cheek teeth alveoli had to be a posterior alveolus of some tooth since enamel is preserved on the posterior edge of the contained root. Granger and Simpson (1929) had interpreted this pair of simply visible alveoli as that of a *Eucosmodon* type M_b. In view of the elongate planar character of the posterior root of a *Eucosmodon* M_b, this simply cannot be. The size and proportions of this ramus agree more closely with those of the type and only San Juan Basin specimen of *C. foliatus* than with any other described species. There are pits in the usual encrustation characteristic of San Juan Basin specimens behind the two principal alveoli. These suggest that this was a mature jaw with M_2 present and that the principal alveoli are those of M_1. My interpretation is that this is a specimen of *C. foliatus* with both closely spaced alveoli for M_b broken away. *Catopsalis foliatus* occurs in the Mantua Lentil of the Polecat Bench Formation based on incisors referred to *Eucosmodon gratus* by Jepsen (1940, p. 279, Pl. 1, fig. 3) and personal observations of: PU 14550, left P^4; 16647, right M^1; 16648, right M^2; 18196, right lower incisor; 16649, left M_1; 16650, left M_2; 14494, 11416, right upper incisors. It also occurs in the lower Puercan of the Denver Formation, South Table Mountain, Golden Colorado (UCM 34979). The larger species, *C. utahensis*, appears in the upper Puercan *Taeniolabis* "zone" of Barrel Spring Arroyo (=De-na-zin Wash), San Juan Basin, New Mexico. It seems likely that the type specimen of *C. foliatus* and AMNH 3030 came from the lower Puercan fossil level (*Hemithlaeus* "zone") either from Kimbeto Wash or Tsosie rincon, Betonnie Tsosie Wash.

CATOPSALIS UTAHENSIS Gazin, 1939
Pl. 6.2, figs. 5–7

Referred specimen.—USNM 23273, left lower incisor from the Puercan *Taeniolabis* "zone" of the lower part of the Nacimiento Formation, Barrel Spring Arroyo (De-na-zin Wash), San Juan Basin, New Mexico.

Discussion.—This incisor has a maximum height of 8.0 mm, maximum width of 5.7 mm, and a ventral radius of curvature of about 51 mm. This is much too large to be an incisor of *C. foliatus*, the type of which has a cross section of 6.5 mm by 4.0 mm. Assuming no change in proportions, this suggests that USNM 23273 came from an animal with M_1 length of 13 mm. This is the size of *C. utahensis* (Gazin, 1939) from the early Torrejonian Dragon local fauna. The tooth had not been in wear long. Anteriorly, the enamel extends onto the dorsal surface as is characteristic of unworn taeniolabidoid incisors. In the posterior half of the fragment, the enamel is restricted to the ventral surface and lower two-thirds of the labial surface. The enamel shows many fine transverse growth lines but no sign of the longitudinal fluting characteristic of *Taeniolabis*. The labial surface of the enamel is smooth and polished for about 5 mm from the tip; more posteriorly the surface is rough in the fashion of short, shallow irregular longitudinal wrinkles. The incisor closely resembles that of *C. joyneri*, except for its much larger size.

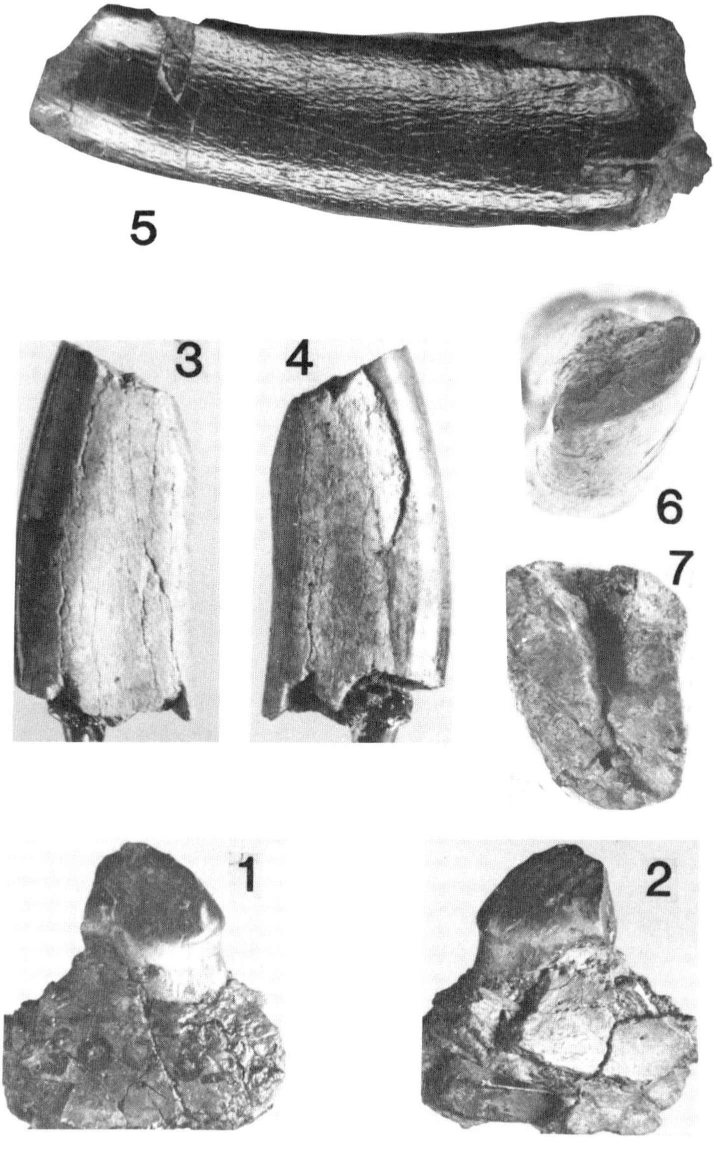

EXPLANATION OF PLATE 6.2

FIGS. *1–4*—*Eucosmodon primus*, AMNH 58085, from Tsosie rincon, right blade in jaw. *1*, labial view; *2*, lingual view, ×3; AMNH 59999 from Tsosie rincon, right lower incisor; *3*, medial view; *4*, labial view, ×6.

5–7—*Catopsalis utahensis*, USNM 23273, left lower incisor from Barrel Spring Arroyo (De-na-zin Wash). *5*, labial view; *6*, cross section at tip; *7*, cross section at root. All ×3.75.

Genus TAENIOLABIS Cope, 1882
TAENIOLABIS TAOENSIS (Cope), 1882
Pl. 6.1, figs. 8–9; Text-figs. 6.14, 6.15

Discussion.—AMNH 729a, collected by the 1892 AMNH expedition from the upper level of the Puercan of "Coal Creek Canyon" is an unworn tip of a *Taeniolabis* right lower incisor. This specimen clearly shows that *Taeniolabis*, like *Meniscoessus* (Clemens 1963, p. 84) and *Catopsalis joyneri*, had a complete enamel cap on the tip of the unworn lower incisor. Five cusps are present in a line on the transverse cutting edge of the incisor; the medial or first cusp projects as far as the third; the second projects farthest. Shallow grooves extend posteriorly between these cusps on both dorsal and ventral surfaces for about 3 mm behind the crest of the second cusp.

AMNH 16321, the best known skull of *T. taoensis*, described by Broom (1914) and Granger and Simpson (1929), has always been restored with a small orbit extending forward to above the front of P^4. This restoration is in error: the restored orbit is really the pocket for the origin of the anterior deep masseter muscle, with the maxillary shelf, the origin of that muscle, broken away. The orbit of *Taeniolabis* is then farther to the rear with its front edge above the middle of M^1 and an apparent orbital diameter of about 35 mm. The large orbit suggests nocturnal or crepuscular habits, a conclusion which also seems likely for the Asiatic forms of this suborder which also have proportionally large eyes. A revised restoration is shown in Text-figure 6.14.

AMNH 752, collected by Wortman and Peterson from the head of "Coal Creek Canyon" in 1892, includes a partial right femur of *T. taoensis*. It shows nothing particularly unusual and is figured (Text-fig. 6.15C) only for the sake of completeness and to demonstrate the size of this largest of all multituberculates. AMNH 3036, the type of *T. taoensis*, collected by Baldwin in 1882, includes some postcranial material as well as the jaws and teeth discussed by others. An astragalus and a pair of

TEXT-FIG. *6.14—Taeniolabis taoensis*, AMNH 16321, skull restoration revised from Granger and Simpson (1929).

TEXT-FIG. *6.15*—*Taeniolabis taoensis*, holotype, AMNH 3036, fragmentary left ulna, dorsal view (*A*) and medial view (*B*); AMNH 752, partial right femur, posterior view (*C*). All × 0.75.

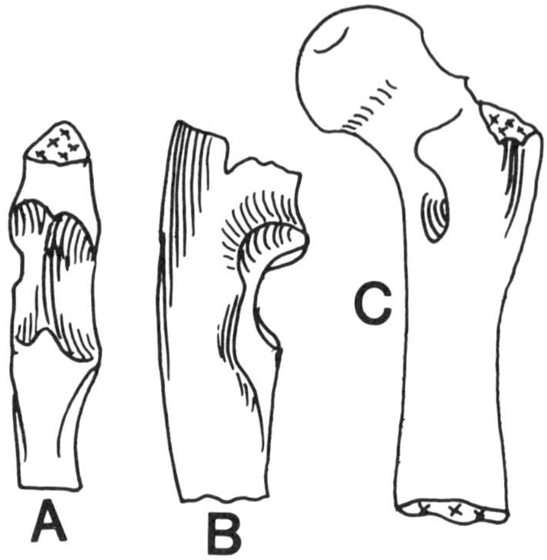

humeri are from some indeterminate placental mammal, but there is a fragmentary left ulna properly referable to this species. As in the case of the femur it serves to show the large size of this species.

SUMMARY

New taxa of multituberculates from the Puercan of the San Juan Basin of New Mexico include *Ptilodus tsosiensis*, *Neoplagiaulax macintyrei*, *Parectypodus vanvaleni* (all *Hemithlaeus* "zone"), and *Catopsalis utahensis* (*Taeniolabis* "zone"). A new skull restoration of *Taeniolabis* and the lower blade of *Kimbetohia* are figured. New taxa from the middle Torrejonian *Deltatherium* "zone" include *Ectypodus szalayi*, *Parectypodus clemensi* and *Mimetodon krausei*. Diagnoses of the genera *Neoplagiaulax*, *Ectypodus* and *Parectypodus* are given. *Charlesmooria childei* Kuehne from England, France, and North America is referred to *Parectypodus*.

ACKNOWLEDGMENTS

I thank R. W. Wilson, W. A. Clemens, Jr., D. E. Russell and M. C. McKenna for permission to study specimens collected by them or under their care. I thank W. R. Hamilton for the photographs of British Museum specimens and the editors and the Albuquerque District of the BLM for aid in the manuscript preparation and revision.

An early version of this paper was submitted and accepted, but not published, in 1967 as part of a multiauthor work. For a variety of reasons the paper did not appear, and I apologize to my colleagues for the long delay in publication.

REFERENCES

Bown, T. M. 1979. Geology and mammalian paleontology of the Sand Creek Facies, lower Willwood Formation (lower Eocene), Washakie County, Wyoming. Geol. Surv. Wyo. Mem. 2, 151 p.

Broom, R. 1914. On the structure and affinities of the Multituberculata. Bull. Am. Mus. Nat. Hist. 33:115–134.

Clemens, W. A. 1963. Fossil mammals of the type Lance Formation Wyoming. Part 1. Introduction and multituberculates. Univ. Cal. Publ. Geol. Sci. 48:1–105.

Cope, E. D. 1888. Synopsis of the vertebrate fauna of the Puerco epoch. Am. Phil. Soc. Trans. 16:298–361.

Delson, E. 1971. Fossil mammals of the early Eocene Powder River local fauna, Eocene of northeast Wyoming. Bull. Am. Mus. Nat. Hist. 146:309–364.

Dorr, J. A. 1952. Early Cenozoic stratigraphy and vertebrate paleontology of the Hoback Basin, Wyoming. Geol. Soc. Am. Bull. 63:59–94.

Gazin, C. L. 1939. Further contribution to the Dragon Paleocene fauna of central Utah. Wash. Acad. Sci. J. 29:273–286.

Granger, W. and G. G. Simpson. 1929. A revision of the Tertiary Multituberculata. Bull. Am. Mus. Nat. Hist. 41:601–676.

Holtzman, R. C. 1978. Late Paleocene mammals of the Tongue River Formation, western North Dakota. North Dakota Geol. Surv. Rept. Inv. 65:1–88.

Jepsen, G. L. 1930a. New vertebrate fossils from the lower Eocene of the Bighorn Basin, Wyoming. Am. Phil. Soc. Proc. 73:117–131.

————. 1930b. Stratigraphy and paleontology of the Paleocene of northeastern Park County, Wyoming. Am. Phil. Soc. Proc. 73:463–528.

————. 1940. Paleocene faunas of the Polecat Bench Formation, Park County, Wyoming. Am. Phil. Soc. Proc. 83:217–341.

Kielan-Jaworowska, Z. and R. E. Sloan. 1979. *Catopsalis* (Multituberculata) from Asia and North America and the problem of taeniolabidid dispersal in the Late Cretaceous. Acta. Paleontol. Polonica 24:187–197.

Krause, D. W. 1977. Paleocene multituberculates (Mammalia) of the Roche Percee local fauna, Ravenscrag Formation, Saskatchewan, Canada. Palaeontographica, Abt. A. 159:1–36.

Kuehne, W. G. 1969. A multituberculate from the Eocene of the London Basin. Proc. Lond. Geol. Soc. 1658:199–202.

Lemoine, V. 1880. Communication sur les Ossements fossil des terrains Tertiare inferieurs. Ass. Franc. Av. Sc., Reims:3–40.

————. 1882. Sur deux *Plagiaulax* Tertiares, recueillis aux environs de Reims. C. R. Acad. Sci. 95:1009–1011.

————. 1885. Etude sur quelques Mammiferes de petite taille de la faune Cernaysienne des environs de Reims. Bull. Soc. Geol. France, ser. 3, 13:203–217.

Marsh. O. C. 1889. Discovery of Cretaceous Mammalia II. Am. J. Sci., Ser. 3, 38:177–180.

Matthew, W. D. and W. Granger. 1921. New genera of Paleocene mammals. Am. Mus. Nat. Hist. Novitates 13:1–7.

McKenna, M. C. 1960a. Fossil Mammalia from the early Wasatchian Four Mile local fauna, Eocene of northwest Colorado. Univ. Cal. Publ. Geol. Sci. 37:1–130.

————. 1960b. A continental Paleocene vertebrate fauna from California. Am. Mus. Nat. Hist. Novitates 2024:1–20.

Rigby, J. K., Jr. 1980. Swain Quarry of the Fort Union Formation, Middle Paleocene (Torrejonian), Carbon County, Wyoming: Geologic setting and mammalian fauna. Evol. Monogr. 3, 179 p.

Russell, D. E. 1964. Les Mammiferes Paleocenes D'Europe. Mus. Nat. D'Hist. Nat. Mem., n.s., Ser. C, 13:1–324.

Schiebout, J. A. 1974. Vertebrate paleontology and paleoecology of Paleocene Black Peaks Formation, Big Bend National Park, Texas. Tex. Mem. Mus. Bull. 24:1–88.

Simpson, G. G. 1935. New Paleocene mammals from the Fort Union of Montana. U.S. Natl. Mus. Proc. 83:221–244.

————. 1936a. Additions to the Puerco fauna, lower Paleocene. Am. Mus. Nat. Hist. Novitates 849:1–11.

————. 1936b. A new fauna from the Fort Union of Montana. Am. Mus. Nat. Hist. Novitates 873:1–27.

————. 1937. The Fort Union of the Crazy Mountain Field, Montana and its mammalian fauna. U.S. Natl. Mus. Bull. 169:1–287.

Sloan, R. E. 1966. Paleontology and geology of the Badwater Creek area, central Wyoming, part 2. The Badwater Multituberculata. Ann. Carn. Mus. 38:309–315.

————. 1979. Multituberculata, pp. 492–498. *In:* R. W. Fairbridge and D. Jablonski (eds.), Encyclopedia of Paleontology. Dowden, Hutchinson and Ross, Stroudsberg, Pa.

Sloan, R. E. and L. Van Valen. 1965. Cretaceous mammals from Montana. Science 148:220–227.

Van Valen, L. and R. E. Sloan. 1966. The extinction of the multituberculates. Syst. Zool. 15:261–278.

West, R. M. 1973. New records of fossil mammals from the early Eocene Golden Valley Formation of North Dakota. J. Mammal. 54:749–750.

Wilson, R. W. 1956. A new multituberculate from the Paleocene Torrejon fauna of New Mexico. Kans. Acad. Sci. Trans. 59:76–84.

7

A review of the Torrejonian (Middle Paleocene) Primates from the San Juan Basin, New Mexico

Glenn C. Conroy
Division of Biology & Medicine,
Brown University
Providence, Rhode Island

INTRODUCTION

Middle Paleocene primates are rare in the San Juan Basin. As presently understood, the primate fauna consists of four genera and five species: *Palaechthon nacimienti*, *P. woodi*, *Torrejonia wilsoni*, *Paromomys* sp., and *Talpohenach torrejonius*. *T. torrejonius* and *Paromomys* sp. are known from only one specimen each whereas *P. woodi* is known from only two (Kay & Cartmill, 1977; Wood et al., 1979; Tomida, this volume). The combined hypodigm published for the remaining two species, *P. nacimienti* and *T. wilsoni*, includes fewer than a dozen specimens, including a crushed skull of *P. nacimienti* (Gazin, 1968; Wilson & Szalay, 1972; Kay & Cartmill, 1974, 1977). Representatives of other Torrejonian age primate families (Plesiadapidae, Carpolestidae, Picrodontidae) known from other sites in western North America are absent or unrecognized in the New Mexico middle Paleocene. There are, however, Tiffanian carpolestids, plesiadapids, and picrodontids from the San Juan Basin (Mason Pocket locality) in southern Colorado.

It is of historical interest to recall that earlier paleontologists found "in the Puerco numerous remains of the Primates, and there is every reason to believe that these animals were both abundant and highly specialized or modernized" (Osborn & Earle, 1895, p. 15). How does one reconcile the known fossil record with this last statement?

By the early 1880's, E.D. Cope, the "discoverer of the Puerco Eocene," had already defined two families of "lemuroids"—Mixodectidae and Anaptomorphidae (Cope, 1883). Cope included within the Mixodectidae no fewer than five genera (*Tricentes*, *Necrolemur*, *Mixodectes*, *Microsyops*, *Cynodontomys*) and within the Anaptomorphidae two genera (*Indrodon* and *Anaptomorphus*). Four of these seven "lemuroid" genera had been recovered from the San Juan Basin (*Tricentes*, *Mixodectes*, *Indrodon*, *Cynodontomys*).

Cope's division of "lemuroids" into two families was based primarily on the number of upper premolars in each, Mixodectidae with three and Anaptomorphidae with two. He had previously removed the Adapidae from the Primates, referring them instead to the insectivorous division of his Bunotheria (allied to tupaiids and erinaceids). This allocation was based primarily on features of the tarsus. At the same time, Cope considered *Chriacus* to be the nearest ally of these Puerco "lemuroids."

Osborn & Earle (1895) formalized Cope's earlier opinions of the affinities of *Chriacus* by defining a third family of "lemuroid primates" from the "Puerco Beds," the Chriacidae. This family included *Chriacus*, *Protochriacus* and *Epichriacus* all from the "lower" beds. These authors also included species of the following genera (*incertae sedis*) within the Primates: *Loxolophus*, *Tricentes*, and *Ellipsodon*. These "lemuroids" were placed in Osborn's Mesoplacentalia, a division of the Eutheria characterized by animals having "a large number of very primitive characters and the apparent incapacity for progressive evolution" (Osborn & Earle, 1895, p. 6).

The year 1895 was the highwater mark for the apparent abundance and diversity of primates from the "Puerco Beds." Only two years hence, Matthew (1897) effectively removed all but one "Puerco Bed" genus (*Indrodon*) from the Primates. *Cynodontomys*, considered a primate by some (e.g. Bown & Gingerich, 1973) is an Eocene form. The other genera, such as *Mixodectes* and *Chriacus*, were referred to Rodentia and Creodonta respectively. When Matthew's (1937) definitive monograph on the Paleocene faunas of the San Juan Basin was published posthumously, primates were conspicuously absent from the faunal lists. Matthew's conclusions were generally accepted until the late 1940's, at least as far as the primates were concerned.

The absence of primates from the early and middle Paleocene faunas of the San Juan Basin was difficult to explain. Paleocene primates had been recovered in some abundance from northern parts of the Rocky Mountain region as early as 1908 (Gidley, 1923; Jepsen, 1930; Simpson, 1937). Torrejonian age primates from these areas include at least eight genera: *Pronothodectes*, *Elphidotarsius*, *Picrodus*, *Paromomys*, *Palenochtha*, *Plesiolestes*, *Palaechthon* and *Torrejonia*. In the absence of any obvious paleoecological barriers to north-south dispersal of early Tertiary primates, one could only speculate that: 1) the primates had not yet been discovered in New Mexico; 2) primates did not inhabit the San Juan Basin in the early and middle Paleocene; and/or 3) the northern Rocky Mountain areas and the San Juan Basin were producing faunas representative of different environmental facies.

Field exploration in the 1940's, particularly by the University of Kansas and the Smithsonian Institution, finally demonstrated that options 1 and 2 were no longer applicable and that option 3 was at least partially correct (see Wilson, 1951, 1956, 1967). In 1948, Robert Wilson and a field party from the University of Kansas discovered primate fossils in the Kutz Canyon area of the San Juan Basin (Wilson, 1951). These specimens were originally cited as "*Palaechthon*? n. sp." and "phenacolemuroid, n. gen. and sp." without description (Wilson, 1956). Most of this material has been subsequently described as a new species of *Palaechthon*, *P. nacimienti* (Wilson & Szalay, 1972; Kay & Cartmill, 1977).

One year after Wilson's original discovery of primates in the middle Paleocene of the San Juan Basin, C.L. Gazin and Franklin Pearce discovered other primate fossils from Torreon Wash (= Arroyo Torrejon), essentially the type locality for the middle Paleocene land mammal "age" as designated by Wood et al. (1941). Gazin (1968) described these specimens as a new genus and species within the Paromomyidae, *Torrejonia wilsoni*. Further specimens referable to this genus were later discovered in the Fort Union (Shotgun Member) and Evanston Formations in Wyoming (Gazin, 1969, 1971).

No further Paleocene primates have been reported from the San Juan Basin in the more than 30 years since Wilson and Gazin's original discoveries. Recent fieldwork (1978-1979) has produced, however, additional primate remains from Torreon Wash. These new specimens, all isolated teeth, have been provisionally referred to *Torrejonia wilsoni* and *Palaechthon woodi* (Wood et al., 1979). Other new discoveries have been reported by Tomida and Taylor (this volume).

There are thus only four genera of middle Paleocene primates generally recognized from the San Juan Basin:

Palaechthon nacimienti Wilson & Szalay, 1972
Palaechthon woodi Gazin, 1971
Torrejonia wilsoni Gazin, 1968
Talpohenach torrejonius Kay & Cartmill, 1977
Paromomys sp. Tomida, this volume

P. nacimienti and *T. torrejonius* are known only from the Kutz Canyon area (Angel's Peak faunule). This local fauna is early Torrejonian in age (*Deltatherium* "zone" of Osborn; see Taylor, this volume). *T. wilsoni* and *P. woodi*, on the other hand, are known from the area around Torreon Wash and are late Torrejonian in age (*Pantolamda* "zone"; see Taylor, this volume, for evidence of *T. wilsoni* in Kutz Canyon).

ABBREVIATIONS

The names of institutions referred to in this paper are abbreviated as follows:

UKMNH	University of Kansas Museum of Natural History, Lawrence.
USNM	U.S. National Museum of Natural History, Washington.
UNM	University of New Mexico, Dept. of Geology, Albuquerque.
MCZ	Museum of Comparative Zoology, Harvard University, Cambridge.

KUTZ CANYON PRIMATES

Remarks.—Kutz Canyon lies some 18 km south of Bloomfield, New Mexico, within the drainage system of the San Juan River. Mammalian fossils were first described from the region by Walter Granger (Granger, 1917; Wilson, 1951). Granger's faunal list included *Tetraclaenodon*, *Mioclaenus*, *Periptychus*, *Anisonchus*, *Protogonodon*, *Tricentes*, *Deltatherium*, and *Psittacotherium*. Granger found no primates.

In 1948, a field party under the direction of Robert Wilson discovered approximately 150 mammalian specimens in the canyon (NW ¼, Sec. 14, T27N, R11W, San Juan County, New Mexico). These were all recovered (with one exception) from one stratigraphic level approximately 50 m below the western rim of the canyon. The one exception was a primate maxillary dentition found some 21 m below the other specimens and about .5 km to the southeast. The bulk of the fossils were recovered from a zone of reddish silt about 1 m thick.

The fauna is a fairly varied one. Wilson (1951) recognized 21 genera and 24 species distributed among 11 to 12 families and 5 to 6 orders. The predominance of ferungulates (91%) suggests a floodplain facies (Wilson, 1951). Wilson saw a greater similarity of the fauna with the *Deltatherium* rather than the *Pantolambda* "zone"; however, he felt these two "zones" were merely facial in character (Wilson, 1951, 1967; Tsentas, this volume).

The primate fauna, although briefly mentioned by Wilson in earlier reports (e.g. 1951, 1956), was first described in 1972 (Wilson & Szalay, 1972). They considered all the material to be *Palaechton nacimienti*, a new species.

Palaechton nacimienti —The type specimen, UKMNH 9559, consists of a left mandible with P_3–M_3 and a right mandible with P_3–M_2. The hypodigm includes several other mandibular fragments with teeth, a palate with right P^3–M^2 and left M^{1-3}, and a crushed skull. One other specimen, UKMNH 7903, was questionably referred to this species, but Kay & Cartmill (1977) have made it the type of a new genus and species, *Talpohenach torrejonius*.

P. nacimienti is the largest of the three known species of *Palaecthon*. The main differences between it and the genotype, *P. alticuspis*, are in the antemolar teeth, particularly in the size relationships of the lower canine and P_2. In *P. alticuspis*, the length of the canine alveolus is less than that of the P_2 alveolus; this relationship is reversed in *P. nacimienti*. The condition in *P. nacimienti* is viewed as being more primitive due to the fact that large canines are characteristic of primitive mammals generally. Other differences between *P. nacimienti* and *P. alticuspis* stressed by Wilson & Szalay (1972) include in the former taxon: P_3 with an incipient talonid; P_4 talonid open lingually; paraconid smaller and metaconid absent on P_4 trigonid (disputed by Kay & Cartmill, 1977); molar paraconids less medially placed.

Kay & Cartmill (1977) published an excellent study on the dental and cranial adaptations of *Palaechthon* which forms the basis for the following summary. The lower incisors of *Palaechthon* and *Plesiadapis* differ in morphology and presumed function (see also Gingerich, 1974). The incisors of *Plesiadapis* were better adapted for shearing or crushing, whereas those of *Palaechthon* were better suited for grasping or stabbing. The mandible of *Palaechthon*, as in most Paleocene primates, had an unfused symphysis.

One of the most important variables in analyzing dietary preferences is body size. Kay (1973) demonstrated that insectivorous primates and marsupials tend to have larger molars than their fruit- and sap-eating relatives based on relationships among body size, tooth size, and dietary preferences. They also tend to have longer molar shearing blades, higher and more acute cusps, and enlarged crushing surfaces on the molars. The estimated body weight for *P. nacimienti* is approximately 100 gm (based on a tupaiid model). Kay & Cartmill's (1977) analyses clearly demonstrated that *P. nacimienti* is most similar to an insect-eating model (its small body size virtually rules out the possibility of the species being a leaf eater).

Palaechthon (and *Plesiadapis*) are unique among mammals in combining small orbits and extreme interorbital breadth (Cartmill, 1974). This morphology, indicative of a large olfactory fossa, presumably functioned as a means for dissipating compressive stresses during incisal biting (Kay & Cartmill, 1977). Data relating to relative orbital size and infraorbital foramen size are consistent with the notion that *Palaechthon* was a terrestrial insectivore (i.e., olfactory sense dominant over vision; full complement of vibrissae).

Talpohenach torrejonius.—One specimen that was tentatively referred to *P. nacimienti* by Wilson & Szalay (1972) was a right maxillary fragment with P^2–M^3 (UKMNH 7903). This particular specimen, redescribed by Kay & Cartmill (1977) as *Talpohenach torrejonius*, new genus, new species, was actually the first mammalian fossil found in the 1948 field season and was recovered some 21 m lower in the stratigraphic section and .5 km southeast of the others. Wilson & Szalay (1972) noted its difference from other maxillary teeth of *P. nacimienti*, namely in having transversely narrower molars with a hypocone extending lingually past the protocone (particularly on M^2). The possession of a protocone on P^3 is a feature which distinguishes this species from other paromomyids. The morphology of the premolars, in particular, rules out a close relationship to either *Plesiolestes* or *Torrejonia* (Kay & Cartmill, 1977).

TORREON WASH PRIMATES

Remarks.—The only other Torrejonian primates from New Mexico published thus far have been a few specimens from Torreon Wash (Gazin, 1968; Wood et al., 1979) in the Deer Mesa and Ojo Encino Mesa quadrangles.

The fauna collected from both flanks of Torreon Wash is the standard reference fauna for the middle Paleocene (Torrejonian) land mammal "age" of North America. Wood et al. (1941) described the characteristic fossils of the region as including: *Anisonchus*, *Periptychus* and *Ptilodus* (of these, only *Ptilodus* is unknown from the Puercan). The index fossils include *Conoryctes*, *Deltatherium*, *Mioclaenus*, *Mixodectes*, and *Triisodon*.

Sinclair & Granger (1914) were the first to note the apparent presence of two faunal "zones" within the Torrejonian (*Deltatherium* and *Pantolambda* "zones") (but see Wilson, 1967; Tsentas, this volume). Neither they, Wortman, nor Baldwin ever discovered primates in the Puercan or Torrejonian of the San Juan Basin.

Palaechton woodi.—*P. woodi* was previously known from only the Shotgun local fauna of the Fort Union Formation, Wyoming (Gazin, 1971) (see Text-fig. 7.1 & 7.2). In his original description of the species, Gazin did not compare it with *P. nacimienti*; most of the comparisons were with *P. alticuspis* and *Navajovius kohlhaasae*. The major distinction between *P. woodi* and *P. nacimienti* appears to be size, with *P. woodi* being about one-third the size of *P. nacimienti*. Further collecting in Torreon Wash in 1978 and 1979 by field parties under the direction of Craig Wood and myself produced two specimens provisionally referred to *P. woodi*. This is the first known appearance of this species in the Paleocene of New Mexico (see Wood et al., 1979). These specimens were

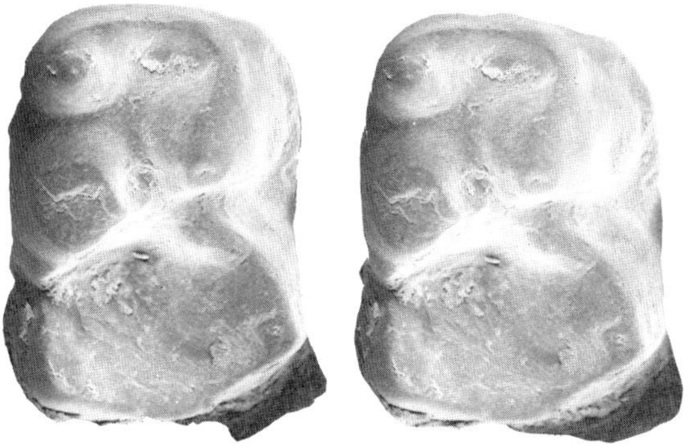

TEXT-FIG. *7.1*—Stereo scanning electron micrograph of M_1, *P. woodi* (MCZ 592), from Shotgun Formation, Wyoming, ×25.

TEXT-FIG. *7.2*—Stereo scanning electron micrograph of M_2, M_3, *P. woodi* (MCZ 2039, 2042), from Shotgun Formation, Wyoming, ×22.

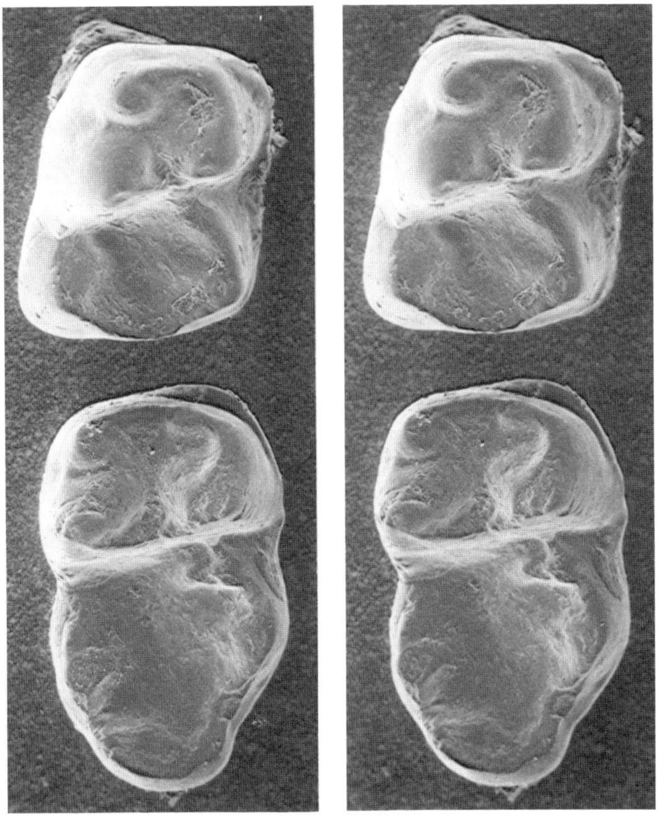

recovered from a new screening locality on the west flank of Torreon Wash (see Tsentas, this volume) in association with a varied fauna which included *Torrejonia* specimens.

The first tooth found, a left M_1 (UNM NP341, Text-fig. 7.3), has a slightly smaller length and width than the type specimen of *P. woodi* (MCZ 18740) (see Gazin, 1971 and Wood et al., 1979 for measurements). The crown of UNM NP341 is preserved intact except for the apex of a well-defined paraconid. The protoconid and metaconid are conical with the metaconid projecting above the level of the protoconid in lateral view and slightly distal to it in occlusal view. The three major cusps of the trigonid are separated from each other by grooves and are all situated around the periphery of the trigonid.

The protolophid (see Van Valen, 1966 for terminology) is bisected by a deep groove forming a V-shaped notch between the protoconid and metaconid. The paralophid runs down the anterior face of the protoconid to the anterolateral corner of the trigonid, at which point it turns medially along the anterior margin of the trigonid to reach the paraconid. The two "arms" of the paralophid run at approximately right angles to one another and form the lateral and anterior margins of the trigonid basin. The trigonid is open medially by the groove separating the paraconid and metaconid and posteriorly by the groove separating the protoconid and metaconid.

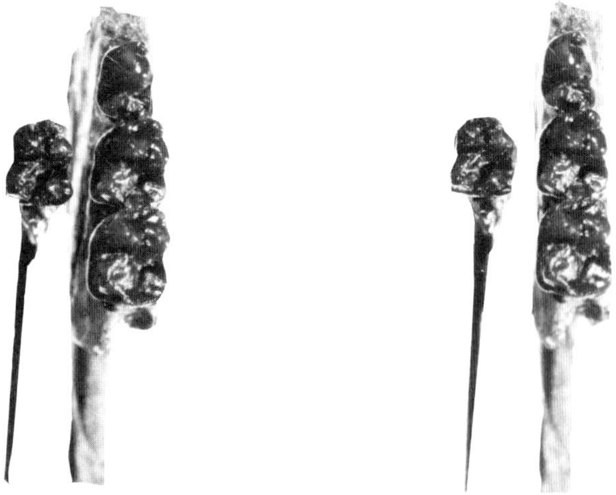

TEXT-FIG. 7.3—Stereophoto of left M_1 (UNM NP341) referred to *P. woodi* from Torreon Wash compared to type specimen (MCZ 18740) of *P. woodi* Gazin, 1971, ×5. Photo courtesy of A. Coleman.

The talonid is broad and deep and carries two major conical cusps, the hypoconid and entoconid. There is no cusplike hypoconulid, but there is a slight swelling of the enamel in that region. Similarly, a swelling along the cristid obliqua represents a rudimentary mesoconid. A deep notch is formed on the lingual side of the talonid basin between the metaconid and entoconid. The cristid obliqua runs from the hypoconid to the vertical posterior face of the trigonid midway between the protoconid and metaconid.

The cingulum commences at the anterolateral side of the tooth and continues around the labial side. It swings around to the posterior aspect of the tooth at the base of the hypoconid and then runs obliquely up the posterior face to end at the swelling in the region of the hypoconulid.

The trigonid cusps are slightly higher than the talonid cusps and are inclined anteriorly at a small angle. Height of the protoconid above the cingulum is 0.8 mm; height of the hypoconid above the cingulum is 0.7mm.

The tooth is hardly worn except for crushing facets on the apices of the major conical cusps. The protoconid and metaconid show the greatest amount of wear, followed by the hypoconid and entoconid.

The morphology of the shearing facets are similar to those described by Kay & Hiiemae (1974) for *Palenochtha.* They demonstrate the distinctiveness of this *P. woodi* molar from those of Cretaceous and early Tertiary insectivores. In primitive tribosphenic molars (e.g. *Gypsonictops, Didelphis*) a principal shearing stroke is that of the paracristid against the metacrista. These two crests, on the lower and upper molars respectively, are obliquely transverse to maximize shear in that direction (Crompton & Hiiemae, 1970). This condition is apparently primitive for both Metatheria and Eutheria (Szalay, 1969).

Another specimen, discovered in 1979, has also been provisionally referred to *P. woodi.* This specimen is a right jaw fragment with M_3 preserved (Text-fig. 7.4). The trigonid has three well-defined conical cusps. The protoconid and metaconid are subequal in height and are separated by a deep groove. The paraconid lies directly anterior to the metaconid on the lingual rim of the trigonid and is separated from it by a shallow groove. Faint ridges run from the bases of the paraconid and metaconid to the base of the protoconid. The paralophid runs anteriorly from the protoconid for a short distance and then turns to form the transverse anterior boundary of the trigonid. This loph ends at the anterior margin of the paraconid.

The trigonid is separated from the talonid by the oblique, vertical, posterior face of the trigonid. The talonid bears four well-defined cusps; hypoconid, entoconid, single hypoconulid (slightly worn), and meso-

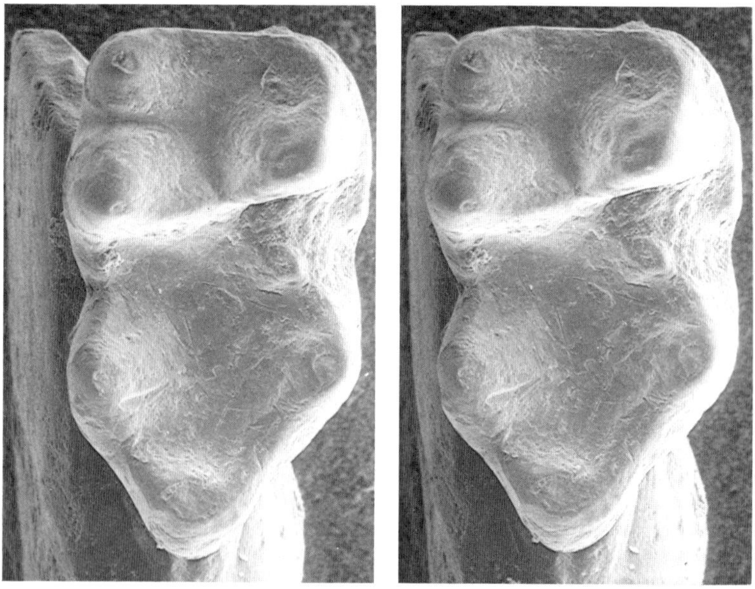

TEXT-FIG. 7.4—Stereo scanning electron micrograph of M_3 (UNM NP342) referred to *P. woodi* from Torreon Wash, × 37.50.

conid on the cristid obliqua. A poorly developed cingulum runs along the labial side of the tooth.

There are several differences between the specimens from Wyoming (Gazin, 1971) and Torreon Wash. Comparisons of UNM NP341 with the type of *P. woodi* (MCZ 18740) and other Wyoming specimens (e.g., MCZ 592) show that the New Mexican specimen differs in that: 1) the cusps of the trigonid are more vertically aligned; 2) trigonid cusps are situated more on the periphery of the trigonid rather than sloping inwards toward each other; 3) the posterior face of the protoconid is relatively broader and flatter.

The M_3 specimen (UNM NP342) from Torreon Wash, tentatively referred to *P. woodi*, is more difficult to assess. Gazin (1971) mentioned only one isolated M_3 specimen (MCZ 2042) from his Shotgun sample with which to compare the New Mexican fossil. The Shotgun specimen is slightly larger than UNM NP342 in all measurements (length 2.0 mm vs. 1.9 mm; width across protoconid and metaconid 0.9 mm vs. 0.8 mm; width across entoconid and hypoconid 1.0 mm vs. 0.9 mm). Besides being slightly smaller in size, the New Mexican specimen also has a narrower hypoconulid region. UNM NP342 differs from *Navajovius* in having a long, narrow hypoconulid region, a well separated entoconid and hypoconulid, and a better developed paraconid. It differs from *Palenochtha* in its more posteriorly projecting hypoconulid region, more conical cusps, and broader trigonid basins.

In general, the Wyoming *P. woodi* material displays slightly more bulbous cusps, a broader M_3 hypoconulid region, and slightly larger size than their New Mexican counterparts. Although these differences might require some taxonomic adjustments in the future, until more material is available, taxonomic splitting would be premature.

Torrejonia wilsoni—Only one year after Wilson's initial discovery of primates in the Torrejonian of New Mexico, Franklin Pearce and C.L. Gazin discovered additional primate remains in the *Pantolambda* "zone" of Torreon Wash which were assigned to *Torrejonia wilsoni*, new genus, new species (Gazin, 1968). Outside of this area the genus is known from the Fort Union (Shotgun Member) and Evanston Formations in Wyoming (Gazin, 1968, 1971). Taylor (this volume) recorded the presence of this taxon at Kutz Canyon.

The type specimen is a left ramus with P_3–M_2 intact (USNM 25255). It was recovered from the east branch of Torreon Wash. The only other specimens of the genus recorded by Gazin from the San Juan Basin include an isolated upper molar (USNM 25256) and an isolated M_2 (USNM 25257), both from the *Pantolambda* "zone" in the west branch of Torreon Wash.

Several other specimens referable to this genus have been recovered in the past two years in Torreon Wash by field parties under the direction of C. Wood and myself. One specimen, a left M_2 (UNM NP340, identified as T-36 in Wood et al., 1979), was found at the same screening locality as the *P. woodi* teeth. These new specimens include an M_1 (UNM B-916, identified as T-37 in Wood et al., 1979), several M_2's (Text-fig. 7.5), and an upper right M^1 (UNM NP087, identified as T-38 in Wood et al., 1979). The upper molar is assigned to this species on the basis of its occlusal relationship to the corresponding lower molar (Wood et al., 1979). This comparison supports Gazin's (1968) identification of USNM 25256 as an upper molar of *Torrejonia*. Since these molars have been ade-

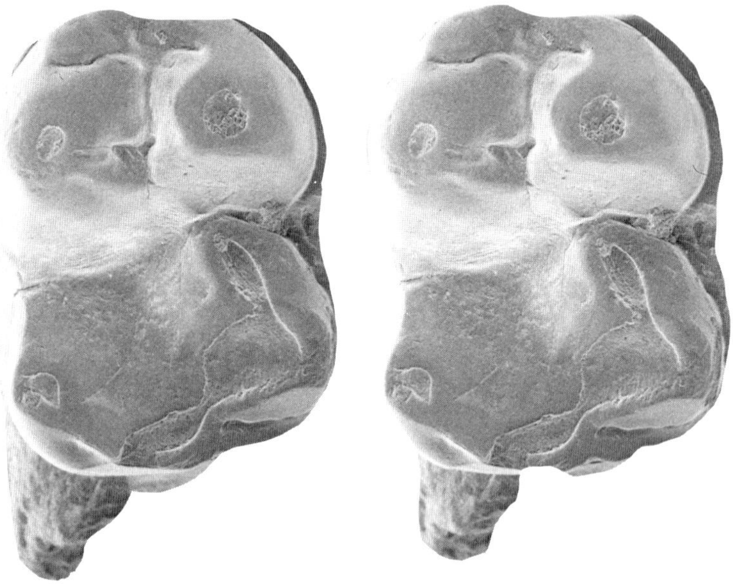

TEXT-FIG. 7.5—Stereo scanning electron micrograph of M_2 referred to *T. wilsoni* (uncatalogued specimen) from Torreon Wash, × 18.75.

quately described by Gazin (1968, 1971), no further elaboration is needed here.

There is some disagreement as to the taxonomic status of *Torrejonia.* Some, like Wilson & Szalay (1972) and Szalay (1973), regard it as synonymous with *Plesiolestes* (Jepsen, 1930), whereas others, notably Kay & Cartmill (1977), consider it to be a valid genus. Part of the problem, no doubt, stems from the fact that detailed comparisons were made mainly with *Paromomys* in the original description. I think (with the benefit of hindsight) that both viewpoints are correct in a certain sense, namely, that *Torrejonia* is certainly distinct from *Paromomys.*

Paromomys has molars with short, closed, and transversely quadrangular trigonid basins (Gidley, 1923; Simpson, 1937). Its molar paraconids are small and, on the M_2's, are almost fused to the metaconids. *Paromomys* is generally considered to be ancestral to the rather specialized *Phenacolemur* (Simpson, 1955). *Torrejonia* does not show any exclusive similarities to this lineage; in fact, Bown & Rose (1976) placed them in different families.

The suggestion that *Torrejonia* should be synonymized with

Plesiolestes is a most interesting one in that the generic distinction of *Plesiolestes* itself is not without its detractors (Simpson, 1937). So far as is presently known, the major distinction between *Plesiolestes* and *Torrejonia* is the highly molariform P_4 of the former. It is odd that such a molariform tooth is absent in the composite drawing of *Plesiolestes sirokyi* (Szalay, 1973); either the P_4 metaconid development is highly variable in *Plesiolestes* species (Bown & Gingerich, 1973), or that specimen cannot belong to *Plesiolestes*.

If *Torrejonia* is not a valid genus, do its affinities lie more with *Plesiolestes* or *Palaechthon*? *Torrejonia*, *Palaechthon*, and *Plesiolestes* are all rather similar except for two features (see chart in Bown & Rose, 1976), the absent or weak P_4 metaconid and the moderate rugosity of the enamel on unworn teeth in *Torrejonia*. In both these features the similarity of *Torrejonia* is with *Palaechthon*, not with *Plesiolestes*. A strong case could be made for lumping all three genera into one with the generic name *Palaechthon* taking priority. In fact, one might suggest that there are fewer morphological distinctions between *Torrejonia* and *Palaechthon* than there are between *P. alticuspis* and *P. nacimienti*. A functional difference between *Torrejonia* and *Plesiolestes* lies in the reduction of molar shearing blades in the former. This presumably indicates that *Torrejonia* had a more mixed diet than *Plesiolestes* (see Kay & Cartmill, 1977). The formal lumping of these three genera should not be undertaken until more material is available from the Torrejonian of New Mexico; otherwise one gets involved in rather sterile exercises in taxonomic juggling. In any event, all the genera so far discovered from the Torrejonian of the San Juan Basin are closely related and clearly belong to one family (either Paromomyidae, *sensu* Simons, 1972 or Microsyopidae, *sensu* Bown & Rose, 1976).

SUMMARY

Fossil primates from Torrejonian deposits of the San Juan Basin, New Mexico, are extremely rare. The situation is all the more remarkable given the extensive number of primates found in other faunas of the same age. It is important to determine whether the reasons for this are paleoecological, zoogeographical, collectors bias, or others. Recent collecting strongly suggests that the paucity of fossil primates in the San Juan Basin in part resulted from the failure to use screen washing techniques (McKenna, 1962). I fully expect the primate faunal lists from the San Juan Basin Torrejonian to expand in future years as these techniques are more extensively employed.

The Torrejonian primates from the San Juan Basin comprise only four genera, all within one family (Paromomyidae). The systematics of this group have not been worked out to everyone's general satisfaction (Simons, 1972). Most recently, Bown & Rose (1976) have drastically reduced to three the number of genera they consider to be in the family (*Paromomys*, *Phenacolemur*, *Ignacius*). All of the genera found in the Torrejonian of the San Juan Basin were removed from the Paromomyidae and placed in the Microsyopidae by them. It is clear that further systematic efforts are needed. It might prove helpful to investigate the enamel prism patterns of these Torrejonian primates with the scanning electron microscope. These prism patterns (and dentinal tubules) can be clearly recognized in Paleocene mammals even without etching the enamel surface (e.g. Text fig. 7.6). These enamel prisms characterize prototherian and therian animals but are not usually found in reptiles or mammal-like reptiles. The earliest known form possessing enamel prisms is the prototherian *Eozostrodon* from the late Traissic (Grine et al., 1979). It is hoped that certain enamel prism patterns can be used for taxonomic purposes in Paleocene primates as they have in studies of Mio-Pliocene primates (Gantt, 1979).

TEXT-FIG. *7.6*—Scanning electron micrograph of enamel prisms in the talonid basin of UNM NP342 (?*P. woodi*) from Torreon Wash, × 1000.

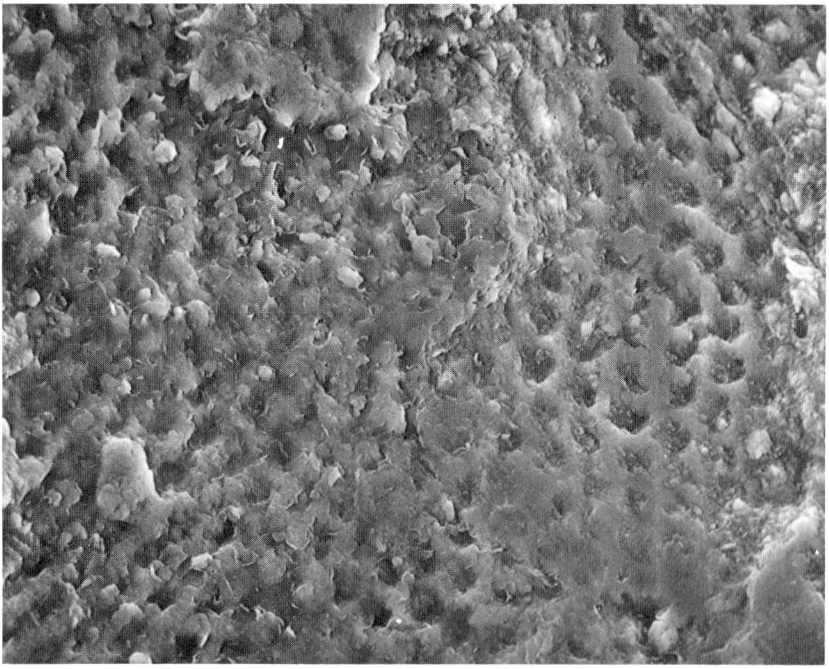

ACKNOWLEDGMENTS

Field work in the San Juan Basin has been supported by grants from Brown University, Providence College, and New York University. Particular thanks are due to the staff of the Bureau of Land Management in New Mexico for all their advice, assistance, and encouragement. My colleague C.B. Wood has played a major role in the discovery and identification of the fossils discussed in this report and of many other Paleocene fossils as well. I thank the field crews from Providence College and New York University for their efforts. J. Froehlich and the editors improved the form of the manuscript greatly. The scanning electron micrographs were produced in the Brown University SEM facility under the technical direction of Ms. S. Koontz.

REFERENCES

Bown, T. and P. Gingerich. 1973. The Paleocene primate *Plesiolestes* and the origin of Microsyopidae. Folia Primatol. 19:1–8.

Bown, T. and K. Rose. 1976. New early Tertiary Primates and a reappraisal of some Plesiadapiformes. Folia Primatol. 26:109–138.

Cartmill, M. 1974. *Daubentonia*, *Dactylopsila*, woodpeckers, and klinorhynchy, p. 655–670. *In*: R. D. Martin, G. Doyle, and A. Walker (eds.), Prosimian Biology, Duckworth. London.

Cope, E. D. 1883. On new lemuroids from the Puerco Formation. Am. Nat., 18:59–61.

Crompton, A. and K. Hiiemae. 1970. Molar occlusion and mandibular movements during mastication in the American opposum, *Didelphis marsupialis*. J. Linnaean Soc. London. 49:21–47.

Gantt, D. 1979. A method of interpreting enamel prism patterns. Scanning Electron Microscopy 2:491–496.

Gazin, C. L. 1968. A new primate from the Torrejon middle Paleocene of the San Juan Basin, New Mexico. Proc. Biol. Soc. Wash. 81:629–634.

————. 1969. A new occurrence of Paleocene mammals in the Evanston Formation, southwestern Wyoming. Smithson. Contrib. Paleobiol. 2:1–17.

————. 1971. Paleocene Primates from the Shotgun Member of the Fort Union Formation in the Wind River Basin, Wyoming. Proc. Biol. Soc. Wash. 84:13–38.

Gidley, J. 1923. Paleocene Primates of the Fort Union, with discussion of the relationships of the Eocene primates. Proc. U.S. Natl. Mus. 63:1–38.

Gingerich, P. 1974. Dental function in the Paleocene primate *Plesiadapis*, p. 531–541. *In:* R. Martin, G. Doyle, and A. Walker (eds.) Prosimian Biology, Duckworth. London.

Granger, W. 1917. Paleocene and lower Eocene mammal horizons of northern New Mexico and southern Colorado. Bull. Am. Mus. Nat. Hist. 32:821–830.

Grine, F., E. Vrba and A. Cruickshank. 1979. Enamel prisms and diphyodonty: linked apomorphies of Mammalia. South African J. Sci. 75:114–120.

Jepsen, G. 1930. Stratigraphy and paleontology of the Paleocene of northeastern Park County, Wyoming. Proc. Am. Phil. Soc. 69:463–528.

Kay, R. 1973. Mastication, molar tooth structure and diet in primates. Unpublished Ph.D. thesis, Yale University, 375 p.

Kay, R. and M. Cartmill. 1974. Skull of *Palaechthon nacimienti*. Nature 252:37–38.

————. 1977. Cranial morphology and adaptations of *Palaechthon nacimienti* and other Paromomyidae (Plesiadapoidea, ?Primates), with a description of a new genus and species. J. Human. Evol. 6:19–53.

Kay, R. and K. Hiiemae. 1974. Jaw movement and tooth use in recent and fossil primates. Am. J. Phys. Anthropol. 40:227–256.

Matthew, W. 1897. A revision of the Puerco fauna. Bull. Am. Mus. Nat. Hist. 9:259–323.

————. 1937. Paleocene faunas of the San Juan Basin, New Mexico. Trans. Am. Phil. Soc. n.s. 30:1–510.

McKenna, M. 1962. Collecting small mammals by washing and screening. Curator 5:221–235.

Osborn, H. and C. Earle, 1895. Fossil mammals of the Puerco Beds. Bull. Am. Mus. Nat. Hist. 7:1–70.

Simons, E. 1972. Primate Evolution, MacMillan. New York, 322 p.

Simpson, G. 1937. The Fort Union of the Crazy Mountain Field, Montana and its mammalian faunas. Bull. U.S. Natl. Mus. 169:1–277.

————. 1955. The Phenacolemuridae, new family of early Primates. Bull. Am. Mus. Nat. Hist. 105:415–441.

Sinclair, W. and W. Granger, 1914. Paleocene deposits of the San Juan Basin, New Mexico. Bull. Am. Mus. Nat. Hist. 33:297–316.

Szalay, F. 1969. Mixodectidae, Microsyopidae and the insectivore-primate transition. Bull. Am. Mus. Nat. Hist. 140:193–330.

————. 1973. New Paleocene primates and a diagnosis of the new suborder Paromomyiformes. Folia Primatol. 19:73–87.

Van Valen, L. 1966. Deltatheridia, a new order of mammals. Bull. Am. Mus. Nt. Hist. 132:1–126.

Wilson, R. 1951. Preliminary survey of a Paleocene faunule from Angel's Peak area, New Mexico. Univ. Kansas Mus. Nat. Hist. Publ. 5:1–11.

————. 1956. A new multituberculate from the Paleocene Torrejon fauna of New Mexico. Trans. Kansas Acad. Sci. 59:76–84.

————. 1967. Fossil mammals in Tertiary correlations. Univ. of Kansas Geol. Dept. Special Publ. 2:590–606.

Wilson, R. and F. Szalay. 1972. New paromomyid primate from Middle Paleocene beds, Kutz Canyon area, San Juan Basin, New Mexico. Am. Mus. Nat. Hist. Novitates 2499:1–18.

Wood, C., G. Conroy and S. Lucas. 1979. New discoveries of fossil primates from the type Torrejonian (middle Paleocene) of New Mexico. Folia Primatol. 32:1–7.

Wood, H., R. Chaney, J. Clark, E. Colbert, G. Jepsen, J. Reeside and C. Stock. 1941. Nomenclature and correlation of the North American continental Tertiary. Bull. Geol. Soc. Am. 52:1–48.

8 Revision of the Middle Paleocene (Torrejonian) Taeniodont (Mammalia) *Psittacotherium* Cope, 1882

Robert M. Schoch
Department of Geology and Geophysics and Peabody Museum of Natural History, Yale University New Haven, Connecticut

INTRODUCTION

Psittacotherium is a middle Paleocene (Torrejonian) representative of the Taeniodonta, an archaic order of mammals first known from the early Paleocene (Puercan) and believed to have disappeared without descendants in the late Eocene (Uintan). *Psittacotherium* was one of the largest animals of the Paleocene, being perhaps the size of a small pig. It possessed large, high, and much compressed claws on the digits and had very powerful limbs, as demonstrated by the short, stout, and massive long bones. The ulna of *Psittacotherium* has a well-developed olecranon process, and the unguals of the manus carry larger claws than those of the pes (cf. Matthew, 1937, figs. 62–67, pls. 59–64).

The skull of *Psittacotherium* (Text-fig. 8.1) is relatively short and wide with a massive mandible, particularly at the symphysis. It has large, subgliriform canines in both the upper and lower jaws, one incisor on each side (upper and lower) and a complete set of cheek teeth, which are usually found worn smooth in such a way that only a band of enamel around the perimeter of the tooth remains. The incisors appear to be miniature counterparts of the canines. *Psittacotherium* and other taeniodonts are rather poorly represented in the fossil record; their fossil remains are generally fragmentary, and often only a few well-worn teeth are found.

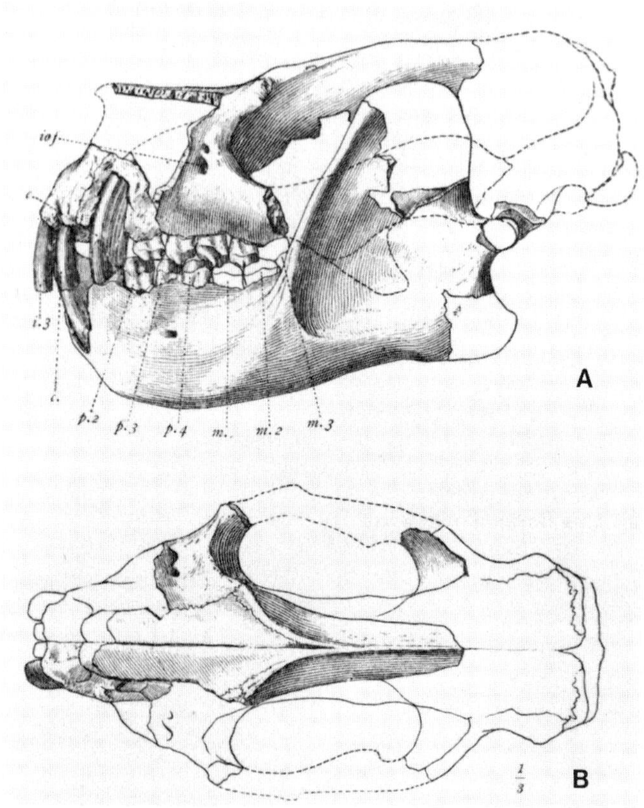

TEXT-FIG. *8.1*—Reconstruction of the skull of *Psittacotherium multifragum* (from Wortman, 1897). *A*, Left lateral view; *B*, Dorsal view.

It has been suggested that progressive taeniodonts, including *Psittacotherium*, were adaptively convergent with the large ground sloths of the later Tertiary and Quaternary (Savage, 1971). Patterson (1949) also speculated that taeniodonts, such as *Psittacotherium*, used their claws and teeth to husk and crack a diet of hard-shelled nuts and fruits. Patterson (1949) attributed the low number of preserved taeniodonts to their supposedly small population sizes rather than to a bias in the fossil record. However, an alternative model of progressive taeniodont adaptations can be proposed, a model which seems to explain better the known record of taeniodonts. Thus, taeniodonts may be rare in the fossil record since they did not live, for the most part, where sediments were being deposited on riverine floodplains, and their remains would not have been readily fossilized and preserved. They may represent one of the first upland radiations of the Mammalia, living in highlands and other areas of denudation, away from most streams and waterways. They used their powerful claws and canines to dig and root for tubers and other succulent subsurface food items, while also eating carrion if they happened to stray across it. The large amount of grit that is an inevitable part of such a diet could account for the extreme wear typical of taeniodont teeth. Feeding on tubers would have freed them from the necessity of always being fairly near water. The taeniodonts were driven to extinction in late Eocene times when other more advanced mammals began to invade the uplands.

One of the best known of the progressive taeniodonts is *Psittacotherium*, first named by E.D. Cope in 1882 on the basis of a fragmentary jaw collected by David Baldwin in the San Juan Basin, New Mexico. Since then, the majority of known remains of *Psittacotherium* have come from New Mexico. Matthew (1937) last reviewed *Psittacotherium* but failed to mention several important specimens, including the type of *Psittacotherium megalodus* Cope, 1887. It is thus the purpose of this paper to revise thoroughly the taxonomy of *Psittacotherium* and to publish photographs of Cope's type specimens for the first time. In this paper AMNH refers to specimens in the vertebrate fossil collections of the American Museum of Natural History, New York.

SYSTEMATIC PALEONTOLOGY

Order TAENIODONTA Cope, 1876
Family STYLINODONTIDAE Marsh, 1875
Subfamily STYLINODONTINAE Schlosser, 1911
Genus PSITTACOTHERIUM Cope, 1882

Synonym.—*Hemiganus* Cope, 28 September, 1882, Am. Nat. v. 16, p. 831 (in part).

Type Species.—*Psittacotherium multifragum* Cope, 25 January 1882, Am. Nat. v. 16, p. 156.

Range.—Middle Paleocene (Torrejonian), Nacimiento Formation, San Juan Basin, New Mexico (Matthew, 1937); Dragon local fauna, North Horn Formation, Dragon Canyon, Emery County, Utah (cf. *Psittacotherium*: Gazin, 1941); Gidley Upper Quarry, upper part of the Lebo Formation, Crazy Mountain Field, Sweetgrass County, Montana (Simpson, 1937); Black Peaks Formation, Big Bend National Park, Brewster County, Texas (*Psittacotherium* or *Lampadophorus*: Schiebout, 1974); Swain Quarry, Fort Union Formation, east flank of the Washakie Basin, Carbon County, Wyoming (Rigby, 1980).

Diagnosis.—Medium-sized stylinodontines (larger than *Wortmania*) distinguished by having greatly enlarged, rooted, sub-gliriform canines with the enamel limited to the anterior face of the tooth and with both crown and root greatly elongated; distinguished from the conoryctines (*Onychodectes*, *Conoryctella*, and *Conoryctes*) and *Wortmania*, in which the canines are comparatively smaller and the entire crown is covered by enamel before wear; distinguished from *Lampadophorus*, *Ectoganus*, and *Stylinodon*, in which the canines are rootless.

PSITTACOTHERIUM MULTIFRAGUM Cope, 1882
Pl. 8.1, 8.2

Type—AMNH 3413 (Pl. 8.1, figs. 1-4) mandible with: left I and C (still showing the original striated surface of the enamel); right P_2, M_1, and M_2, all only moderately worn; root of right I and C and alveoli for left M_2, M_3, and right P_1 and M_3.

Horizon and locality of the type.—Presumably Torrejonian-aged strata of the Nacimiento Formation near Huerfano Peak, San Juan Basin, New Mexico (Matthew, 1937).

Synonyms.—The following are considered here to be synonyms of *P. multifragum*: 1) *Psittacotherium aspasiae* Cope, 20 February 1882, Paleont. Bull. No. 34 and Proc. Am. Phil. Soc., v. 20, p. 192. Type: AMNH 3416 (Pl. 8.1, figs. 5-6), posterior portion of left horizontal mandibular ramus with alveoli for M_1, M_2 and containing the partially erupted M_3. M_2 fell out of the jaw post-mortem and is attached to the outside of the ramus. M_2 is badly crushed, and only part of the talonid remains (hypoconid and accessory cuspules). M_2 shows only slight amounts of wear on the hypoconid. Horizon and locality of the type: presumably Torrejonian-aged strata of the Nacimiento Formation, San Juan Basin, New Mexico (Matthew, 1937).

2) *Hemiganus vultuosus* Cope, 28 September 1882, Am. Nat., v. 16, p. 831. Type: AMNH 3390 (Pl. 8.2, figs. 1-8), left lower C on which the enamel striations are worn smooth, a lower right(?) I, an upper right C and an upper cheek tooth, all extremely worn, which presumably came from the same individual. Horizon and locality of the type: presumably Torrejonian-aged strata of the Nacimiento Formation, San Juan Basin, New Mexico (Matthew, 1937).

3) *Psittacotherium megalodus* Cope, May 1887, Am. Nat., v. 21, p. 469. Type: AMNH 3418 (Pl. 8.2, figs. 9-12), anterior portion of right mandibular ramus with root of right C and alveoli for P_1, P_2, P_3, P_4, M_1, and M_2. Isolated right P_2, which shows only a slight amount of wear, presumably from the same individual. Horizon and locality of the type: presumably Torrejonian-aged strata of the Nacimiento Formation, San Juan Basin, New Mexico (Matthew, 1937).

Diagnosis.—Same as that for the genus.

Discussion.—Wortman (1897) believed that *P. multifragum*, *P.*

aspasiae, *P. megalodus*, and *Hemiganus vultuosus* represented a single species. Matthew (1937) synonymized *Hemiganus vultuosus* with *P. multifragum* but kept *P. aspasiae* as a valid species without discussing it or referring any additional material to it. He failed to discuss the taxonomic status of *P. megalodus*.

I have examined all of the AMNH *Psittacotherium* specimens and find that they represent one species that has little morphological variation except in size. Lengths and widths of eight lower canines (among the most common and, thus, directly comparable of the remains of *Psittacotherium*) were measured at the base of the enamel (Table 8.1). The mean anterior-posterior length was 28.0 mm and varied from 25.4 mm to 31.2 mm with a standard deviation of 1.75. The mean buccal-lingual width was 17.8 mm and varied from 15.9 mm to 19.5 mm with a standard deviation of 1.34. The type of *P. multifragum* falls near the small end of the range, whereas the types of *H. vultuosus* and *P. megalodus* fall near the large end of the range. This is also reflected in the size of P_2, which is approximately 20% smaller in the type of *P. multifragum*, as compared to the P_2 associated with the type of *P. megalodus* (Table 8.1). These differences in size may be due to sexual dimorphism or individual variation within the species. The coefficient of variation for canine length is 6.3; that for canine width is 7.5. These fall slightly on the high side of the range of coefficients of variation that Simpson et al. (1960) cite as typical for most size related parameters in a single species. With further collecting, *Psittacotherium* may prove to contain more than one species; however, given the overall similarity not only of the canines, but of all known *Psittacotherium* remains, it is most reasonable to assign all specimens to one species at the present time.

The type of *P. aspasiae* (Pl. 8.1, figs. 5–6) is a subadult, accounting for the small size of the mandibular ramus, which does not include a lower canine. Nonetheless, it is referable to the same species as the type of *P. multifragum* on the basis of the molars, which are identical in morphology and size of the M_2 (Pl. 8.1, figs. 3–6 and Table 8.1). Thus it is concluded that the genus *Psittacotherum* is presently known from only one valid species, *Psittacotherium multifragum* Cope, 1882.

ACKNOWLEDGMENTS

I thank Drs. R. H. Tedford and M.C. McKenna (both AMNH) for permission to examine specimens in their care. I am grateful to Spencer Lucas for his help, discussions, and encouragement throughout this project and for his review of the manuscript. Finally, I thank the editors of this volume for allowing me to contribute this paper. This research was supported in part by the National Science Foundation. Contributions to the Evolution of the Taeniodonta, Number 1.

Table 8.1—Measurements of *Psittacotherium* specimens in AMNH (in mm, L = Length, W = Width).

AMNH No.	C_1 L	C_1 W	P_2 L	P_2 W	M_1 L	M_1 W	M_2 L	M_2 W	M_3 L	M_3 W
3413[a]			7.1	13.0	10.4	10.1	9.3	9.0		
3416[b]							9.2	9.0	9.3[c]	8.3[c]
3390[d]	28.9	18.2								
3418[e]	27.8[f]	16.0[f]	9.8	16.4						
757	26.6	19.1								
3392	28.7	17.0								
3393	28.5	18.2								
16660	27.0	18.3								
16661	25.4	15.9								
88383	31.2	19.5								
Mean	28.0	17.8								
Standard Deviation	1.75	1.34								
Coefficient of Variation	6.3	7.5								

[a]Type of *P. multifragum*. The canine is not exposed at the base of the enamel, but what is exposed of the canine is the same size as AMNH 16661.
[b]Type of *P. aspasiae*.
[c]Rough estimates on partially erupted M_3.
[d]Type of *Hemiganus vultuosus*.
[e]Type of *P. megalodus*.
[f]Estimates from the broken root of the canine.

EXPLANATION OF PLATE 8.1

FIGS. *1–6*—Type specimens of *Psittacotherium multifragum* and *P. aspasiae*. *1*, Left lateral view of anterior portion of mandible, AMNH 3413, *P. multifragum*, ×.5; *2*, Occlusal view of anterior portion of mandible, AMNH 3413, *P. multifragum*, ×.5; *3*, Internal view of right ramus, AMNH 3413, *P. multifragum*, ×.75; *4*, Occlusal view of right ramus, AMNH 3413, *P. multifragum*, ×.75; *5*, Lateral view of left ramus with attached M_2, AMNH 3416, *P. aspasiae*, ×1; *6*, Occlusal view of left ramus, AMNH 3416, *P. aspasiae*, ×1.

1

2

3

4

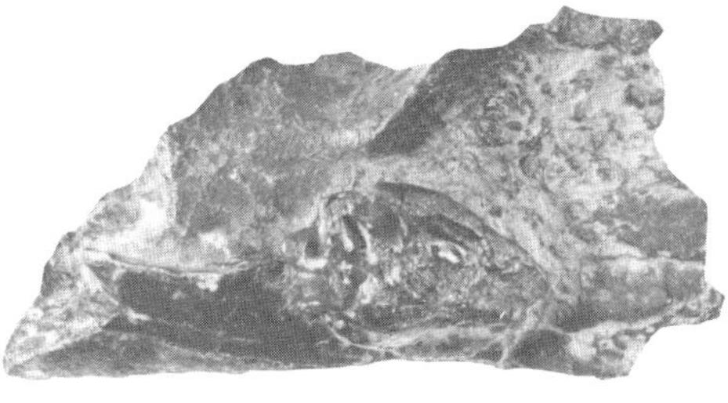

5

6

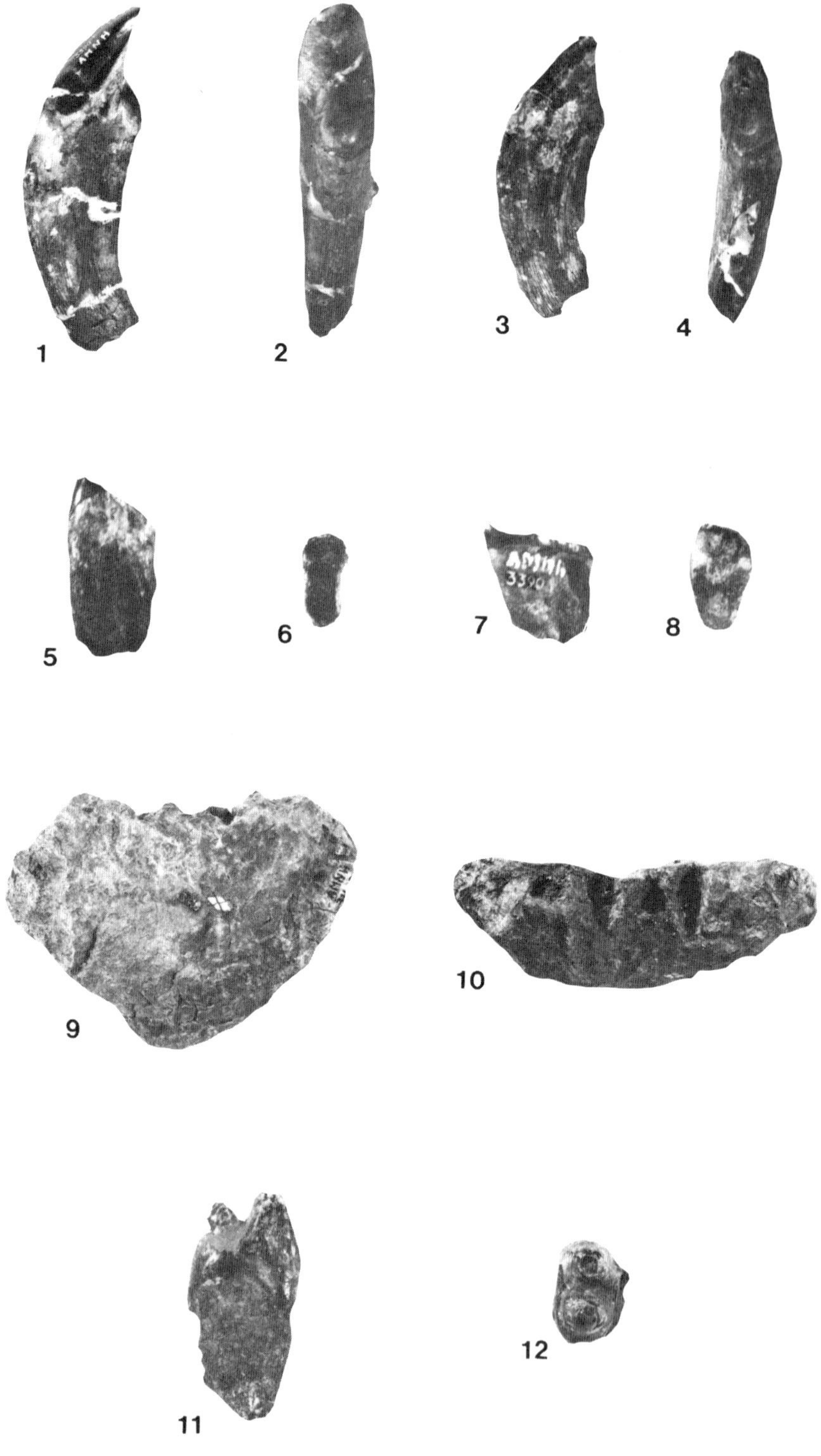

REFERENCES

Gazin, C. L. 1941. The mammalian faunas of the Paleocene of central Utah with notes on the geology. Proc. U.S. Natl. Mus. 91:1–53.

Matthew, W. D. 1937. Paleocene faunas of the San Juan Basin, New Mexico. Trans. Am. Phil. Soc. n. s. 30:1–510.

Patterson, B. 1949. Rates of evolution in taeniodonts p. 243–278. *In* Jepsen, G. L., G. G. Simpson and E. Mayr (eds.), Genetics, paleontology, and evolution. Princeton Univ. Press.

Rigby, J. K., Jr. 1980. Swain Quarry of the Fort Union Formation, middle Paleocene (Torrejonian), Carbon County, Wyoming: Geologic setting and mammalian fauna. Evol. Monogr. 3, 179 p.

Savage, D. E. 1971. Taeniodonta. McGraw-Hill encyclopedia of science and technology 13:408.

Schiebout, J. A. 1974. Vertebrate paleontology and paleoecology of Paleocene Black Peaks Formation, Big Bend National Park, Texas. Bull. Texas Memorial Museum 24:1–88.

Simpson, G. G. 1937. The Fort Union of the Crazy Mountain Field, Montana, and its mammalian faunas. Bull. U.S. Natl. Mus. 169:1–287.

Simpson, G. G., A. Roe and R. C. Lewontin. 1960. Quantitative Zoology. Harcourt, Brace, New York, 440 p.

Wortman, J. L. 1897. The Ganodonta and their relationships to the Edentata. Bull. Am. Mus. Nat. Hist. 9:59–110.

EXPLANATION OF PLATE 8.2

FIGS. *1–12*—Type specimens of *Hemiganus vultuosus* and *Psittacotherium megalodus*. *1*, Buccal view of left lower canine, AMNH 3390, *H. vultuosus*, ×.5; 2, Posterior view of left lower canine, AMNH 3390, *H. vultuosus*, ×.5; 3, Buccal view of right upper canine, AMNH 3390, *H. vultuosus*, ×.5; *4*, Posterior view of right upper canine, AMNH 3390, *H. vultuosus*, ×.5; *5*, Lateral view of lower incisor, AMNH 3390, *H. vultuosus*, ×1; *6*, Occlusal view of lower incisor, AMNH 3390, *H. vultuosus*, ×1; *7*, Lateral view of upper cheek tooth, AMNH 3390, *H. vultuosus*, ×1; *8*, Occlusal view of upper cheek tooth, AMNH 3390, *H. vultuosus*, ×1; 9, Lateral view of right ramus, AMNH 3418, *P. megalodus*, ×.5; *10*, Occlusal view of right ramus, AMNH 3418, *P. megalodus*, ×.5; *11*, Posterior view of right P_2, AMNH 3418, *P. megalodus*, ×1; *12*, Occlusal view of right P_2, AMNH 3418, *P. megalodus*, ×1.

Part 3
Local Faunas and Biostratigraphy

9

The Alamo Wash Local Fauna: A new look at the old Ojo Alamo Fauna

Thomas M. Lehman
Department of Geological Sciences
University of Texas,
Austin, Texas

INTRODUCTION

The Late Cretaceous and early Tertiary vertebrate faunas of the San Juan Basin have been studied intermittently since 1874 when Edward Drinker Cope first prospected the badlands west of the Jemez Mountains. Among Cope's remarkable collections, which were mostly of early Tertiary mammals, were a few dinosaur teeth which David Baldwin had sent him from the area around Ojo Alamo. At that time, Ojo Alamo was no more than a spring with a large cottonwood tree. However, by the turn of the century a small store built there for trading with the Navajo Indians took the same name. Baldwin had, in passing, discovered an important new locality for Late Cretaceous vertebrates; however, almost thirty years passed before the region was again visited by paleontologists.

Ojo Alamo lies at the confluence of a series of washes which are tributaries of the Chaco River (Text-fig. 9.1). The exposures at the heads of Hunter Wash, Willow Wash, Alamo Wash, and De-na-zin Wash are referred to here collectively as the Ojo Alamo region (Text-fig. 9.2). Many collections of Cretaceous vertebrates have been made from this region, but poor collecting practices and the constant shuffling of stratigraphic nomenclature have prevented an adequate understanding

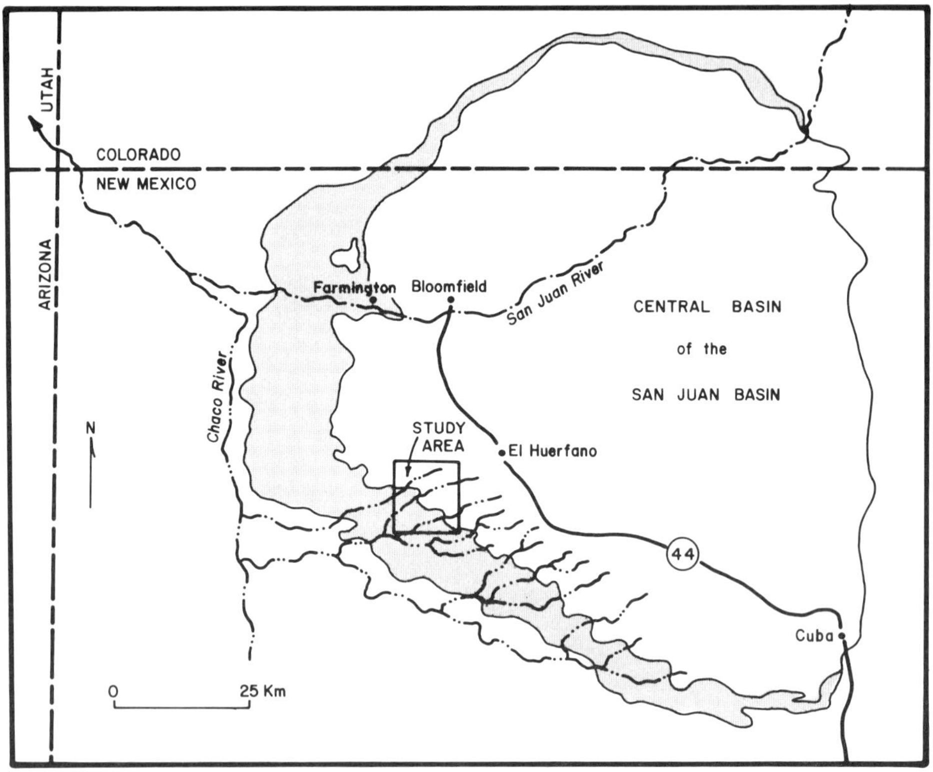

TEXT-FIG. *9.1*—Location map showing the combined outcrop belt of the Fruitland Formation and Kirtland Shale (shaded) and the location of the Ojo Alamo region enlarged in Text-fig. 9.2.

of their proper association and correlation. Only the uppermost dinosaur-bearing horizon exposed in the vicinty of the old Ojo Alamo Trading Post is considered in the present study. It is hoped that this study will help clarify some of the present misunderstandings about the correlation of this horizon and its vertebrate fauna with those of better known regions of North America.

The uppermost dinosaur-bearing horizon in the Ojo Alamo region is the Naashoibito Member of the Kirtland Shale; its vertebrate fauna may be referred to as the Alamo Wash local fauna. While both the rock-stratigraphic and the biostratigraphic terminology applied to this horizon have had a long and checkered history, I believe that these terms are appropriate for future work. Specimens from the Naashoibito Member, which have been adequately described in the literature, and those in the collection of the University of New Mexico Geology Department are reviewed here in an attempt to provide an accurate description of the Alamo Wash local fauna.

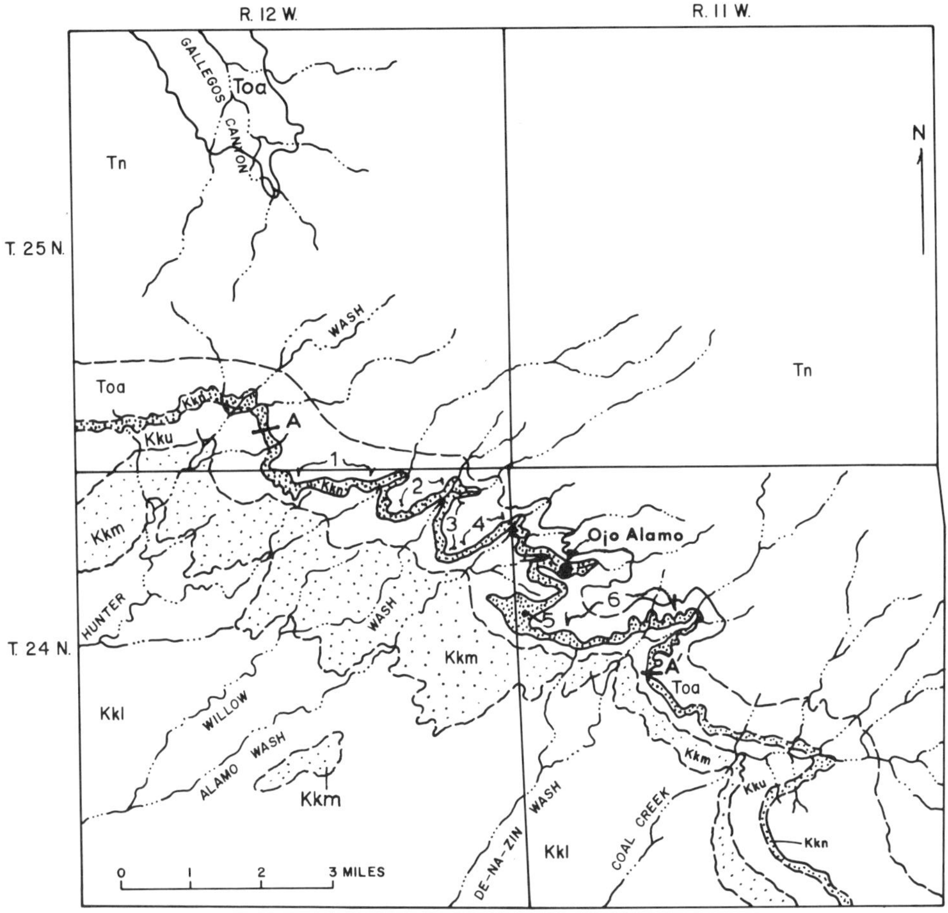

TEXT-FIG. 9.2—Geologic map of the Ojo Alamo region after Baltz et al. (1966). The location of the old Ojo Alamo trading post is indicated. The Alamo Wash local fauna has been collected from the portion of the outcrop between points A and A′. The six major collecting areas are: 1) north side of the south branch of Hunter Wash, 2) north side of Willow Wash, 3) south side of Willow Wash, 4) north side of Alamo Wash, 5) "Rick's site" (UNM locality BUNM-77-675), and 6) north side of De-na-zin Wash. Abbreviations are those used in Text-fig. 9.3.

HISTORY OF STUDY

George H. Pepper of the Hyde Exploring Expedition rediscovered Baldwin's dinosaur-bearing strata near Ojo Alamo in 1902 (Brown, 1910). His collection sparked the interest of New York's American Museum of Natural History, which, in 1904, dispatched the renowned dinosaur collector, Barnum Brown, to investigate the area. Brown, and later Walter Granger of the same institution with W.J. Sinclair of Princeton University, made the first important collections from these strata, which Brown (1910) appropriately named the "Ojo Alamo Beds." Brown (1910; see also Sinclair & Granger, 1914) and Oliver P. Hay (1908, 1910) studied these early collections.

Field parties from the United States Geological Survey soon followed. In 1908 J. H. Gardner and J. W. Gidley, and in 1915 C. M. Bauer with J. B. Reeside secured a place for the Survey in the description of the area's geology. Among other things, Bauer (1916) redefined the American Museum workers' Ojo Alamo Beds to include the Fruitland Formation, Kirtland Shale, and Ojo Alamo Sandstone. His partner Reeside continued work in this area into the mid 1920's. In the meantime, Charles W. Gilmore (1916, 1919, 1922) studied the collections of Cretaceous vertebrates acquired by these groups.

Beginning in 1921 and continuing through 1924, Charles H. Sternberg assembled a large collection of dinosaur and turtle specimens from the Cretaceous strata of the San Juan Basin. Some of these came from the area around Ojo Alamo. His collections were sold to many institutions and were studied by H. F. Osborn (1923), C. Wiman (1930, 1931, 1933), Gilmore (1935), J. H. Ostrom (1961, 1963), and N. Mateer (1976; this volume). A 1929 field party from the United States National Museum collected additional specimens, which were described by Gilmore (1935). Except for brief collecting trips by the University of Oklahoma in 1940 and 1941, this marked the end of the first era of collecting at Ojo Alamo, and little else was done until the 1960's.

Collections made in 1962 (Museum of Natural History, University of Kansas) and throughout the 1970's (University of Arizona, Museum of Northern Arizona, and University of New Mexico) have yet to be fully described. Parts of these collections were obtained from the Ojo Alamo region. Recent work by W. A. Clemens (1973) and J. G. Armstrong-Ziegler (1978) has concentrated on specimens collected from the Fruitland Formation and from the lower part of the Kirtland Shale, west of the Ojo Alamo region. Jon Powell (1968) identified a turtle collected by the University of Arizona from this region; however, it is the only specimen from the collections of this institution mentioned in the literature to date. This study, and work by Kues, et al. (1980) make the first contributions to the literature from the UNM collections.

STRATIGRAPHY

Rock-stratigraphy.—The stratigraphic nomenclature of the rocks straddling the Cretaceous–Tertiary boundary in the San Juan Basin has been a source of much confusion to rock-stratigraphers and paleontologists alike. This problem has been discussed at length by many authors (e.g. Baltz et al., 1966; Powell, 1973; Fassett, 1973; Clemens, 1973), and although disagreement remains, the stratigraphic nomenclature adopted for this report is that proposed by Baltz et al. (1966). Their revised nomenclature was based on a thorough study of the field relationships of strata exposed in the Ojo Alamo region and provides a logical association of these strata (Text-fig. 9.3).

The Naashoibito Member of the Kirtland Shale comprises strata that were formerly included as the lower part of Bauer's (1916) Ojo Alamo Sandstone. Recently, however, several authors (e.g. Baltz et al., 1966; Fassett and Hinds, 1971; O'Sullivan et al., 1972; Lucas & Rigby, 1979) have demonstrated the presence of a widespread, regionally angular,

TEXT-FIG. 9.3—Stratigraphic nomenclature chart for rocks exposed in the Ojo Alamo region (after Baltz et al, 1966) showing the biostratigraphic units used in this report. Abbreviations are those used in Text-fig. 9.2.

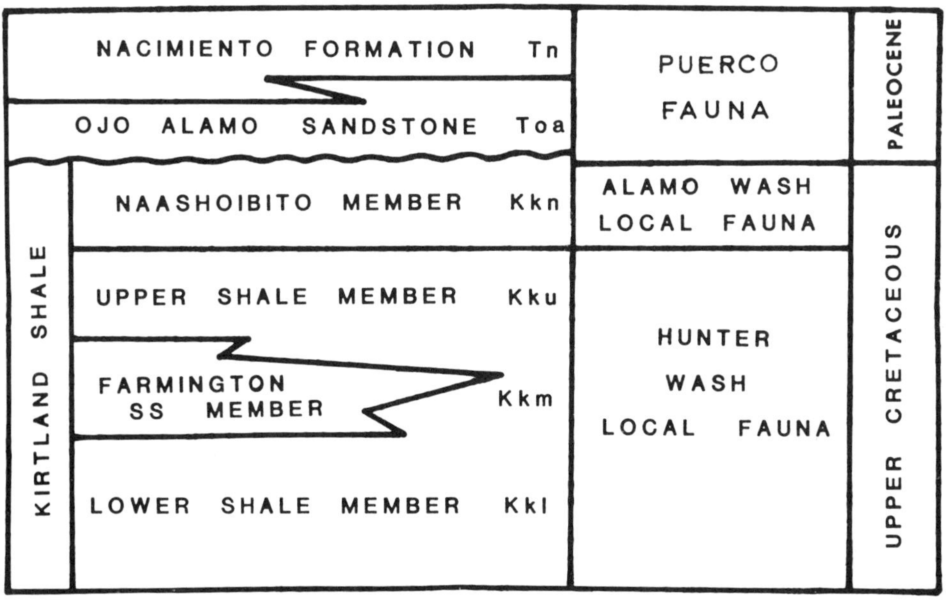

unconformity at the base of the upper conglomerate of Bauer's (1916) Ojo Alamo Sandstone. Further, on the basis of palynology (Anderson, 1960), vertebrate fauna (Rigby & Lucas, 1977), and intertonguing relationships (Baltz et al., 1966), this upper conglomeratic sandstone is demonstrably Paleocene in age. These facts, together with the existence of lithologic and paleontologic similarities, suggest that the Naashoibito beds should, indeed, be included with those of the Kirtland Shale, and that the term "Ojo Alamo Sandstone" should now be used in the restricted sense proposed by Baltz et al. (1966) to include only the upper conglomeratic sandstone of Bauer (1916).

Throughout the Ojo Alamo region the Naashoibito consists of a basal conglomeratic sandstone overlain by a distinctive series of interfingering gray, purple, and maroon shales, and soft white sandstones (Text-fig. 9.4). Baltz et al.(1966) demonstrated the lateral continuity and persistence of the lower conglomeratic sandstone throughout the Ojo Alamo region. However, lack of exposures on the divides between the major wash systems north and south of the Hunter Wash–De-na-zin Wash confluence make it difficult, if not impossible, to physically trace the Naashoibito into other areas. Moreover, the absence in other areas of the distinctively colored shale and soft sandstone characteristic of the Naashoibito, as well as the lack of any persistent conglomeratic sandstone, suggests that the Naashoibito may be restricted to the Ojo Alamo region.

Lithologically, the Naashoibito resembles the McDermott Member of the Animas Formation. This unit is exposed along the northwest side of the San Juan Basin and occupies a stratigraphic position comparable to that of the Naashoibito (Reeside, 1924). It contains conglomeratic sandstones and purple and maroon shales which are strikingly similar to those of the Naashoibito. In addition, both the McDermott and Naashoibito contain varying amounts of detrital silicic and intermediate

TEXT-FIG. 9.4—*A*, view to north across head of De-na-zin Wash near Barrel Spring showing the upper shale member (Ku) and Naashoibito member (Kn) of the Kirtland Shale, pinnacled sands of the Ojo Alamo Sandstone (Toa), and variegated shales of the Nacimiento Formation (Tn); *B*, view to east of Naashoibito Member on the divide between Alamo and De-na-zin washes. The basal conglomeratic sandstone of the Naashoibito forms the small ledges in the foreground (*A*).

A

B

volcanic rocks. It seems possible, therefore, that the Naashoibito represents an erosional remnant of the McDermott south of the San Juan River.

Biostratigraphy.—The vertebrate fauna of the Naashoibito was for a long time called the "Ojo Alamo fauna," due largely to the work of C. W. Gilmore (1916, 1919, 1922). The name has persisted in the literature. However, with the separation of the strata that bear this fauna from the Ojo Alamo Sandstone (restricted) and the inclusion of these strata with the Kirtland Shale, another name should be used for this vertebrate assemblage. The unfortunate consequences of constructing a biostratigraphic name from a rock-stratigraphic name are clearly evident.

Clemens (1973) proposed the term "Alamo Wash local fauna" for the fossils collected from the Naashoibito and upper shale members of the Kirtland "in the vicinity of the old Ojo Alamo trading post." Powell (1973) utilized the name "Alamo local fauna" for this same faunal unit, and more recently Lindsay et al. (1978) used the term "Ojo Alamo local fauna." Neither of these terms is as desirable as Clemens' original designation, which is based on a geographic name. In addition, Clemens' term removes the confusion that might result from association with the Ojo Alamo Sandstone. Hence, "Alamo Wash local fauna" is the term used here.

In contrast to Clemens' use of the term, however, its use here is restricted solely to the Naashoibito Member of the Kirtland Shale. Clemens (1973) hesitated to do this because it was uncertain whether sauropod dinosaurs were restricted to that unit. The occurrence of the sauropod dinosaur *Alamosaurus sanjuanensis* Gilmore is significant because its remains do not occur outside of the Naashoibito within the San Juan Basin. At the time of Clemens' work the type locality of *A. sanjuanensis* was the only sauropod locality known. Recent work has resulted in the discovery of five additional localities, all within the Naashoibito (e.g. Kues et al., 1980). Moreover, these localities occur throughout the Naashoibito outcrop area from the south branch of Hunter Wash to the north side of De-na-zin Wash (Text-fig. 9.2). Thus, the proposed restriction is a logical one for the collections currently available. Therefore, the Hunter Wash local fauna has been extended to include the upper shale member of the Kirtland (Text-fig. 9.3).

The Alamo Wash local fauna has been collected exclusively from the Hunter Wash–De-na-zin Wash confluence. Major collections of this fauna are housed in the American Museum of Natural History, National Museum of Natural History, the University of Uppsala, Sweden, and the University of New Mexico.

ABBREVIATIONS

The abbreviations used in the text and figures are as follows:

AMNH American Museum of Natural History

USNM United States National Museum of Natural History.

UNM University of New Mexico, Geology Department. BUNM is the prefix given to the University of New Mexico collecting localities; B- and FKK- are prefixes given to specimens in the UNM collection.

THE ALAMO WASH LOCAL FAUNA

No articulated skeletons have yet been recovered from the Naashoibito, although associated skeletal remains are common. As a result, the Alamo Wash local fauna is known entirely from isolated and usually fragmentary specimens and single bones. In addition, existing collections are biased toward larger vertebrates because as yet only small-scale screen washing projects have been attempted. In the following annotations for the Alamo Wash faunal list (Table 9.1), the previous record of each group is treated first, followed by brief discussions of the new material in the University of New Mexico collection.

Lepisosteidae and Cyprinidae.—Isolated lozenge-shaped ganoid scales were first reported from the Naashoibito by Gilmore (1916). The UNM collection contains a large number of these scales, and, according to the usual practice, they are referred here to the genus *Lepisosteus*. A single pharyngeal tooth (UNM FKK-036) indicates the presence of cyprinid fishes in the Alamo Wash local fauna.

Baenidae, Trionychidae, and Dermatemydidae.—Study of the diverse turtle fauna of the Naashoibito began with the description of four new species by Hay (1908). The type specimens of these species (*Thescelus rapiens* AMNH 6066, *Aspideretes vorax* AMNH 6140, *A. fontanus* AMNH 6070, and *A. austerus* AMNH 6068) were all collected in the vicinity of Ojo Alamo Trading Post. No precise stratigraphic information was given by Hay; nor was any given by Brown (1910), who collected the specimens. Fortunately, Gilmore (1916) later determined from Brown that all of the specimens had been collected from what is now the Naashoibito.

In 1909, Gardner and Gidley collected the type specimens of two additional species, *Basilemys nobilis* (USNM 6555) and *Adocus vigoratus* (USNM 6554), from the Naashoibito near Ojo Alamo. These were described by Hay in 1910. Gilmore (1916) added an occurrence of *Compsemys* sp., which can also be referred to the Naashoibito.

Table 9.1—Vertebrates of the Alamo Wash local fauna. Sources for identifications are given in text.

Class Osteichthyes
Subclass Actinopterygii
Order Lepisosteiformes
Family Lepisosteidae
Lepisosteus sp.
Order Cypriniformes
Family Cyprinidae
gen. et. sp. indet.
Class Reptilia
Subclass Anapsida
Order Testudines
Family Baenidae
Thescelus insiliens
Neurankylus eximius
"Baena" nodosa
Compsemys sp.
Family Trionychidae
Aspideretes vorax
Aspideretes fontanus
Aspideretes austerus
Family Dermatemydidae
Adocus vigoratus
Basilemys nobilis
Hoplochelys cf. *H. bicarinata*
Subclass Archosauria
Order Saurischia
Family Tyrannosauridae
?*Albertosaurus* sp.
Family Saurornithoididae
gen. et. sp. indet.
Family Titanosauridae
Alamosaurus sanjuanensis
Order Ornithischia
Family ?Hypsilophodontidae
gen. et. sp. indet.
Family Hadrosauridae
Hadrosaurus navajovius
Parasaurolophus tubicen
Family Ceratopsidae
Pentaceratops sternbergii
aff. *Torosaurus utahensis*
Family Nodosauridae
?*Panoplosaurus* sp.
Order Crocodylia
Family Crocodylidae
Crocodylus sp.
?*Allognathosuchus* sp.
Order Goniopholididae
Goniopholis sp.

In addition, two specimens described by Wiman (1933), for which only "Ojo Alamo Sandstone" was given as provenience, can now be shown to have come from the Naashoibito. These specimens were collected by C.H. Sternberg in 1921. A letter from Sternberg to Wiman dated 14 November, 1921, indicates that the specimens were collected from the Barrel Spring area on De-na-zin Wash "from between the conglomerates [of the] Ojo Alamo Sandstone," and thus from what is now the Naashoibito (Mateer, this volume). One of the specimens was referred to *Neurankylus baueri* ("Exemplar No. 3" of Wiman, 1933), the other to *Neurankylus* sp.

Most recently, Powell (1968) reported a single specimen of *Baena nodosa* from the Naashoibito. Gaffney (1972) demonstrated that *Thescelus rapiens* is a junior synonym of *Thescelus insiliens*, and that *Neurankylus baueri* is a junior synonym of *Neurankylus eximius*. He also questioned the assignment of *Baena nodosa* to the genus *Baena*. The faunal list (Table 9.1) is amended accordingly. A revision of the dermatemydids and trionychids will doubtless reveal similar synonymous relationships; however, the original references to the species of these families are retained here until such a study has been conducted.

The UNM collection contains an additional record in the form of a single, complete entoplastron (Text-fig. 9.5). The shape of the bone and the configuration of the sulci agree with the entoplastron of *Hoplochelys bicarinata* (Gilmore, 1919). Identification of such scant material is tenuous, but the specimen resembles no other species reported from the San Juan Basin. *Hoplochelys* is common in the Nacimiento Formation, occurring as low as 50 ft (15 m) above its base (Gilmore, 1919). If the present identification is correct, the range of *Hoplochelys* must be extended downward into the Upper Cretaceous.

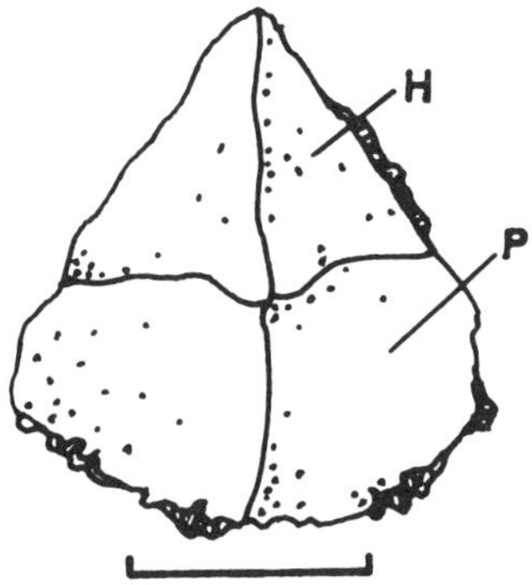

TEXT-FIG. 9.5—*Hoplochelys* cf. *H. bicarinata* (UNM FKK-002). Entoplastron showing sulci for humeral (H) and pectoral (P) scutes. Length of reference line = 2 cm.

Tyrannosauridae.—The presence of carnivorous dinosaurs in what is now the Kirtland Shale was first recognized by Barnum Brown (1910, p.268). Brown noted that carnosaur remains "were numerous; chiefly disassociated limb bones, lacking articular ends but characterized by hollow shafts, and vertebrae representing a dinosaur as large as *Albertosaurus* of the lower Edmonton series." Detached carnosaur teeth were present in Brown's collection, as well as in E. D. Cope's collection made from the same area (Brown, 1910), but it is not possible to determine if any of them came from the Naashoibito Member.

Sinclair and Granger (1914) collected teeth of carnivorous dinosaurs from both the Naashoibito and upper shale members of the Kirtland, and Brown referred some of the teeth in their collection to ?*Deinodon* sp. However, the stratigraphic provenience of these teeth was not given. Thus, the occurrence of *Deinodon* in the Naashoibito, long propagated in the literature, is dubious. The point is moot because the genus *Deinodon* has never been satisfactorily defined and is now considered a *nomen dubium* (cf. Russell, 1970).

Although it is likely that many of the carnosaur remains mentioned above came from the Naashoibito, the first undoubted record of their occurrence here was provided by Gilmore (1916). Isolated teeth were collected from several localities. In addition, several fragmentary vertebrae of a "large carnivorous dinosaur of the proportions of *Tyrannosaurus*" were collected from the Naashoibito (Gilmore, 1919). Gilmore did not, however, attempt to refer any of the specimens to a genus.

A single, complete, right fourth metatarsal (on loan from the collection of Ronald P. Ratkevich of Alamogordo, New Mexico) is tentatively referred here to *Albertosaurus* sp.(Text-fig. 9.6). The drawing and transverse sections shown here may be compared directly with those given by Lambe (1917) and Parks (1928) for *Albertosaurus*. Russell (1970), however, indicated that the metatarsus of *Daspletosaurus* is similar to, but slightly shorter than, comparable specimens of *Albertosaurus*.

A large number of detached tyrannosaurid teeth are present in the UNM collection. While such teeth are, in most cases, not identifiable at the generic level, they may have serration counts which are diagnostic for groups of related genera. Serration counts recorded for the UNM collection of tyrannosaurid teeth indicate the presence of animals more nearly related to *Albertosaurus* or *Daspletosaurus* (with about 9 to 15 serrations per 5 mm) than to *Tyrannosaurus* (with about 6 to 9 serrations per 5 mm; measurements from Ostrom, 1969 and Russell, 1970).

Saurornithoididae.—The UNM collection contains several small, detached teeth which are referable to this family (Text-fig. 9.7). That

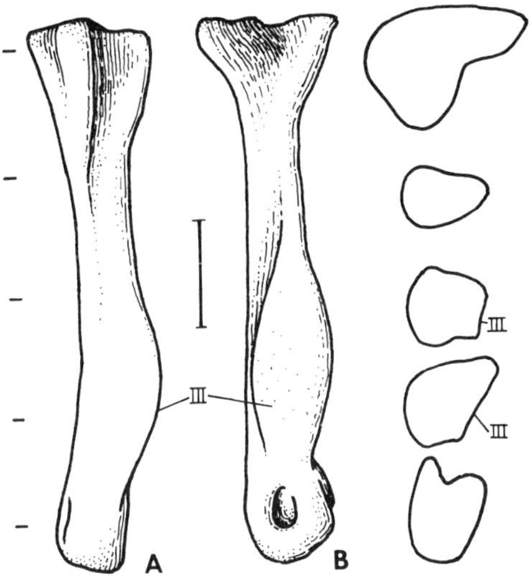

TEXT-FIG. 9.6—?*Albertosaurus* sp. (specimen borrowed from Ronald P. Ratkevich of Alamogordo, New Mexico), right metatarsal IV in anterior (*A*) and medial (*B*) views, with outlines of transverse sections enlarged slightly more. Location for sections is given at left. III = surface for application of metatarsal III. Length of reference line = 10 cm.

TEXT-FIG. 9.7—Saurornithoididae (UNM FKK-014), tooth in lingual (*A*), anterior (*B*), posterior (*C*), and labial (*D*) views. F = wear facet. Length of reference line = 5 mm.

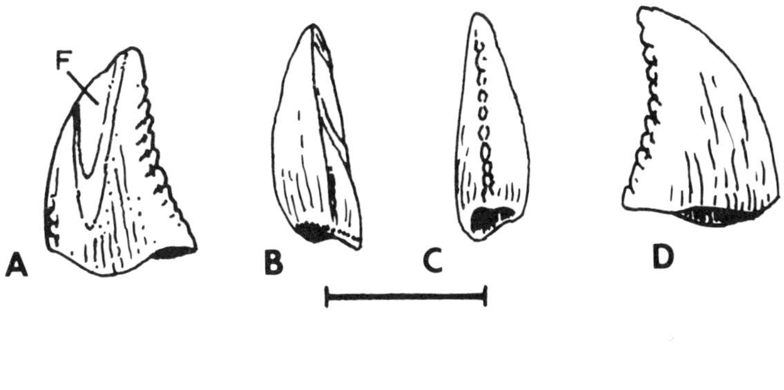

these were functional teeth is indicated by the presence in several specimens of distinct lateral wear facets. The teeth agree in size, shape, serration count, and absence or restriction of the anterior serrations to the base of the crown, with those described for *Saurornithoides junior* (Barsbold, 1974). In fact, were it not for the present taxonomic uncertainties affecting *Saurornithoides* and several related genera (e.g. *Stenonychosaurus*, *Chirostenotes*, *Troödon*), I would not hestitate to refer these specimens to that genus. For the present, however, it seems best to refer the UNM specimens to the family Saurornithoididae only.

Another specimen, a complete first metatarsal from the right foot of a small theropod (on loan from the collection of Neal LaFon of Albuquerque, New Mexico), is comparable in every way to that bone in deinonychosaurs (Text-fig. 9.8). It differs from the same element in *Dromaeosaurus* and *Deinonychus* in that the articular contact for metatarsal II is more expanded and the distal end more compressed. The

TEXT-FIG. 9.8—Deinonychosauria (specimen borrowed from Neal LaFon of Albuquerque, New Mexico), right metatarsal I in lateral (*A*), posterior (*B*), medial (*C*), and inferior (*D*) views. Length of reference line = 2 cm.

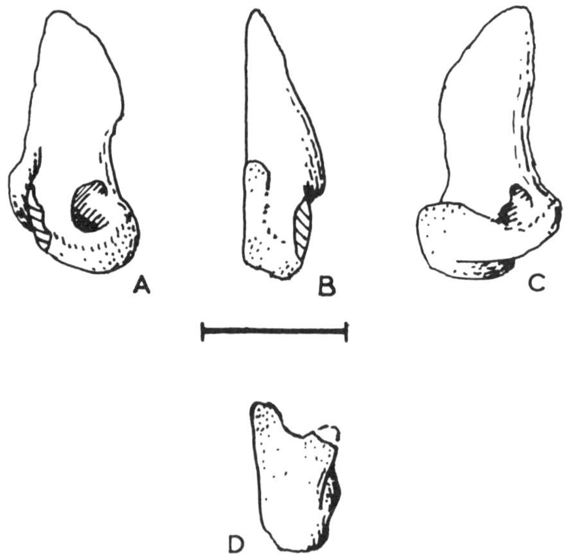

specimen resembles more closely the first metatarsal of *Stenonychosaurus inequalis* (Russell, 1969). It is not, however, possible to assign this specimen to either of the deinonychosaur families—Dromaeosauridae or Saurornithoididae.

Titanosauridae.—The only sauropod presently known from the Naashoibito is *Alamosaurus*, a genus established by C. W. Gilmore (1922) based on specimens collected by J. B. Reeside in 1921. The type materials were collected from the Naashoibito and consist of a left scapula and a right ischium (USNM 10486 and 10487). In 1921 C. H. Sternberg collected a sauropod posterior sacral vertebra with the first two caudals, a fragment of a right ilium, and some poorly preserved cervical vertebrae from the same locality which had yielded Gilmore's type. This material was deposited in the Paleontological Institute of Uppsala, Sweden, where it remained undescribed until recently (Mateer, 1976). A distal caudal centrum and a caudal neural spine (USNM 15658) had been collected at the same locality in 1916 by J. B. Reeside, and these specimens were referred to *Alamosaurus* thirty years later by Gilmore (1946). The locality was revisited in 1974 by David Thomas, who collected a sauropod femur and a tooth (UNM FKK-033 and FKK-029, respectively). The femur is too badly crushed to be of much diagnostic value, and it lacks most of the articular surfaces; however, the preserved portion is 1500 mm long. It is possible that all of these specimens pertain to a single individual.

Also present in the UNM collection are a distal caudal centrum (UNM B-522) and a dozen shed teeth which are probably referable to *Alamosaurus sanjuanensis*. The teeth have been described and illustrated elsewhere (Kues et al., 1980).

Hypsilophodontidae.—A single unusual scapula from the Naashoibito suggests the possible presence here of hypsilophodont dinosaurs (Text-fig. 9.9). Though fragmentary, the specimen exhibits several peculiarities, most important of which is the absence of an acromion process. In addition, the bone is stoutly constructed with a greatly thickened and blunt surface for articulation with the coracoid, which was evidently free or not firmly united to the scapula. The distal end of the blade, though somewhat fragmented, is thickened, as if for attachment to a suprascapular. While the UNM specimen is much larger, it agrees in shape and structure with the scapulae of the hypsilophodont, *Parksosaurus warreni* (Parks, 1926).

Hadrosauridae.—Large detached hadrosaur teeth are present in the Cope collection made from the Ojo Alamo area (Brown, 1910); however, it is not certain whether any of these came from the Naashoibito. In 1904, the holotype of "*Kritosaurus*" *navajovius* Brown (AMNH 5799) was collected near Ojo Alamo. Again no stratigraphic provenience was given (Brown, 1910); however, Brown later informed C. W. Gilmore that the specimen came from the upper shale member of the Kirtland (Gilmore, 1916).

The first hadrosaur remains definitely referable to the Naashoibito were collected by Sinclair and Granger (1914) "a few feet above the conglomerate separating the two horizons at which dinosaur bones were found." This specimen (AMNH 5797) consists of a maxillary and part of a lower jaw, identified by Brown as *Kritosaurus navajovius* (Sinclair & Granger, 1914). This specimen represents the only known occurrence of "*Kritosaurus*" in the Naashoibito. Recently, Baird and Horner (1977) suggested that "*Kritosaurus*" is congeneric with the eastern genus *Hadrosaurus*. In light of this, "*Kritosaurus*" is here considered a junior synonym, and the faunal list (Table 9.1) is amended accordingly.

In 1916 Gilmore added several hadrosaur occurrences for the Naashoibito, but none of the specimens was generically determinable. In 1921 C. H. Sternberg collected a fragmentary hadrosaur skull from what was then considered the Ojo Alamo Sandstone. A letter dated 14 November 1921 from Sternberg to Wiman indicates that the specimen was collected from the Naashoibito near Barrel Springs on De-na-zin Wash (N. Mateer, this volume). Wiman (1931) described the specimen as the type of a new species, *Parasaurolophus tubicen*.

Previous faunal lists for the Naashoibito Member have not included *Parasaurolophus tubicen*. This is probably due to an oversight by Gilmore (1935), who referred to *P. tubicen* another specimen collected by N. H. Boss in 1929 from the Kirtland Shale. Thereafter Gilmore included *P. tubicen* in his faunal lists for the Kirtland but neglected to include it in his Ojo Alamo (Bauer's definition) faunal lists. This oversight was propagated in later lists, probably because Gilmore's work was more widely known than Wiman's. Moreover, with the description by Ostrom (1963) of *P. cyrtocristatus* from the Fruitland Formation, it is evident that Gilmore's specimen (USNM 13492) might well be referred to either species. The USNM specimen is, however, from an animal substantially larger than the type of *P. cyrtocristatus* and more in accord with the size one would expect for *P. tubicen*.

The UNM collection does not contain any generically determinate hadrosaur remains; however, several specimens are worthy of comment. One of these, a very large left humerus (Text-fig. 9.10 A), indicates the presence of large hadrosaurines in the Naashoibito. A reasonable estimate of its length, as restored in figure 9.10, would be about 740 mm. The only hadrosaur humeri reported which are larger than this belong to a form related to *Lambeosaurus* from the El Gallo Formation of Baja California (Morris, 1972). Another specimen, consisting of a fragmentary right tibia (Text-fig. 9.10 B), would be well over 1000 mm

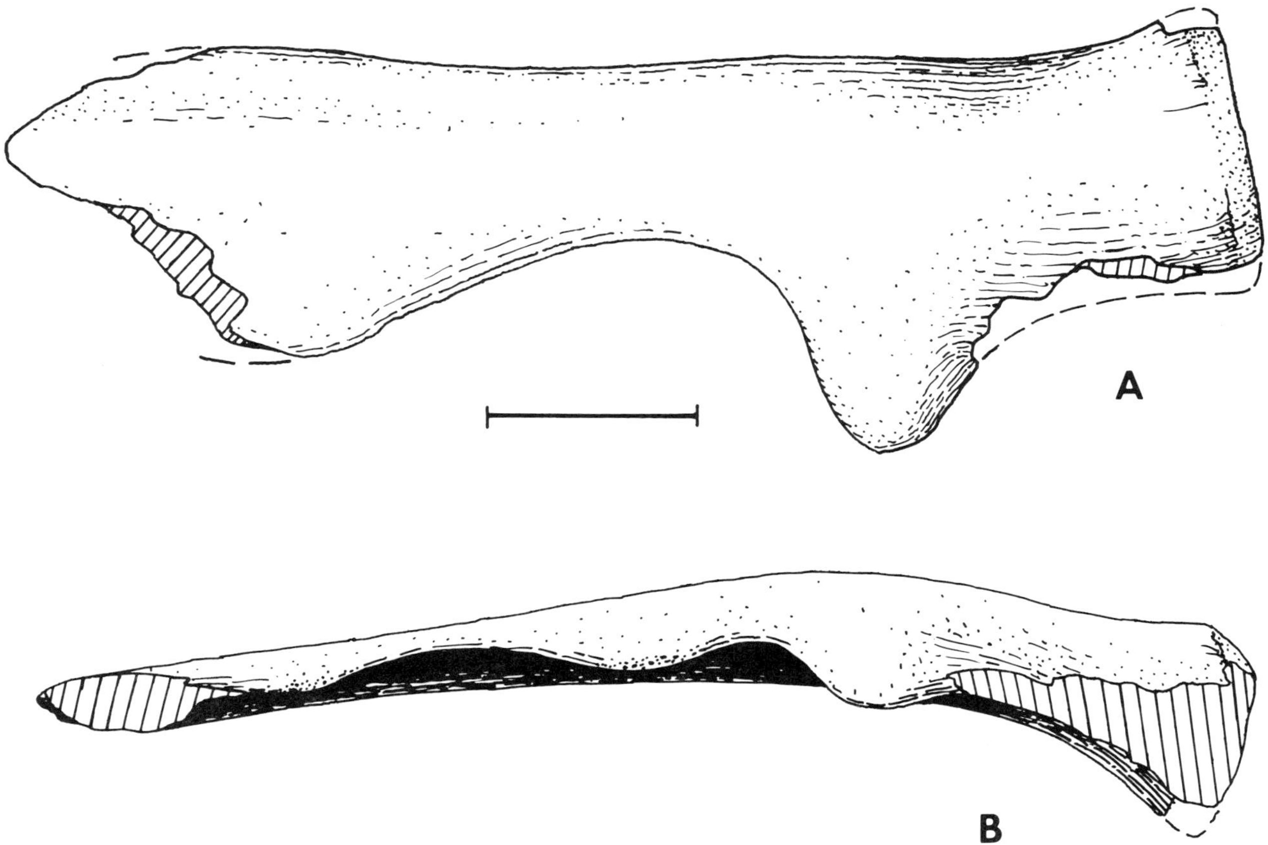

TEXT-FIG. 9.9—?Hypsilophodontidae (UNM B-709) right scapula in lateral (*A*) and posterior (*B*) views. Length of reference line = 10 cm.

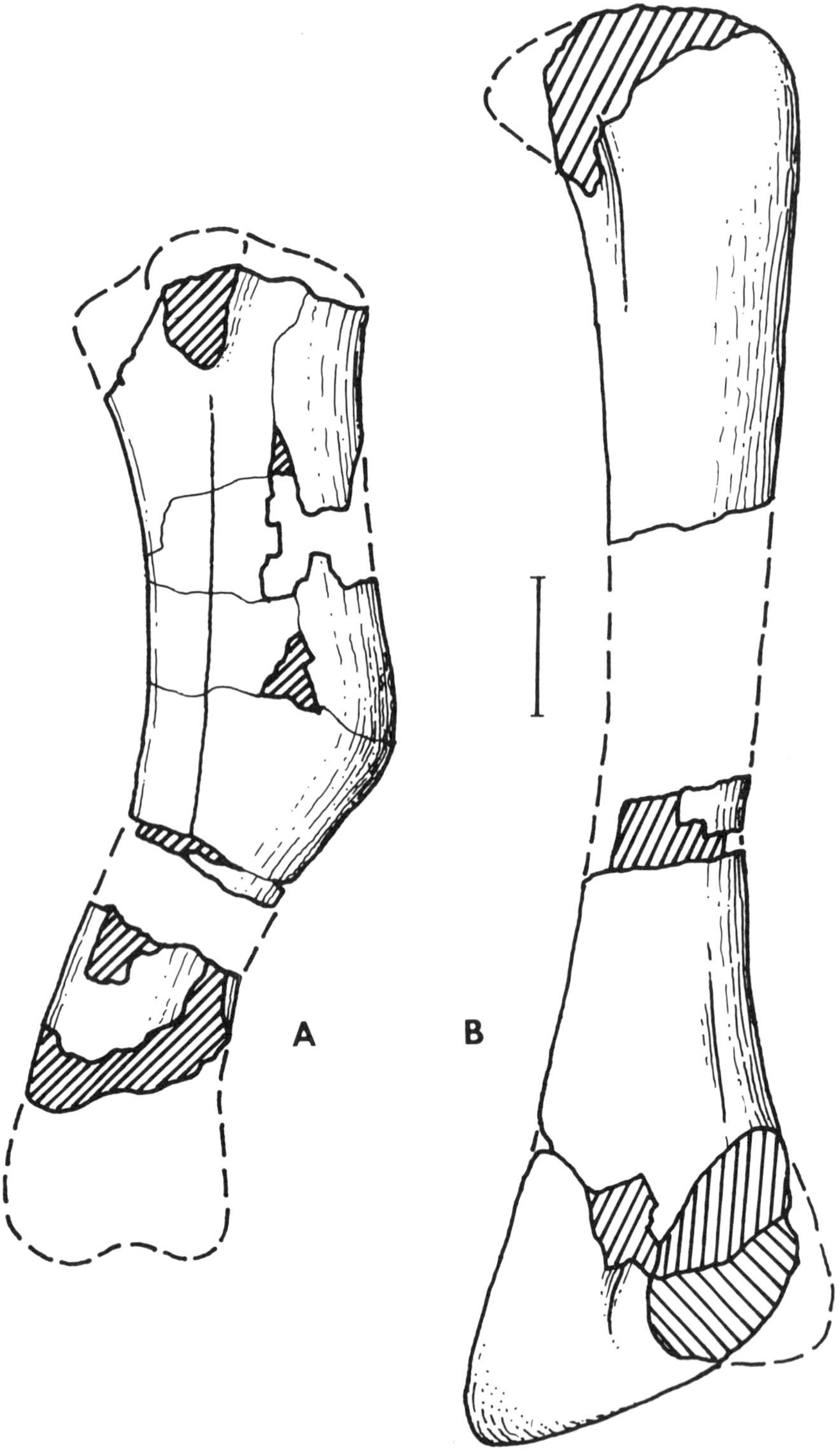

TEXT-FIG. *9.10*—Hadrosauridae *A*, left humerus (UNM B-706); *B*, right tibia (UNM B-517), both in anterior view. Length of reference line = 10 cm.

in length. Thus the hadrosaur to which it belonged was as large as any of the largest North American forms.

Ceratopsidae.—None of the ceratopsian material from the Ojo Alamo area reported by Brown (1910) can be referred with certainty to the Naashoibito Member because the stratigraphic information supplied by Brown is inadequate. Although Gilmore (1916) reported some ceratopsian frill fragments and teeth from what is now the Naashoibito, he could not add any determinable specimens. Gilmore did, however, assign Brown's specimen of *Monoclonius*? sp. (AMNH 5798) to the Ojo Alamo Sandstone (Bauer's definition). If Gilmore's assignment is correct, this specimen should now be referred to the Naashoibito. Gilmore (1916) did not give his reasons for making this assigment, but perhaps he had questioned Brown regarding the stratigraphic level at which the specimen was collected. Gilmore (1916) did determine from Brown the horizons at which the type specimens of *Kritosaurus navajovius* and the four species of turtles described by Hay (1908) were collected, and thus it is likely that he did the same for the *Monoclonius*? specimen. For this reason, the specimen (AMNH 5798) is retained as part of the Alamo Wash local fauna, though it is here removed from the genus *Monoclonius*.

Specimen AMNH 5798 consists of a part of a supraorbital horncore, 120 mm long. Brown based his identification of the specimen on its resemblance to *Monoclonius recurvicornis*, but judged it to be distinct. *M. recurvicornis* was, however, subsequently removed from *Monoclonius* and referred to *Ceratops* by Hatcher, though Lull suggested it might represent a new genus (Hatcher et al., 1907). Later, Sternberg (1938) referred the species to *Centrosaurus*, adding at the same time that "the shape and proportion of the horn cores are quite different from those of any known specimens of *Centrosaurus*." Evidently, *M. recurvicornis* cannot be convincingly referred to *Monoclonius*, *Centrosaurus*, or *Ceratops* as they are currently defined. Therefore, AMNH 5798 cannot be assigned to *Monoclonius*. Clearly, this specimen cannot be generically identified, nor should it or any other isolated horn core be, given the present poor understanding of the ontogenetic development and intraspecific variation of most members of the Ceratopsia.

Brown (1910) noted the occurrence of another ceratopsid in the area around Ojo Alamo. He wrote (p. 269): "The squamosal fragments of another individual observed in the field were much thinner than that bone in the genus *Triceratops* but similarly marked by deep vascular grooves. The horn and other skull fragments were apparently from a mature animal representing a ceratopsian genus smaller than *Triceratops* or *Torosaurus*, but the remains are too fragmentary for characterization."

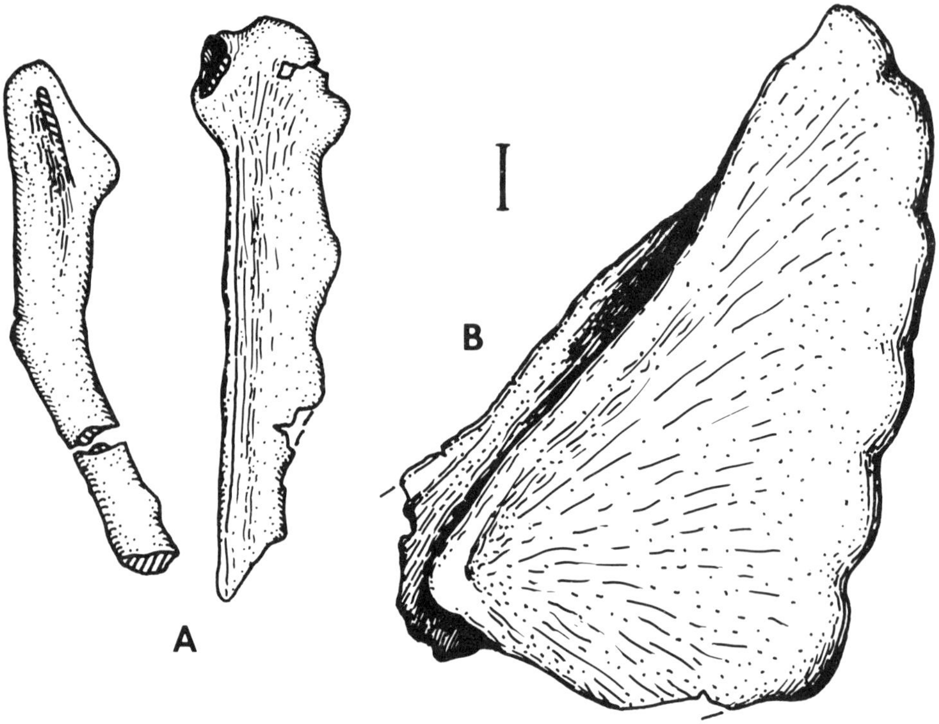

TEXT-FIG. 9.11—*Pentaceratops* cf. *P. sternbergii* (UNM FKK-035), fragmentary frill of juvenile specimen. *A*, posterior and lateral parts of parietal in dorsal view; *B*, right squamosal in medial view. Length of reference line = 5 cm.

A similar situation confronted Gilmore nine years later. Describing a fragmentary skull which had been collected on De-na-zin Wash, he wrote:

> From the great number of fragments a considerable part of the posterior end of what appears to be the median or dermo-supraoccipital was pieced together. Although this large fragment is not sufficiently characteristic to indicate the generic affinities, it is of interest as showing the presence here of a ceratopsian which is distinct from either *Triceratops*, *Ceratops*, or *Monoclonius*. This fragment shows a fenestrated frill in which the openings are apparently smaller than in either of the genera mentioned above. In all probability it represents an undescribed form. (Gilmore, 1919, p. 65).

It is possible that the specimens Brown and Gilmore described belong to *Pentaceratops*, a genus not known at the time of their work but later found in the Fruitland Formation and lower part of the Kirtland Shale. A ceratopsian supraorbital horn core, braincase, and squamosal fragment (USNM 12741), referred by Gilmore to *Pentaceratops sternbergii*,

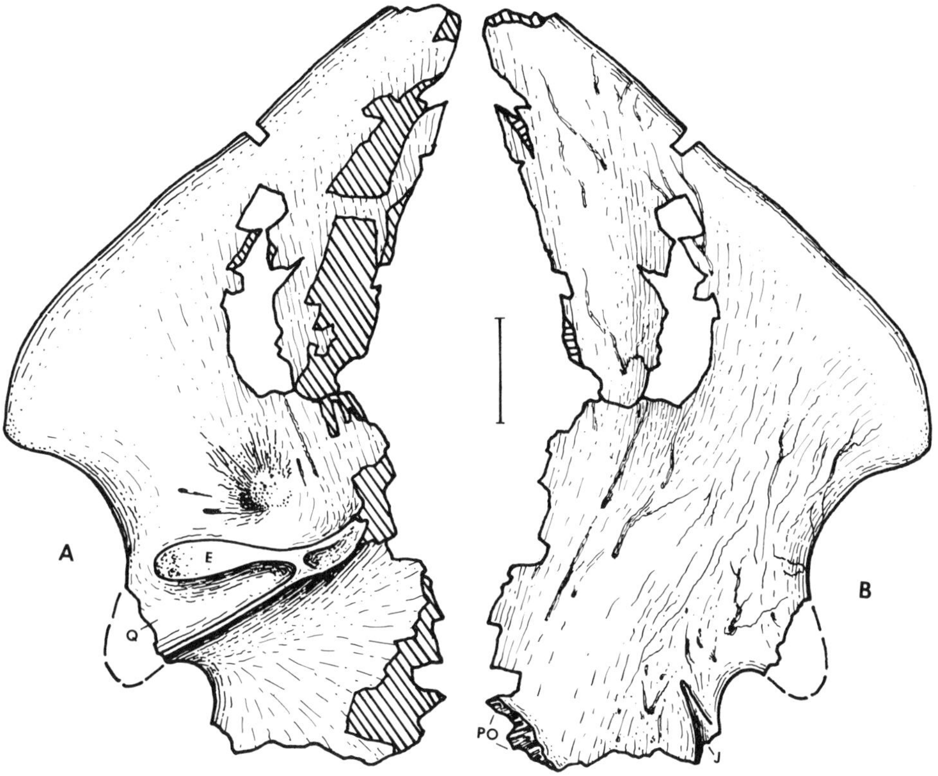

TEXT-FIG. 9.*12*—aff. *Torosaurus utahensis* (UNM B-628), left squamosal in medial (*A*) and lateral (*B*) views, showing: E, buttress for articulation with exoccipital; Q, pit for process of quadrate; PO, sutural contact for postorbital; and J, groove for process of jugal. Length of reference line = 10 cm.

was collected in 1929 from the purple shales of the McDermott Formation on Hunter Wash (T.25N., R.12W.). This specimen has not been previously mentioned in the literature. Baltz et al. (1966) have shown, however, that the beds mapped in this area by Reeside (1924) as part of the McDermott Formation belong, in fact, to the Naashoibito. Thus, *Pentaceratops* does occur in the Naashoibito, and it is likely that the specimens Brown and Gilmore described are referable to this genus as well.

Specimens in the UNM collection, while no less fragmentary, include the remains of at least five skulls and numerous isolated jaws and teeth. One specimen (Text-fig. 9.11), consisting of part of a right squamosal and fragments of the parietal, is identified here as a juvenile *Pentaceratops.* The small squamosal is similar in size and morphology to that of *Eoceratops canadensis* (Lambe, 1915). The parietal fragments indicate, however, that this individual possessed a slender, medially in-

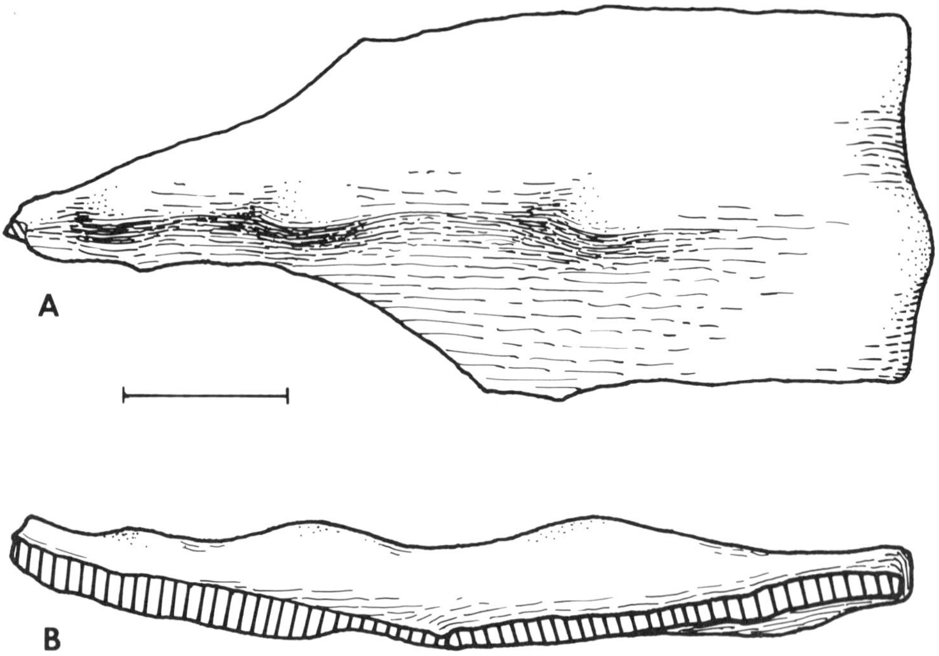

TEXT-FIG. 9.*13*—aff. *Torosaurus utahensis* (UNM FKK-013), postero-medial part of parietal in dorsal (*A*) and lateral (*B*) views. Length of reference line = 10 cm.

dented parietal bar, and large parietal fenestrae, as seen in *Chasmosaurus* and *Pentaceratops*. Instead of referring the specimen to *Eoceratops*, which may itself represent the juvenile form of some other ceratopsid (e.g., *Chasmosaurus*), it seems reasonable to assume that the small size of the present specimen is a juvenile character, and thus not taxonomically significant.

Part of another squamosal (Text-fig. 9.12) represents an additional and much larger ceratopsian in the Naashoibito, but the remains are not sufficient for complete characterization. The thick, broadly triangular squamosal has no undulations or epoccipitals along its outer border and is well invested with vascular sulci on its external surface. The internal surface is smooth with the exception of a large roughened boss just posterior to the exoccipital buttress. This feature may be pathologic. Several other specimens in the UNM collection are referable to this form; however, they are equally fragmentary. One of these (Text-fig. 9.13), consisting of the postero-medial part of a parietal, exhibits additional diagnostic characters. The internal surface of the parietal is flat, while the central axis is raised in a longitudinal series of shallow undulations, which give the bone a triangular transverse section. The bone thins rapidly away from the midline and carries no trace of vascular channels. The combination of features seen in these specimens resembles condi-

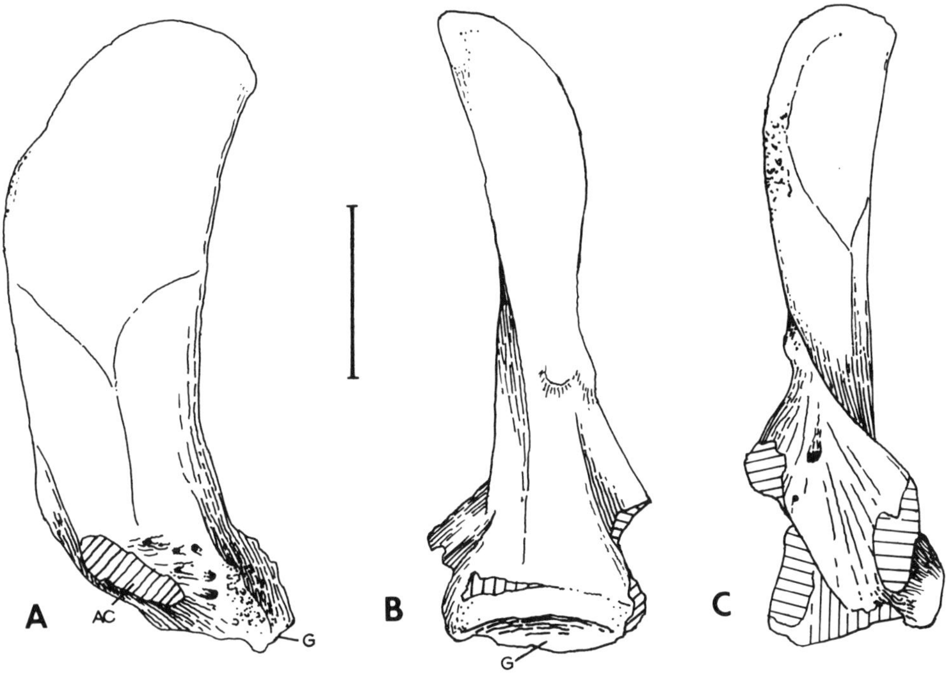

TEXT-FIG. 9.14—?*Panoplosaurus* sp. (USNM 8571), left scapula in lateral (*A*), inferior (*B*), and superior (*C*) views showing: AC, acromion process; G, part of glenoid cavity. Length of reference line = 10 cm.

tions in *Torosaurus utahensis* (Gilmore, 1946; Lawson, 1976), and they are, thus, tentatively referred.

Nodosauridae.—In 1919, Gilmore described the first nodosaurid remains from the Naashoibito. No ankylosaurid remains have yet been recovered. Included in Gilmore's (1919) collection is a nearly complete left scapula (USNM 8571) and two additional specimens, each consisting of a number of suturally united or associated dermal plates (USNM 8610 and 8611). Gilmore (1919, p. 66) wrote that the scapula "undoubtedly represents an undescribed form" and that the dermal plates "in all probability pertain to the same kind of an animal."

Additional figures are given here of the scapula (USNM 8571) described by Gilmore (Text-fig. 9.14). This specimen certainly pertains to one of the large, advanced nodosaurids of the "North American lineage" (*sensu* Coombs, 1978) and is tentatively referred to ?*Panoplosaurus* sp. It is indistinguishable from the scapula associated with *Panoplosaurus* (=*Edmontonia*) *rugosidens* (L. S. Russell, 1940). Postcranial armor fragments in the UNM collection support the occurrence of nodosaurids in the Naashoibito. However, these isolated keeled plates, compressed conical plates, and small ossicles are generically indeterminate.

Goniopholididae and Crocodylidae.—The "crocodile" tooth mentioned by Brown (in Sinclair & Granger, 1914) is the first recorded from the Kirtland Shale. Its stratigraphic position was not given, nor was it assigned a catalog number. Gilmore (1916) reported shed crocodilian teeth and osteoderms from two localities, both of which can now be referred to the Naashoibito. However, they serve only to indicate the presence of crocodilians in the Naashoibito Member. Later, Gilmore (1946) referred some of these specimens to *Crocodylus* sp., but no figures or catalog numbers have ever been reported for any of these specimens.

The occurrence of *Crocodylus* in Cretaceous deposits is subject to some question, particularly since all such occurrences are based on very fragmentary material. Steel (1973, p. 60), however, remarked that while "reports of the genus in Mesozoic deposits should perhaps be treated with caution, the persistence of such a relatively unspecialized crocodilian genus throughout the Cenozoic appears to be entirely feasible." Gilmore's identification of *Crocodylus* sp. will be retained here until such time as the material may be re-examined or the occurrence otherwise substantiated.

The UNM collection includes several large, stout, blunt crocodilian teeth with regularly spaced longitudinal costae (Text-fig. 9.15 A). These teeth may be assigned to *Goniopholis* based upon their similarity to a tooth from the Kirtland Shale, figured by Wiman (1932, pl. 5, fig. 2,3).

A number of small, low-crowned, "button-shaped" teeth—elongate and subovate in occlusal view—are present in the UNM collection (Text-fig. 9.15 B). Each tooth is covered with fine anastomosing pustulose ridges, which converge upward along a more prominent ridge. This ridge bisects the long axis of the tooth. Here again, while the specimens are meager, they are probably sufficient to indicate the presence of *Allognathosuchus* in the Naashoibito. If correct, this would be the first occurrence of this genus in Cretaceous deposits, although it is very abundant in the Paleocene deposits of the overlying Nacimiento Formation.

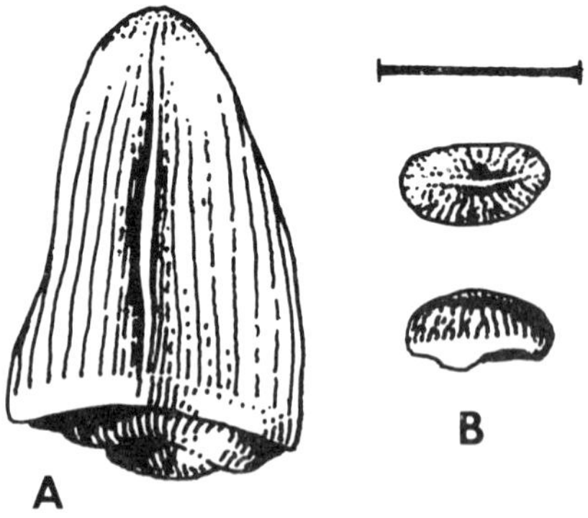

TEXT-FIG. 9.15—*A*, *Goniopholis* sp. (UNM B-675), tooth in lateral view; *B*, ?*Allognathosuchus* sp. (UNM B-657), tooth in occlusal and lateral view. Length of reference line = 1 cm.

It is advisable, however, to regard its occurrence in the Naashoibito as questionable until better specimens become available. Additional crocodilian remains, consisting of numerous osteoderms, cranial fragments, and teeth, are present in the UNM collection. None of this material, however, is sufficient to permit generic or even familial identification.

AGE AND CORRELATION OF THE ALAMO WASH LOCAL FAUNA

The Naashoibito Member is the highest dinosaur-bearing horizon on the western side of the San Juan Basin; therefore it has an important bearing on the interpretation of local stratigraphic relationships and regional correlations. In fact, it is the importance of the Alamo Wash local fauna in working out the temporal sequence of dinosaur communities in western North America that led to the present study (e.g. Russell, 1975 refers to "the Fruitland-Kirtland Problem").

The first step in clarifying the relationships of the Alamo Wash local fauna is to compare it with the fauna of the underlying parts of the Kirtland Shale and Fruitland Formation—the Hunter Wash local fauna. The following comparison chart was constructed for this purpose (Table 9.2). Only the larger vertebrates (turtles, dinosaurs, and crocodilians) will be considered here, for, as noted above, only small-scale screen washing projects have so far been attempted in the Naashoibito; and the extensive sample of microvertebrates obtained from the Hunter Wash local fauna (Armstrong-Ziegler, 1978) has not been duplicated in the overlying deposits. One locality in the Naashoibito (locality 5, Text-fig. 9.2) has, however, proven to be productive. Further work there should increase the number of micro-vertebrate taxa known from the Alamo Wash local fauna.

Before proceeding, some points regarding the Hunter Wash local fauna (Table 9.2) should be clarified. *Pentaceratops fenestratus* Wiman (1930) is not included here, as it may be considered a pathologic and aged individual of *P. sternbergii* (Lull, 1933; but see Mateer, this volume). Similarly, *Parasaurolophus tubicen* is omitted from the Hunter Wash local fauna because the specimen on which its occurrence is based (USNM 13492; Gilmore, 1935), though it is much larger, could easily pertain to *P. cyrtocristatus*. The questionable occurrence of *Euoplocephalus* sp. is based on the identification of a right humerus (USNM 8360) by Barnum Brown (unpublished note on specimen label). Finally, identifications of the generically indeterminate ornithomimid and saurornithoidid are based on numerous specimens in the UNM collection.

Examination of Table 9.2 reveals strong resemblances between the

Table 9.2—Comparison chart for the larger vertebrates of the Hunter Wash local fauna and the Alamo Wash local fauna. Sources for identifications of the Hunter Wash local fauna are: Gilmore (1916, 1919, 1935), Wiman (1931, 1933), Armstrong-Ziegler (1978). Sources for the Alamo Wash local fauna are cited in the text. A question mark indicates that the occurrence of a taxon is for some reason questionable.

	Hunter Wash local fauna	Alamo Wash local fauna
Family Baenidae		
"Baena" ornata	X	
"Baena" nodosa	X	X
Boremys pulchra	X	
Thescelus insiliens	X	X
Neurankylus eximius	X	X
Compsemys sp.	X	X
Family Trionychidae		
Aspideretes ovatus	X	
Aspideretes vorax	X	X
Aspideretes fontanus		X
Aspideretes austerus		X
Trionyx sp.	X	
Plastomenus robustus	X	
Platypeltis sternbergi	X	
Family Dermatemydidae		
Adocus bossi	X	
Adocus kirtlandius	X	
Adocus vigoratus		X
Basilemys nobilis	X	X
Hoplochelys cf. *H. bicarinata*		X
Family Tyrannosauridae		
Albertosaurus sp.	X	?
Family Coeluridae		
Paronychodon lacustris	X	
Family Ornithomimidae		
gen. et. sp. indet.	X	
Family Dromaeosauridae		
gen. et. sp. indet.	?	
Family Saurornithoididae		
gen et. sp. indet.	X	X
Family Titanosauridae		
Alamosaurus sanjuanensis		X
Family Hypsilophodontidae		
gen et. sp. indet.		?
Family Hadrosauridae		
Hadrosaurus navajovius	X	X
Parasaurolophus cyrtocristatus	X	
Parasaurolophus tubicen		X
Family Ceratopsidae		
Chasmosaurus sp.	X	
Monoclonius sp.	?	
Pentaceratops sternbergii	X	X
aff. *Torosaurus utahensis*		X

Family Pachycephalosauridae		
gen. et. sp. indet.	X	
Family Ankylosauridae		
Euoplocephalus sp.	?	
Family Nodosauridae		
Panoplosaurus sp.		?
Family Crocodylidae		
Brachychampsa sp.	X	
Crocodylus sp.	X	X
Leidyosuchus sp.	X	
Thoracosaurus sp.	?	
Allognathosuchus sp.		?
Family Goniophoilididae		
Goniopholis kirtlandicus	X	
Goniopholis sp.		X

Hunter Wash local fauna and Alamo Wash local fauna. Many species carry over from one fauna to the other. Most of the apparent differences may reasonably be attributed to preservational or collecting biases. Several of these differences, however, reflect a real distinction between the two faunas, and these may be used as criteria for their separation.

Most important is the introduction of *Alamosaurus sanjuanensis* in the Alamo Wash local fauna. The fragmentary appendicular remains of this large animal are numerous in the Naashoibito Member, and its shed teeth are locally abundant in channel deposits (Kues et al., 1980). Its complete absence in the underlying parts of the Kirtland Shale and Fruitland Formation is therefore significant, though admittedly a conclusion based on negative evidence. Likewise, the presence of the ceratopsian, aff. *Torosaurus utahensis* in the Alamo Wash local fauna is significant, for its remains, which are quite distinct from those of *Pentaceratops*, would certainly have been reported had they been found in the Hunter Wash local fauna. Also important is the fragmentary occurrence of two "Paleocene-aspect" reptiles in the Alamo Wash local fauna—the turtle *Hoplochelys* and the crocodilian ?*Allognathosuchus*.

Though not considered in the comparison chart, the complete absence of amiid fishes and the batoid chondrichthyian *Myledaphus* from the Alamo Wash local fauna may be significant because their remains are abundant in the Hunter Wash local fauna, and screen washing techniques are not needed in order to find them.

The Hunter Wash local fauna is, on the basis of contemporaneous marine invertebrates (Cobban, 1973), pollen (Fassett and Hinds, 1971; Tschudy, 1973), and vertebrates (Gilmore, 1935; Armstrong-Ziegler, 1978) demonstrably Campanian in age. The age of the Alamo Wash local fauna is more difficult to determine owing to a mixing of Campanian dinosaurs (*Hadrosaurus*, *Parasaurolophus*, and *Pentaceratops*) with Maestrichtian dinosaurs (*Alamosaurus* and *Torosaurus*). Because no invertebrates or pollen suites have yet been collected from the Naashoibito, the correlation must rest entirely upon the vertebrate evidence.

The age of the Alamo Wash local fauna may be interpreted in two ways. First, it may be Campanian in age and thus approximately equivalent temporally to, or by principle of superposition, slightly younger than the Hunter Wash local fauna. If this interpretation holds, the differences between the faunas are wholly ecologic and probably attributable to facies differences (as argued by S. Lucas, this volume). The absence of *Alamosaurus* and *Torosaurus* in other Campanian deposits is explained by suggesting that these animals were restricted to an inland environment, rarely preserved, while the more typical Campanian dinosaurs were eurytopic and thus found in inland as well as the more common coastal floodplain faunas.

An alternative interpretation, first advanced by Sloan (1970) and later reinforced by Lawson (1976), holds that the Alamo Wash local fauna is Maestrichtian. I favor this interpretation because of the strong resemblance of the Alamo Wash local fauna to assemblages from the lower part of the North Horn Formation of Utah (Gilmore, 1946), and the lower part of the Tornillo Group of west Texas (Lawson, 1976). All three faunas are characterized by the presence of *Alamosaurus* and *Torosaurus*.

Torosaurus is known elsewhere only from the Maestrichtian Lance and Hell Creek formations. In addition, Lawson (1976) has demonstrated that the Maestrichtian dinosaur, *Tyrannosaurus rex*, is present in the lower part of the Tornillo Group. The presence of "Paleocene-aspect" reptiles (*Hoplochelys* and *Allognathosuchus*) in the Alamo Wash local fauna also points to a latest Cretaceous age. Thus, a combination of evidence suggests that the "*Alamosaurus* community" of Sloan (1970), which includes the Alamo Wash local fauna, is indeed Maestrichtian in age.

The occurrence of Campanian-aspect dinosaurs in the Alamo Wash local fauna, therefore, requires explanation. Perhaps the dinosaurs *Hadrosaurus* and *Parasaurolophus* survived into Maestrichtian time in the southern part of the Western Interior while they were replaced in the north by the more characteristic Maestrichtian form, *Anatosaurus*. This suggestion is strengthened by the recent discovery of *Parasaurolophus* in the Maestrichtian Kaiparowits Formation of Southern Utah (Weishampel and Jensen, 1979). *Pentaceratops*, which was apparently endemic to the San Juan Basin region, survived into the Maestrichtian as well, while in the north its ecologic equivalents (*Chasmosaurus* and perhaps *Anchiceratops*) were replaced by *Triceratops*. In addition, the nodosaurid ?*Panoplosaurus* evidently persisted well into Maestrichtian time in the south, while it lasted only through the early Maestrichtian in the north (Coombs, 1978).

Torosaurus inhabited both northern and southern realms during the Maestrichtian, though it is represented by a different species in each; *T. latus* in the north and *T. utahensis* in the south (Lawson, 1976). Likewise, the evidence suggests that *Tyrannosaurus rex* inhabited both realms (Lawson, 1976). On the other hand, *Alamosaurus* apparently migrated into the southern realm during the Maestrichtian (probably from South America) but did not penetrate into the north (Sloan, 1970). Fragmentary titanosaurid remains found in the Maestrichtian Evanston Formation of southwest Wyoming may represent the northernmost occurrence of *Alamosaurus* (D. Lawson, written comm. to S. Lucas, 1978).

Thus it appears that during Maestrichtian time two distinct faunal realms were present in the Western Interior of North America (Text-fig. 9.16). In the north was the "*Triceratops* community" of Sloan (1970), characterized by *Anatosaurus*, *Triceratops*, and *Torosaurus latus*. In the south was the "*Alamosaurus* community" characterized by *Alamosaurus*, *Pentaceratops*, and *Torosaurus utahensis*, as well as the persistence of the Campanian-aspect dinosaurs, *Hadrosaurus* and *Parasaurolophus*.

The "geographic distribution and lithologic association" of *Torosaurus* prompted Lawson (1976) to infer that the *Triceratops* community was characteristic of coastal environments, while the *Alamosaurus* community was an inland fauna living "hundreds of kilometers from the coast." This may well have been the case. This inland depositional setting might explain the absence of amiid fishes and the batoid *Myledaphus* from the Alamo Wash local fauna and other examples of the *Alamosaurus* community. They are seemingly ubiquitous in coastal depositional settings of the Upper Cretaceous.

The two faunal realms are separated somewhere in the area of western Colorado, as the southernmost reliably reported occurrence of *Triceratops* is in the vicinity of Denver (Lull, 1933). Paleogeographic reconstructions of western North America for Maestrichtian time indicate that the rising San Juan, Nacimiento, and Sangre de Cristo mountain ranges produced major topographic highs in northern New Mexico and southern Colorado (Text-fig. 9.16; McGookey et al., 1972). These features were not present during Campanian time. It seems possible then that growing geographic barriers were at least partly responsible for the separation of the two realms.

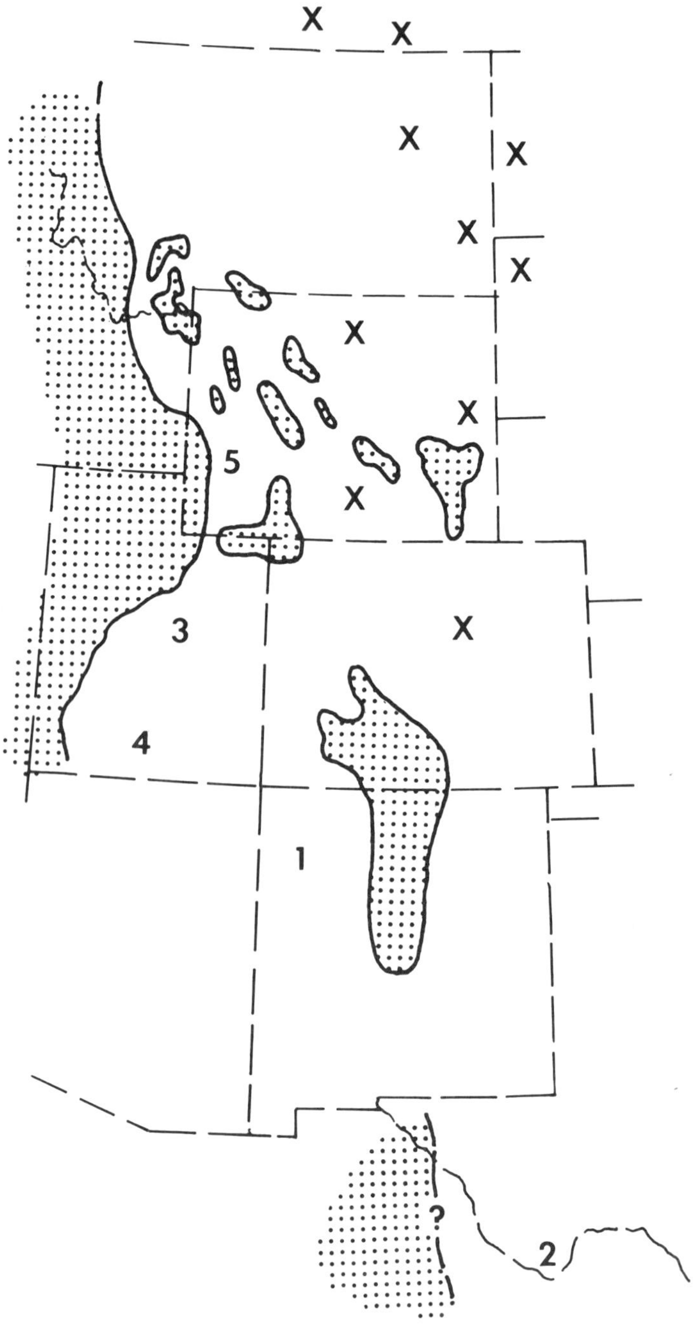

TEXT-FIG. *9.16*—Paleogeographic map of western North America during Later Maestrichtian time, after McGookey et al. (1972), showing the location of major topographic highs. Localities 1–5 are sites of known or possible representatives of the *Alamosaurus* community (Sloan, 1970) including: 1. Alamo Wash local fauna, 2. fauna of the lower part of the Tornillo Group, 3. fauna of the lower part of the North Horn Formation, 4. fauna of the Kaiparowits Formation, and 5. fauna of the Evanston Formation. X indicates the location of major collections of the *Triceratops* community.

ACKNOWLEDGMENTS

I would like to thank Howard T. Lehman and Neal LaFon for their help with the fieldwork, David Thomas for the donation of specimens in his collection to the University of New Mexico, and Wann Langston, Jr., John McIntosh, Donald Baird, John Horner, Walter Coombs, Jr., Bob Sullivan, Spencer Lucas, and J. K. Rigby, Jr. for their help in identifying fragmentary specimens and for much helpful information. Lastly, I would like to thank Dr. Wann Langston, Jr. for reviewing this paper and offering much helpful criticism.

REFERENCES

Anderson, R. Y. 1960. Cretaceous-Tertiary palynology, eastern side of the San Juan Basin, New Mexico. N. Mex. Bur. Mine and Mineral Resources Mem. 6, 59 p.

Armstrong-Ziegler, J. G. 1978. An aniliid snake and associated vertebrates from the Campanian of New Mexico. J. Paleont.52:480–483.

Baird, D. and J. Horner. 1977. A fresh look at the dinosaurs of New Jersey and Delaware. New Jersey Acad. Sci. Bull. v. 22, no. 2: 50.

Baltz, E. H., S. R. Ash and R. Y. Anderson. 1966. History of nomenclature and stratigraphy of rocks adjacent to the Cretaceous–Tertiary boundary western San Juan Basin, New Mexico. U.S. Geol. Surv. Prof. Paper 524-D, 23 p.

Barsbold, R. 1974. Saurornithoididae, a new family of small theropod dinosaurs from central Asia and North America. Results Polish–Mongol. Palaeont. Exped., Pt. V, Palaeont. Pol. 30:5–22.

Bauer, C. M. 1916, Stratigraphy of a part of the Chaco River valley. U.S. Geol. Surv. Prof. Paper 98-P:271–278.

Brown, B. 1910. The Cretaceous Ojo Alamo Beds of New Mexico, with description of the new dinosaur genus *Kritosaurus*. Am. Mus. Nat. Hist. Bull. 28:267–274.

Clemens, W. A. 1973. The roles of fossil vertebrates in interpretation of Late Cretaceous stratigraphy of the San Juan Basin, New Mexico. Four Corners Geol. Soc. Mem.: 154–167.

Cobban, W. A. 1973. Significant ammonite finds in uppermost Mancos Shale and overlying formations between Barker Dome, New Mexico and Grand Junction, Colorado. Four Corners Geol. Soc. Mem.: 148–153.

Coombs, W. P. 1978. The families of the ornithischian dinosaur order Ankylosauria. Paleontology 21:143–170.

Fasset, J. E. and J. Hinds. 1971. Geology and fuel resources of the Fruitland Formation and paleontologist should be friends. Four Corners Geol. Soc. Mem.: 123–130.

Fasset, J. D. and J. Hinds. 1971. Geology and fuel resources of the Fruitland Formation and Kirtland Shale of the San Juan Basin, New Mexico and Colorado. U.S. Geol. Surv. Prof. Paper 676, 76 p.

Gaffney, E. S. 1972. The systematics of the North American family Baenidae (Reptilia, Cryptodira). Amer. Mus. Nat. Hist. Bull. 147:241–320.

Gilmore, C. W. 1916. Vertebrate faunas of the Ojo Alamo, Kirtland and Fruitland Formations, U.S. Geol. Surv. Prof. Paper 98-Q:279–308.

————. 1919. Reptilian faunas of the Torrejon, Puerco, and underlying Upper Cretaceous formations of San Juan County, New Mexico. U.S. Geol. Surv. Prof. Paper 119, 71 p.

———. 1922. A new sauropod dinosaur from the Ojo Alamo Formation of New Mexico. Smithsonian Misc. Colln. 72, no. 14, 9 p.

———. 1935. On the Reptilia of the Kirtland Formation of New Mexico, with descriptions of new species of fossil turtles. U.S. Natl. Mus. Proc. 83:159–188.

———. 1946. Reptilian fauna of the North Horn Formation of central Utah. U.S. Geol. Surv. Prof. Paper. 210-C:29–53.

Hatcher, J. B., O. C. Marsh and R. S. Lull. 1907. The Ceratopsia. U. S. Geol. Surv. Monograph 49, 198 p.

Hay, O. P. 1908. The fossil turtles of North America. Carnegie Inst. Wash. Publ. no. 75, 568 p.

———. 1910. Descriptions of eight new species of fossil turtles from west of the one-hundredth meridian. Proc. U. S. Natl. Mus. 38:307–326.

Kues, B. S., T. Lehman and J. K. Rigby, Jr. 1980. The teeth of *Alamosaurus sanjuanensis*, a late Cretaceous sauropod. J. Paleont. 54:864–869.

Lambe, L. M. 1915. On *Eoceratops canadensis*, gen. nov., with remarks on other genera of Cretaceous horned dinosaurs. Can. Nat. Mus. Bull. 12:1–49.

———. 1917. The Cretaceous theropodous dinosaur *Gorgosaurus*. Can. Geol. Surv. Mem. 100, 84 p.

Lawson, D. A. 1976. *Tyrannosaurus* and *Torosaurus*, Maestrichtian dinosaurs from Trans-Pecos Texas. J. Paleont. 50:158–164.

Lindsay, E. H., L. L. Jacobs and R. F. Butler. 1978. Biostratigraphy and magnetostratigraphy of Paleocene terrestrial deposits, San Juan Basin, New Mexico. Geology 6:425–429.

Lucas, S. G. and J. K. Rigby, Jr. 1979. Comment on 'Biostratigraphy and magnetostratigraphy of Paleocene terrestrial deposits, San Juan Basin, New Mexico.' Geology 7:323–325.

Lull, R. S. 1933. Revision of the Ceratopsia or horned dinosaurs. Peabody Mus. of Nat. Hist. Mem. v. 3, pt. 3, 175 p.

Mateer, N. J. 1976. New topotypes of *Alamosaurus sanjuanensis* Gilmore (Reptilia: Sauropoda). Geol. Inst. Univ. Uppsala N.S. 6:93–95.

McGookey, D. P., J. D. Haun, L. A. Hale, H. G. Goodell, D. G. McCubbin, R. J. Weimer and G. R. Wulf. 1972. Cretaceous System, in Geologic atlas of the Rocky Mountain region. Rocky Mountain Assoc. of Geologists. A. B. Hirschfeld Press, Denver, Colo.

Morris, W. J. 1972. A giant hadrosaurian dinosaur from Baja California. J. Paleont. 46:777–779.

Osborn, H. F. 1923. A new genus and species of Ceratopsia from New Mexico, *Pentaceratops sternbergii*. Amer. Mus. Nat. Hist. Novitates no. 93, 3 p.

Ostrom, J. H. 1961. A new species of hadrosaurian dinosaur from the Cretaceous of New Mexico. J. Paleont. 35:575–577.

———. 1963. *Parasaurolophus cyrtocristatus*, a crested hadrosaurian dinosaur from New Mexico. Fieldiana Geology. 14:143–168.

———. 1969. Osteology of *Deinonychus antirrhopus*, an unusual theropod from the Lower Cretaceous of Montana. Peabody Mus. Nat. Hist. Bull. 30, 165 p.

O'Sullivan, R. B., C. A. Repenning, E. C. Beaumont and H. G. Page. 1972. Stratigraphy of the Cretaceous rocks and the Tertiary Ojo Alamo Sandstone, Navajo and Hopi Indian Reservation, Arizona, New Mexico and Utah. U.S. Geol. Surv. Prof. Paper 521-E, 65 p.

Parks, W. A. 1926. *Thescelosaurus warreni*, a new species of orthopodous dinosaur from the Edmonton Formation of Alberta. Univ. Toronto Studies, Geol. Ser. no. 21, 42 p.

———. 1928. *Albertosaurus arctunguis*, a new species of theropodous dinosaur from the Edmonton Formation of Alberta. Univ. Toronto Studies, Geol. Ser. no. 25, 42 p.

Powell, J. S. 1968. Reptilian fossils and geology of uppermost Cretaceous deposits of the San Juan Basin, New Mexico, abs. in Geol. Soc. Amer. Cordilleran Section, 64th annual meeting, p. 97.

————. 1973. Paleontology and sedimentation models of the Kimbeto Member of the Ojo Alamo Sandstone. Four Corners Geol. Soc. Mem.: p. 111–122.

Reeside, J. B., Jr. 1924. Upper Cretaceous and Tertiary formations of the western part of the San Juan Basin of Colorado and New Mexico. U.S. Geol. Surv. Prof. Paper 134, 70 p.

Rigby, J. K., Jr. and S. Lucas. 1977. Fossil mammals from the Ojo Alamo Sandstone. New Mexico Geol. Soc. Guidebook, 28th Field Conference, Supplement: 55–56.

Russell, D. A. 1969. A new specimen of *Stenonychosaurus* from the Oldman Formation (Cretaceous) of Alberta. Can. Jour. Earth Sci. 6:595–612.

————. 1970. Tyrannosaurs from the Late Cretaceous of western Canada, Can. Nat. Mus. of Natural Sci. Publ. in Paleo. no. 1, 30 p.

Russell, L. S. 1940. *Edmontonia rugosidens* (Gilmore) an armored dinosaur from the Belly River Series of Alberta. Univ. of Toronto Studies Geol. Ser. no. 43, 27 p.

————. 1975. Mammalian faunal succession in the Cretaceous system of western North America. Geol. Assoc. Canada Spec. Paper no. 13, p. 137–161.

Sinclair, W. J. and W. Granger. 1914. Paleocene deposits of the San Juan Basin, New Mexico. Amer. Mus. Nat. Hist. Bull. 33:297–316.

Sloan, R. E. 1970. Cretaceous and Paleocene terrestrial communities of western North America. North Amer. Paleont. Convention, Chicago, Proc. E:427–453.

Steel, R. 1973. Crocodylia. Handbuch der palaeoherpetologie, part 16, Gustav Fischer Verlag, Stuttgart, 116 p.

Sternberg, C. M. 1938. *Monoclonius* from southeastern Alberta compared with *Centrosaurus*. J. Paleont. 12:284–286.

Tschudy, R. H. 1973. The Gasbuggy core—a palynological appraisal. Four Corners Geol. Soc. Mem.: 131–143.

Weishampel, D. B. and J. A. Jensen. 1979. *Parasaurolophus* (Reptilia:Hadrosauridae) from Utah. J. Paleont. 53:1422–1427.

Wiman, C. 1930. Über Ceratopsia aus der Oberen Kreide in New Mexico. Nova Acta Reg. Soc. Sci. Upsaliensis, ser. 4, v. 7, no. 2, 19 p.

————. 1931. *Parasaurolophus tubicen* n. sp. aus der Kreide in New Mexico. Nova Acta Reg. Soc. Sci. Upsaliensis, ser. 4, v. 7, no. 5, 11 p.

————. 1932. *Goniopholis kirtlandicus*, n. sp. aus der Oberen Kreide in New Mexico. Uppsala Univ. Geol. Inst. Bull. 23: 181–190.

————. 1933. Über Schildkröten aus der Oberen Kreide in New Mexico. Nova Acta Reg. Soc. Sci. Upsaliensis, ser. 4, v. 9, no. 5, 35 p.

10 "Dragonian" fossils from the San Juan Basin and status of the "Dragonian" Land Mammal "Age"

Yukimitsu Tomida
Department of Geosciences, University of Arizona
Tucson, Arizona

INTRODUCTION

Gazin (1938, 1939, 1941) first described the Dragon fauna from the North Horn Formation of Dragon Canyon in central Utah, and concluded that the Dragon fauna is intermediate in age between the Puerco and Torrejon faunas in the San Juan Basin. Wood et al. (1941), based on Gazin's work, proposed the Dragonian Provincial (Land Mammal) "Age" between the Puercan (early Paleocene) and the Torrejonian (middle Paleocene) Land Mammal "Ages." However, since no other faunas representing the same age and containing the same unique assemblage as the Dragon fauna have been found, the recognition of the "Dragonian" Land Mammal "Age" has been questioned by several vertebrate paleontologists (e.g. Evernden et al., 1964; Tedford, 1970).

Magnetostratigraphic study has been undertaken to better correlate the Dragon fauna in the North Horn Formation with the faunal sequence in the San Juan Basin, where the topotypic faunas of the Puercan and Torrejonian Land Mammal "Ages" occur (Tomida & Butler, 1980). Tomida & Butler (1980) demonstrated that the Dragon fauna occurs slightly earlier than the topotypic Torrejonian fauna in the San Juan Basin. They also reported new occurrences of a few mammalian taxa from the Dragon fauna indicating closer affinity of the Dragon fauna to the Torrejonian than previously known. However, the Dragon fauna still includes a few genera characteristic of the Puercan Land Mammal "Age" and a few genera restricted to the Dragon fauna, as well as several genera characteristic of the Torrejonian Land Mammal "Age." Thus, they concluded that the Dragon fauna occurs in a time interval between the type Puercan and Torrejonian faunas, but in addition to the chronologic assignment of the "Dragonian," discovery and study of faunas in other areas (especially the San Juan Basin) within the same time interval is necessary for the complete assessment of the "Dragonian" Land Mammal "Age."

In the San Juan Basin, mammalian fossils have been almost totally unknown from the stratigraphic interval between the Puercan *Taeniolabis* zone and the Torrejonian *Deltatherium* zone. The exceptions are: (1) three specimens of *Periptychus carinidens* from the upper part of Sinclair & Granger's (1914) Barrel Spring Arroyo section, and (2) one specimen of *P. carinidens* from the upper part of their Kimbeto section. Although Sinclair & Granger (1914) placed the lower boundary of the Torrejon "Formation" at the level of the first occurrence of *P. carinidens*, all other Torrejonian fossils have been found at higher levels (Sinclair & Granger, 1914; Taylor & Butler, 1980).

Magnetostratigraphic study in the North Horn Formation identified the stratigraphic interval of the San Juan Basin in which "Dragonian" fossils should be found. This interval, previously considered barren, was investigated by the University of Arizona field party in 1978. Several new localities were discovered in that interval. Fossils from those localities provide an improved understanding of the "Dragonian" Land Mammal "Age." Thus, the purpose of the present paper is to report those fossils found from the "Dragonian" time interval in the San Juan Basin and to discuss the status of the "Dragonian" Land Mammal "Age."

POSITION OF THE "DRAGONIAN" IN THE MAGNETIC POLARITY TIME SCALE

A composite magnetic polarity sequence was constructed from two paleomagnetic sections in the Paleocene part of the North Horn Formation, including the Wagonroad and Dragon faunal levels (Tomida & Butler, 1980; left column in Text-fig. 10.1). The Wagonroad fauna is characterized by the presence of such genera as *Taeniolabis*, *Oxyclaenus*, *Ectoconus*, *Carsioptychus*, and *Desmatoclaenus* and is considered late Puercan in age. The Dragon fauna includes several characteristic Torrejonian taxa, such as *Aphronorus*, cf. *Palaechthon*, *Protictis haydenianus*, *Mimotricentes*, *Chriacus* cf. *C. baldwini*, and *Periptychus*. The Dragon fauna also includes two genera, *Loxolophus* and *Oxyclaenus*, that are characteristic of the Puercan, as well as two

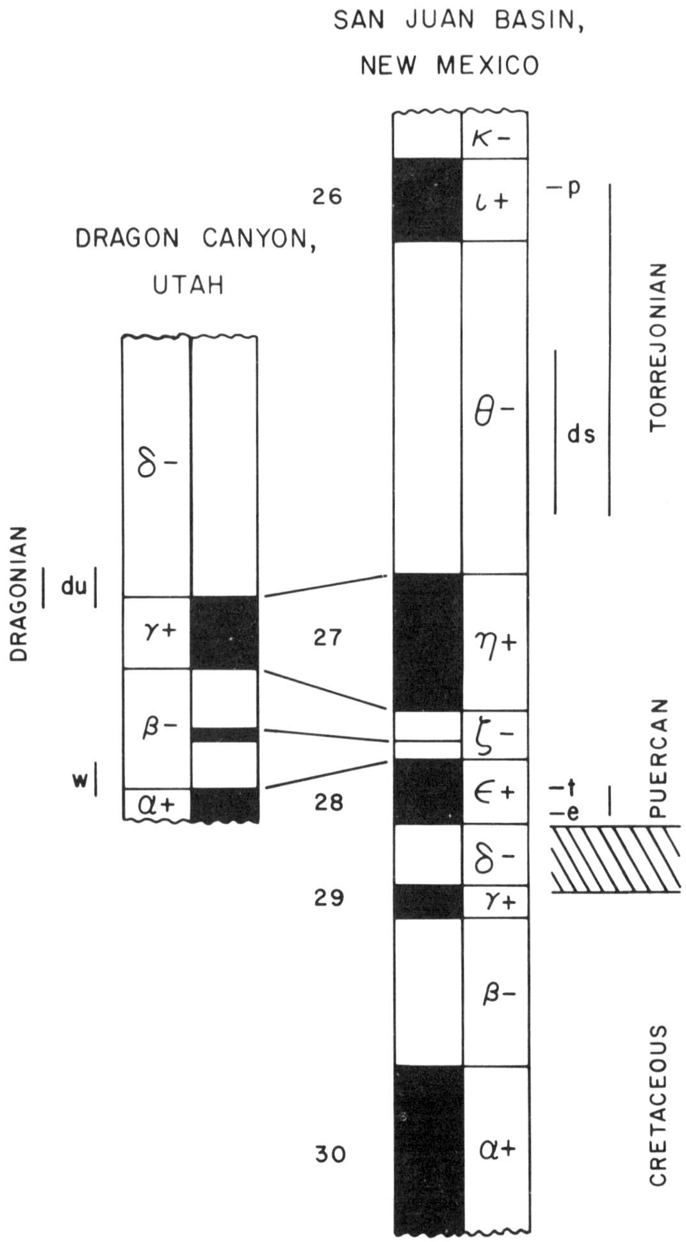

TEXT-FIG. *10.1*—Correlation of Dragon Canyon section with San Juan Basin section (Tomida & Butler, 1980). Black and white bars designate normal and reversed polarities, respectively. Stratigraphic intervals of faunal levels are shown by vertical and horizontal lines adjacent to the magnetozone designations. Letters adjacent to those lines indicate: ds, *Deltatherium* zone; du, Dragon faunal level; e, *Ectoconus* zone; p, *Pantolambda* zone; t, *Taeniolabis* zone; w, Wagonroad faunal level. Numbers between the magnetic polarity columns are the magnetic anomaly designations of the normal polarity intervals in the polarity time scale with which the Dragon Canyon and San Juan Basin magnetozones are correlated.

genera, *Dracontolestes* and *Conoryctella*, that are presently restricted to the Dragon fauna. Thus, the Dragon fauna is considered somewhat older than Torrejonian in age on the basis of its faunal composition (Tomida & Butler, 1980).

The composite magnetic polarity sequence from Dragon Canyon was correlated with the San Juan Basin magnetic polarity sequence (Lindsay et al., 1978, 1981) on the basis of the similarity in patterns of magnetic polarity zones and restriction of the faunal composition (Text-fig. 10.1). As shown in Text-figure 10.1, the Wagonroad faunal level (indicated by line 'w') is slightly younger than and possibly partly contemporaneous with the known Puercan faunal level, and the Dragon faunal level (indicated by line 'du') is slightly older than previously known Torrejonian faunal levels in the San Juan Basin.

Based on these correlations, the Dragon fauna is contemporaneous with the deposition of the upper part of normally magnetized sediments correlative with magnetic anomaly 27 and the base of the reversely magnetized sediments immediately above it. Thus, in the San Juan Basin, "Dragonian" fossils should be found within or slightly above the normally magnetized stratigraphic interval correlative with magnetic anomaly 27.

"DRAGONIAN" FOSSILS IN THE SAN JUAN BASIN

In 1977 a few *Periptychus carinidens* jaws, turtle shell fragments, and gar fish scales were found in the stratigraphic interval correlative with anomaly 27. These specimens thereby more or less represent "Dragonian" time in the San Juan Basin. This interval was intensively investigated by the University of Arizona field party in the following year, and several new localities were discovered. Those localities are scattered in three general areas: the lower Kutz Canyon area, near the mouth of the West Fork (UALP [University of Arizona, Laboratory of Paleontology] Locs. 7896, 7897, and 7899); the upper Kimbeto area, about 6.5 km north of the village of Kimbeto (UALP Locs. 7784, 77113, and 77114); and the upper Tsosie Arroyo area, about 3 km northeast of the "Tsosie" locality (Sinclair & Granger, 1914) (UALP Locs. 7782 and 7783). Stratigraphic positions of all of the localities are plotted relative to magnetic polarity columns (Text-fig. 10.2).

As shown in Text-figure 10.2, the normal polarity zones B+ in the Kutz Canyon section, C+ in the Kimbeto Arroyo section, and the upper part of C+ in the Tsosie Arroyo section are correlated with anomaly 27 of the magnetic polarity time scale (LaBrecque et al., 1977), with some reservation on the Tsosie Arroyo section. Thus, all localities mentioned above are correlated with the "Dragonian" time interval.

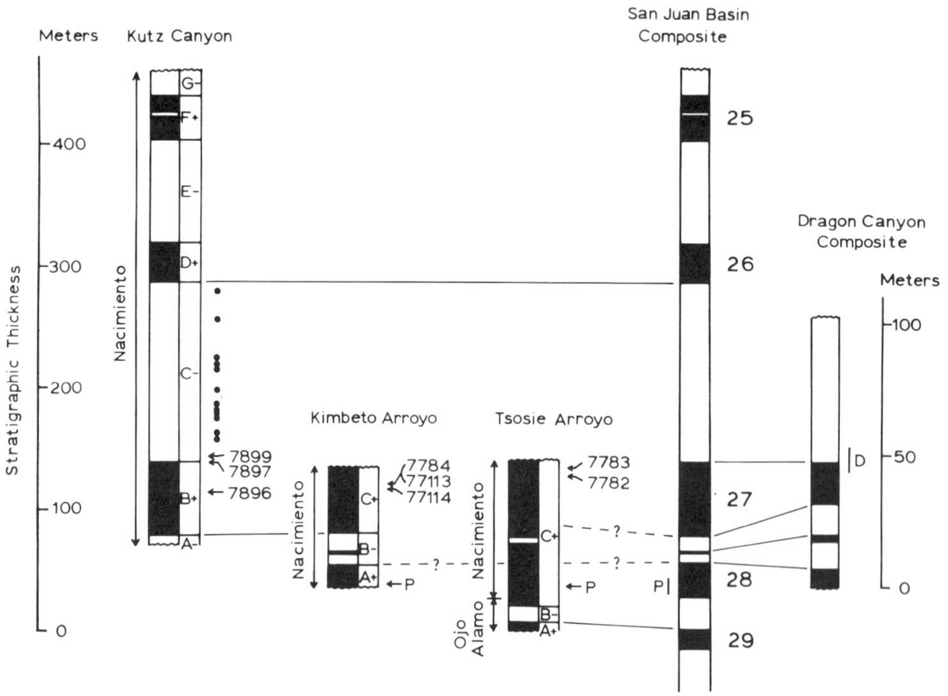

TEXT-FIG. *10.2*—Stratigraphic positions of the localities (7897, 77114, 7782, etc.) described in this paper relative to magnetic polarity columns. P indicates the Puercan faunal level, while D indicates the Dragon faunal level. Solid circle indicates the stratigraphic position of other localities in Kutz Canyon. Solid line connecting the sections indicates the best correlation, whereas broken line indicates questionable correlation between the sections. Other symbols as in Text-fig. 10.1.

Detailed data on each locality are as follows:

UALP Loc. 7896—At the level of SJ276 paleomagnetic sample site, which is about 25 m below the top of B+ zone in the Kutz Canyon section.

UALP Loc. 7897—Approximately at the level of SJ286 paleomagnetic sample site, which is about 1 m below the top of B+ zone in the Kutz Canyon section.

UALP Loc. 7899—Approximately at the level of SJ293 paleomagnetic sample site, which is about 4.5 m above the B+ zone in the Kutz Canyon section.

UALP Locs. 7784 and 77113—Both approximately at the level of SJ665 paleomagnetic sample site, which is about 13 m below the top of the Kimbeto Arroyo section.

UALP Loc. 77114—Approximately at the level of SJ664 paleomagnetic sample site, which is about 18 m below the top of the Kimbeto Arroyo section.

UALP Loc. 7782—Approximately at the level of SJ554 paleomagnetic sample site, which is about 13 m below the top of the Tsosie Arroyo section.

UALP Loc. 7783—Approximately at the level of SJ556 paleomagnetic sample site, which is about 6 m below the top of the Tsosie Arroyo section.

Mammalian fossils discovered from these localities are listed in Table 10.1. Taxonomic identifications of the fossils follow:

Table 10.1—Distribution of taxa by UALP (University of Arizona, Laboratory of Paleontology) locality.

	Upper Tsosie Wash		Upper Kimbeto Arroyo			Lower Kutz Canyon		
	7782	7783	7784	77113	77114	7896	7897	7899
Primates								
Paromomyidae								
cf. *Paromomys* sp.				X				
Carnivora								
Miacidae								
Protictis cf. *P. haydenianus*							X	
Condylarthra								
Arctocyonidae								
cf. *Loxolophus kimbetovius*	X							
Triisodon cf. *T. antiquus*							X	
Hyopsodontidae								
cf. *Mioclaenus turgidus*							X	
Periptychidae								
Periptychus carinidens	X	X	X		X			X
Anisonchus sectorius						X		

cf. *Paromomys* sp.
Pl. 10.1, fig. 1

Material.—UALP 12016: isolated left M_2, locality 77113.

Discussion.—Although the lower molars of the Paleocene paromomyid and microsyopid (classification of Bown & Rose, 1976) primates are less diagnostic than upper dentitions, the following observations on UALP 12016 indicate close affinity to the genus *Paromomys* (Bown & Rose, 1976, Table 1): 1) valley between protoconid and metaconid nearly filled; 2) trigonid strongly inclined anteriorly; 3) talonid slightly wider than trigonid; 4) no distinct hypoconulid; instead of that a weak ridge (basin rim) connects the entoconid and hypoconid; 5) paraconid very close to metaconid. Measurements (in mm) are: length 2.5, width 1.9.

Protictis cf. *P. haydenianus*
Pl. 10.1, fig. 2

Material.—UALP 12005: left maxillary fragment with P^4 and labial part of M^1, locality 7897.

Discussion.—On P^4 the protocone is somewhat crushed and removed posteriorly, and the metastyle blade is slightly bent labially at the position of the notch (terminology after MacIntyre, 1966). On M^1, only the parastyle, stylar shelf, stylar cingulum, and metastyle region are preserved. Despite the poor state of preservation, there is no doubt that this specimen is a miacid because of the unique morphology of P^4 and M^1. UALP 12005 obviously differs from *Ictidopappus* in morphology (deep concavity of the labial outline on M^1 and the peculiar shearing facet in *Ictidopappus*) and in size. The general morphology of UALP 12005 indicates closest resemblance to *Protictis.* UALP 12005 is within the known size range of *P. haydenianus* (MacIntyre, 1966) and distinctly larger than other known species of the genus. However, on P^4 of UALP 12005 the metastyle blade is elongated posteriorly and lower in height than *P. haydenianus.* Measurements (labial length only, in mm) are: P^4 = ca. 8.0, M^1 = 6.5.

cf. *Loxolophus kimbetovius*
Pl. 10.1, fig. 4

Material.—UALP 10960: left partial jaw with M_1–M_3, locality 7782.

Discussion.—A formal reason for the synonymy of *Protogonodon* with *Loxolophus* has not been published (see Russell, 1967). Van Valen (1978) simply mentioned this synonymy. However, I agree with the synonymy for the following reasons: 1) overall similarity of their dentition, especially bulbous and flat-appearing molars; and 2) *L. hyattianus—L. priscus—P. protogonoides—P. pentacus—P. kimbetovius* form a linear sequence in size.

Although UALP 10960 has slightly less bulbous molars and a rather unusually large M_3, the specimen resembles *L. kimbetovius* in the M_1 with the talonid and trigonid of equal width, and in the M_2 with the talonid only slightly wider than the trigonid. Species of the genus exhibit a wide range of variation in size and massiveness. Measurements of the teeth (length and width, in mm) are: M_1 = 10.0 x 7.3, M_2 = 10.5 x 8.9, M_3 = 13.5 x 8.6.

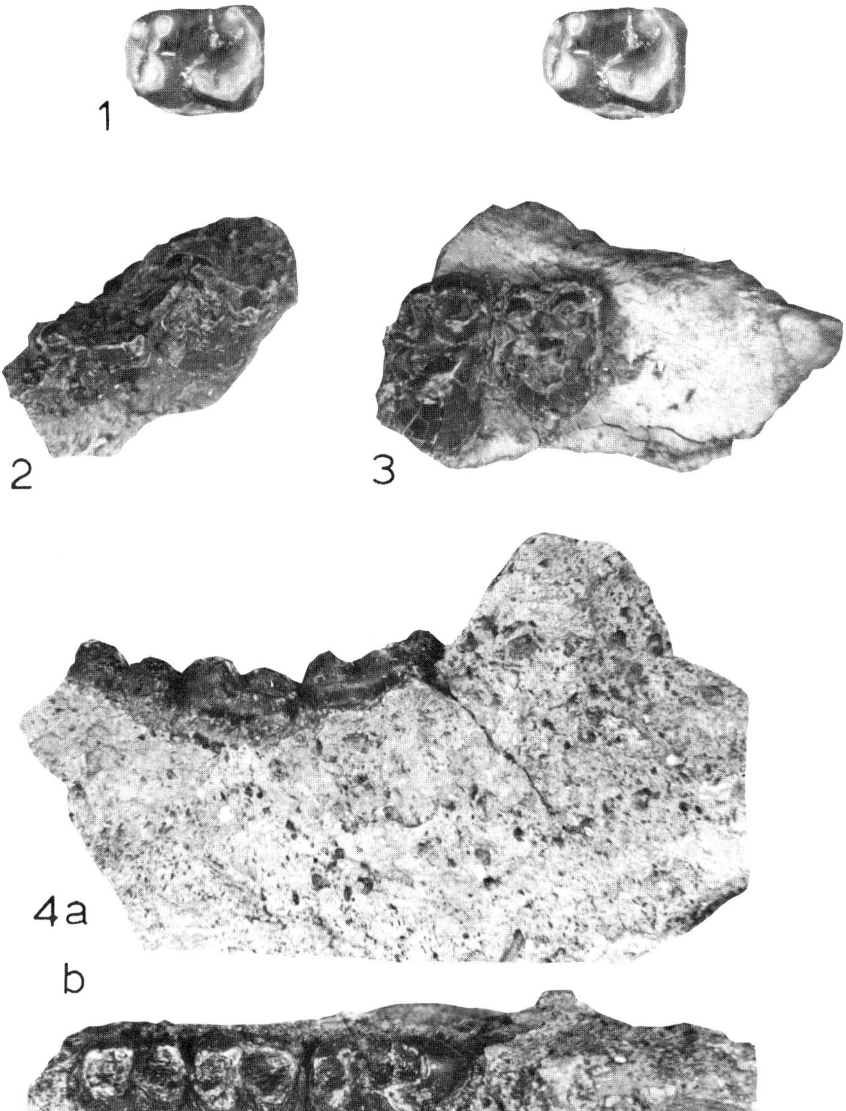

EXPLANATION OF PLATE 10.1

FIG. *1*—cf. *Paromomys* sp., UALP 12016, left M_2 occlusal view, stereophotograph, ×8.1.

2—*Protictis* cf. *P. haydenianus*, UALP 12005, left P^4 and labial portion of M^1 occlusal view, ×2.5.

3—cf. *Mioclaenus turgidus*, UALP 12004, right M^{1-2} occlusal view, ×2.5.

4—cf. *Loxolophus kimbetovius*, UALP 10960, left partial jaw with M_{1-3}: *a*, labial view, b, occlusal view, both ×1.5.

Triisodon cf. *T. antiquus*
Pl. 10.2

Material.—UALP 12001: partial skull with right P^3–P^4 and M^2–M^3, left jaw fragment with M_1–M_2, right jaw with a few teeth almost completely covered by concretion, and a number of cranial fragments all covered by concretion; UALP 12002: maxillary fragment with right M^2, labial wall broken; UALP 12003: isolated, antero-posteriorly compressed right M^1(?); all from locality 7897.

Discussion.—Van Valen (1978) synonymized *T. antiquus* with *T. quivirensis* without explanation. However, *T. quivirensis* has a much reduced paraconid on the lower molars and is slightly smaller than *T. antiquus*. For these reasons, I prefer to retain both species until an intensive study of the genus is made.

In UALP 12001, the paraconid is fairly well developed on M_{1-2}, and the upper dentition most closely resembles in morphology AMNH (American Museum of Natural History) 3175 which is referred to *T. antiquus*. UALP 12001 is about the same size as the type of *T. antiquus* (AMNH 3174). However, there are a few morphological differences between UALP 12001 and *T. antiquus*. In UALP 12001, 1) the postcingulum is less developed on M^2; 2) P^3 and P^4 are transversely wider relative to their length; 3) no distinct entoconid is developed on M_{1-2}, and 4) the protoconid is located more labially, so that the labial wall of the trigonid is almost vertical. Since UALP 12002 and 12003 are poorly preserved, detailed comparison cannot be made, except by their size. I refer them tentatively to the same taxon as UALP 12001. Measurements of the teeth are given in Table 10.2.

Table 10.2—Measurements (in mm) of teeth of *Triisodon* cf. *T. antiquus* from UALP Loc. 7897.

Specimen Number		P^3	P^4	M^1	M^2	M^3	M_1	M_2
UALP 12001	L	10.8	13.2		15.0	9.5	ca 16.5	18.5
	W	9.7			21.0	16.3	9.5	10.4
UALP 12002	L				16.8			
	W				22.6			
UALP 12003	L							
	W			ca 19				

cf. *Mioclaenus turgidus*
Pl. 10.1, fig. 3

Material.—UALP 12004: right M^{1-2} connected by concretion, locality 7897.

Discussion.—Although both molars are somewhat crushed, especially M^1, basic characters are still visible. UALP 12004 closely resembles *M. turgidus*, the only known species of the genus, in general morphology and in size. Direct comparison was made with several specimens from other Kutz Canyon localities (see Taylor, this volume) and localities north of Aztec, New Mexico. The major differences are: 1) the metacone on M^2 is more labially located on UALP 12004 than on any other specimens that I have examined, so that the labial outline of the tooth is nearly parallel to that of M^1, and 2) both conules, the parastyle, and the mesostyle are somewhat weakly developed on UALP 12004. Measurements of the teeth (length and width, in mm) are: M^1 = 6.1 x 7.5, M^2 = ? x 9.8.

Periptychus carinidens
Pl. 10.3, figs. 1–3

Material.—UALP 10959: left partial jaw with roots of P_2–M_3, locality 7782; UALP 10961: right jaw with broken P_1, P_{2-3}, broken P_4–M_3, locality 7783; UALP 10962: right jaw fragment with P_{2-3}, locality 7784; UALP 10963: left partial jaw with broken P_1–M_1, locality 7784; UALP 11997: isolated right P_1, locality 77114; UALP 12000: isolated right M^3 and fragments of molar and premolars, locality 7782; UALP 12011: almost complete left jaw with partly broken C-M_3, partial right jaw with root of P_1 and partly broken P_2–M_3, left and right partial maxillae with M^{1-2} on both, fragments of other teeth, and several postcranial fragments, locality 7899.

Discussion.—Simpson (1959) clearly showed the synonymy of *P. rhabdodon* with *P. carinidens.* Thus, only one species of the genus is presently recognized in the San Juan Basin. Another species, *P. gilmorei*, is known from the Dragon fauna (Gazin, 1941). However, *P. gilmorei* is easily distinguished from *P. carinidens* by having much wider teeth, especially the upper dentition, and by having the fourth premolars only slightly larger than molars. These fourth premolars are distinctly larger in *P. carinidens.* Moreover, in *P. carinidens* the protostyle and hypocone on the upper molars are less lingually located, and the talonid of P_4 is slightly better developed (Gazin, 1941). All specimens listed above show almost identical characters of *P. carinidens*, and the size of all measurable teeth is well within the known range of *P. carinidens* (Simpson, 1959). Measurements of the teeth are given in Table 10.3.

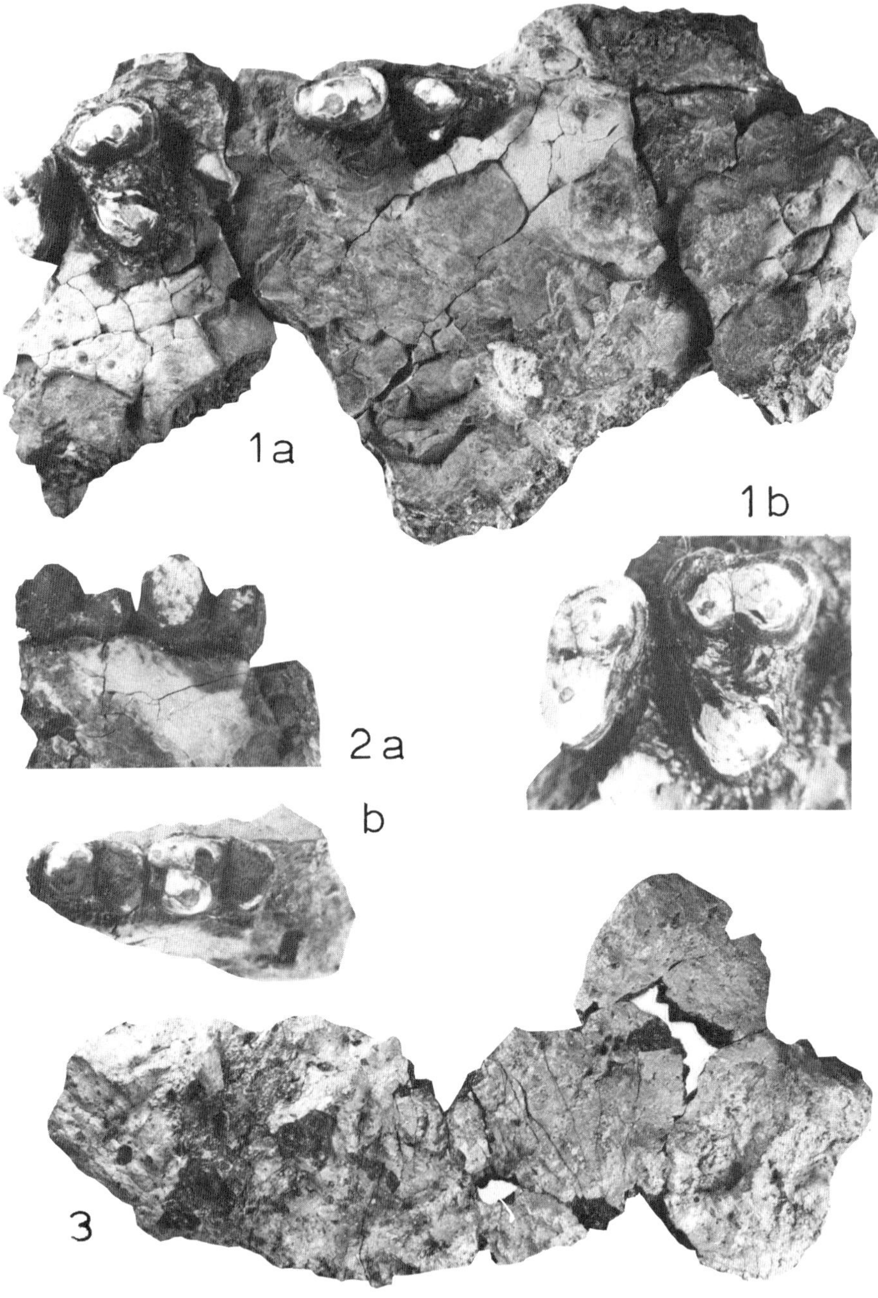

EXPLANATION OF PLATE 10.2

FIGS. *1–3*—*Triisodon* cf. *T. antiquus*, UALP 12001. *1*, partial skull with right P^{3-4} and M^{2-3}: *a*, occlusal view, ×1; *b*, right M^{2-3}, ×1.5; *2*, left jaw fragment with M_{1-2}, ×1: *a*, labial view; *b*, occlusal view; *3*, right jaw covered by concretion, ×0.5.

Table 10.3—Measurements (in mm) of teeth of *Periptychus carinidens* from various localities described in this paper.

Specimen Number		C	P_1	P_2	P_3	P_4	M_1	M_2	M_3
UALP 10961	L			11.8	13.5	13.2	12.3	11.2	
	W		6.9	9.3	10.7	ca 11			
UALP 10962	L			10.2	11.1				
	W			7.7	9.7				
UALP 10963	L		ca 6	11.7	13.3	ca 13	ca 11		
	W		ca 5	ca 8	ca 10		8.1		
UALP 10964	L						10.1	8.4	10.8
	W						7.7	8.3	7.9
UALP 11997	L		5.7						
	W		4.8						
UALP 12011	L	8.6	7.2	12.3	13.0	13.4	11.8	10.9	ca 14
	W	8.4	5.8	8.8	10.7	10.9	10.2	10.3	9.6

		M^1	M^2	M^3
UALP 12000	L			8.0
	W			9.4
UALP 12011	L	10.8	10.5	
	W	14.0	13.9	

Anisonchus (Anisonchus) sectorius
Pl. 10.3, fig. 4

Material.—UALP 12010: left partial jaw with P_4–M_2, locality 7896.

Discussion.—It is fairly difficult to distinguish A. (*Anisonchus*) from A. (*Mithrandir*) by the morphological characters of lower molars alone, but I agree with Van Valen (1978) that the hypoconid on the lower molars of A. (*Mithrandir*) is less lingual than in A. (*Anisonchus*). Nevertheless, UALP 12010 can be distinguished from A. (*M.*) *gillianus*, A. (*M.*) *oligistus*, and A. (*M.*) *onostus* by its larger size. The lower dentition of A. (*M.*) *fortunatus* is unknown.

The lower molars of the Dragon species of the subgenus A. (*Anisonchus*), A. (*A.*) *dracus*, are about the same size as those of A. (*A.*) *sectorius*, but in A. (*A.*) *dracus*, the cristid obliqua appears distinctly lower than the entocristid (Gazin, 1941). In UALP 12010, this condition is reversed. UALP 12010 shows no significant difference from several other specimens of A. (*A.*) *sectorius* from other Kutz Canyon and Torrejon Arroyo localities I have examined, except for the slightly greater width of P_4 in UALP 12010. Measurements of the teeth (length and width, in mm) are: P_4 = 6.4 x 4.4, M_1 = 5.8 x 4.3, M_2 = 5.4 x 4.4.

Of the seven taxa identified above, six (the exception is cf. *Loxolophus kimbetovius*) were restricted previously to the Torrejonian. The genera *Protictis*, *Loxolophus*, *Periptychus*, and *Anisonchus* are recorded from the Dragon fauna, although Dragon species of those genera (other than *Protictis*) are different from those of the typical Torrejonian.

DISCUSSION

As shown in Text-figure 10.1, the "Dragonian" interval can be interpreted as the time interval represented by approximately the younger half of magnetic anomaly 27 and the basal part of the reversed polarity zone immediately above it. This is fairly close to, but distinct from, the previously known duration of the Torrejonian (Lindsay et al., 1978) and obviously distinct from the known duration of the Puercan.

Text-figure 10.3 summarizes the known ranges of those genera described in the present paper and several other common and/or important genera known from the Puercan to Torrejonian in the San Juan Basin and/or Dragon Canyon. Of those taxa known from the "Dragonian" time interval in the San Juan Basin and/or Dragon Canyon, *Palaechthon*, *Paromomys*, *Triisodon*, *Haploconus*, *Periptychus*, *Mioclaenus*, *Anisonchus sectorius*, *Protictis haydenianus*, *Chriacus*, *Protoselene*, and *Mimotricentes* are characteristic and/or common Torrejonian taxa. Since *Jepsenia* and *Oxytomodon* were synonymized with *Litaletes* and *Litomylus* (both known from northern Torrejonian localities; Russell,

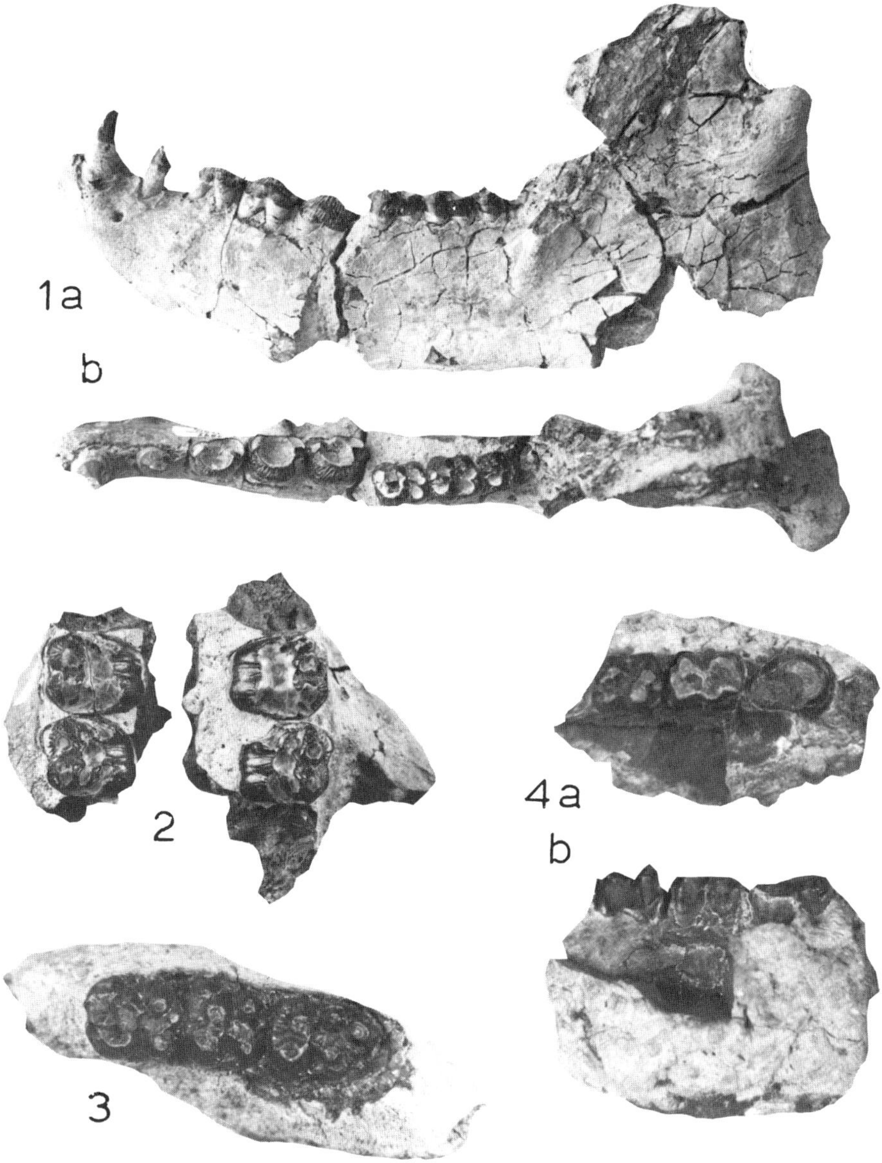

EXPLANATION OF PLATE 10.3

FIGS. *1–3*—*Periptychus carinidens*. *1* and *2*, UALP 12011. *1a*, *b*, left jaw with C–M_3, labial and occlusal views, ×0.6; *2*, right and left maxillary fragments with M^{1-2} occlusal view, × 1; *3*, UALP 10964, left partial jaw with M_{1-3} occlusal view, × 1.5.

4—*Anisonchus sectorius*, UALP 12010, left partial jaw with P_4–M_2: *a*, occlusal view; *b*, lingual view, both × 1.9.

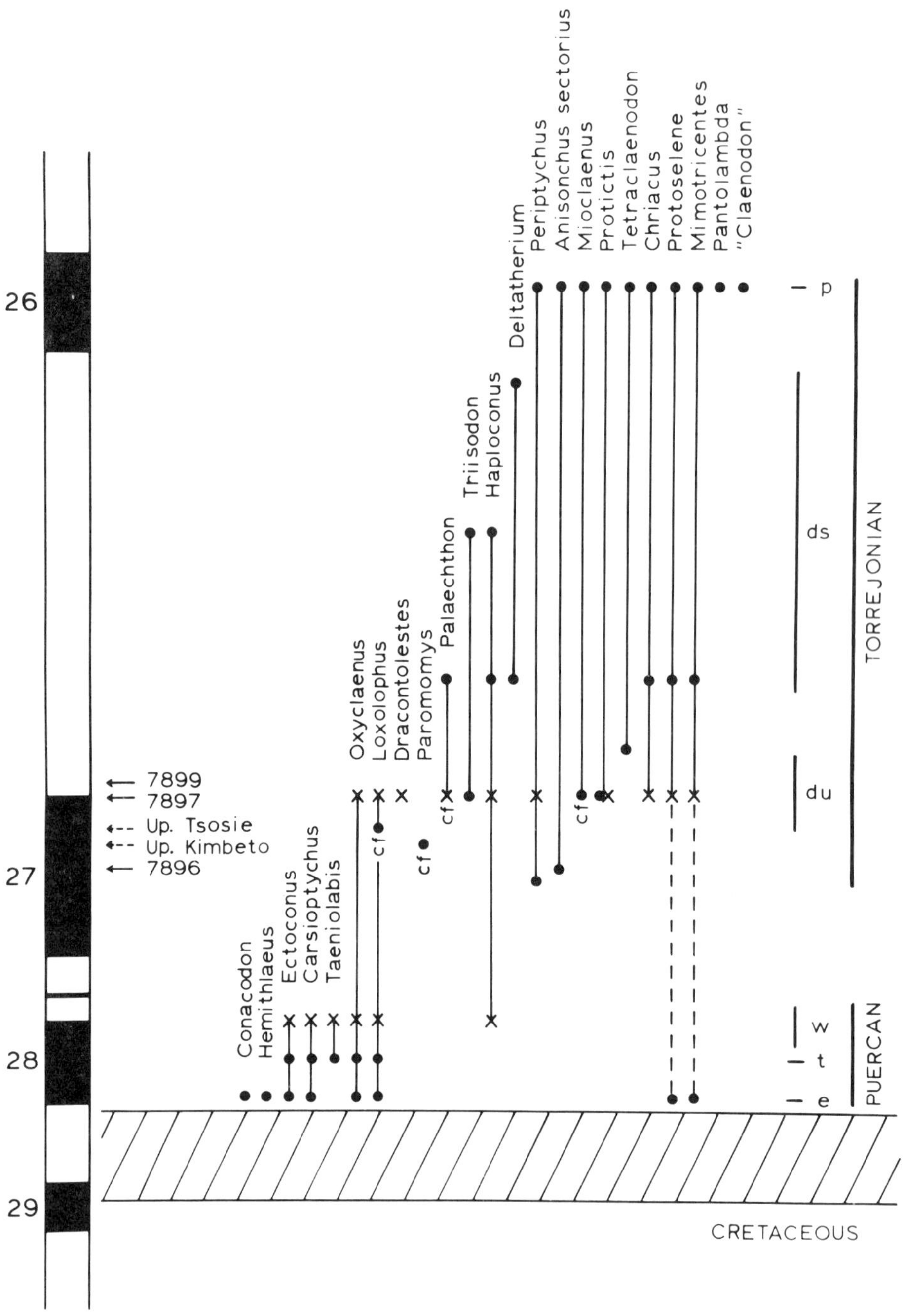

TEXT-FIG. *10.3*—Range chart of some common and/or important genera (including all taxa identified in this paper) from the San Juan Basin and Dragon Canyon placed in a composite magnetic column. Ranges are occurrences in the San Juan Basin and Dragon Canyon only. Solid circle indicates occurrence in the San Juan Basin, X indicates occurrence in Dragon Canyon, and cf indicates comparable or questionable assignment to that genus. Other symbols as in Text-fig. 10.1.

1967) respectively by Van Valen (1978), only *Dracontolestes* and *Conoryctella* (both poorly known) are presently restricted to the Dragon fauna. However, *Loxolophus* and *Oxyclaenus*, both common in Puercan faunas, are also recorded from the "Dragonian" time interval. Probably these are the latest known occurrences of these genera. The genera *Deltatherium* and *Tetraclaenodon* have not been recorded from this interval. The earliest known record of *Tetraclaenodon puercensis* in the San Juan Basin, as reported by Lindsay et al. (1981) and by Taylor (this volume), occurs lower than the known range of the genus *Deltatherium*. The earliest *Tetraclaenodon* record is from UALP Loc. 75139, which is 1.5 km southwest from and about 16 m stratigraphically higher than UALP Loc. 7897 (Text-fig. 10.2). Magnetostratigraphically this locality is rather close to the highest locality in Dragon Canyon (Text-fig. 10.3), but *Tetraclaenodon* has not been recorded from Dragon Canyon. *T. puercensis* is one of the most common species in the San Juan Basin Torrejonian. However, it has not been recovered from any of the localities described in this paper, which are relatively evenly distributed throughout the "Dragonian" interval. Thus, the time interval represented by the "Dragonian" is characterized by the occurrence of *Oxyclaenus*, *Loxolophus*, *Chriacus*, *Triisodon*, cf. *Mioclaenus*, *Periptychus*, *Mimotricentes*, *Protoselene*, *Protictis*, cf. *Palaechthon*, and cf. *Paromomys*, and by the absence of *Deltatherium* and *Tetraclaenodon*.

Thus, the "Dragonian" time interval is more closely related to the Torrejonian both temporally (Text-fig. 10.1) and faunally than previously known. The fauna of this interval is dominated by Torrejonian genera (although it lacks such characteristic Torrejonian genera as *Deltatherium* and *Tetraclaenodon*), and lacks its own characteristic, widespread genera. Therefore, the "Dragonian" Land Mammal "Age" is best treated as earliest Torrejonian, and its status as a distinct land mammal "age" should be abandoned. The old "Dragonian" is, however, distinct from the *Deltatherium* zone because of the presence of *Loxolophus* and *Oxyclaenus* and the absence of *Deltatherium* and *Tetraclaenodon*. I propose a new zone, named the *Periptychus-Loxolophus* zone, for the "Dragonian" time interval. The name signifies that the common Torrejonian genus *Periptychus* and the common Puercan genus *Loxolophus* occur together in this zone (both in the San Juan Basin and in Dragon Canyon), which makes this zone unique.

The *Pantolambda* zone was first described by Osborn and Matthew (1909) as a "life zone." Sinclair and Granger (1914) discovered the lower fossiliferous "horizon" at the type Torrejon and found that the genus *Deltatherium* was confined to that "horizon." Osborn (1929) defined the *Pantolambda* and *Deltatherium* zones on the basis of observations of Sinclair & Granger (1914), although Matthew (1937) and

Wilson (1951) questioned the significance of the time difference between these zones. Thus, the *Pantolambda* and *Deltatherium* zones were originally defined as biostratigraphic units. However, Taylor & Butler (1980) tied these zones into the magnetic polarity time scale. With an independent (from biochronologic or biostratigraphic) chronologic ordering (e.g., magnetostratigraphic) these zones became equivalent to the concept of chronozones (chronostratigraphic units). The magnetostratigraphic correlation of strata in the San Juan Basin and Dragon Canyon allowed comparison of faunal levels in those areas relative to an independent chronologic ordering. This technique helped to define the time interval dealt with in the present paper. Thus, the *Periptychus-Loxolophus* zone is not only a biostratigraphic assemblage zone in the two areas where it is recognized, but it is also a chronozone (chronostratigraphic unit) since its limits and duration can be recognized on the basis of magnetic polarity zonation in the absence of fossils. This zone could be recognized as a concurrent-range zone, but does not have to be since it is based, in part, on magnetostratigraphic correlation.

In practice, however, most vertebrate paleontologists have been using the *Pantolambda* and *Deltatherium* zones as well as the Puercan *Ectoconus* and *Taeniolabis* zones as subdivisions of land mammal "ages," or biochronologic units (e.g. Sloan, 1970, p. 438). R. Sloan (written comm.: handouts distributed at the Society of Vertebrate Paleontology Annual Meeting in Pittsburgh, 1979) divided the Paleocene land mammal ages into a number of zones including three zones for the Torrejonian. I believe the zones proposed by Sloan are biochronologic units, based on the stage of evolution of mammalian fossils. Sloan's earliest subdivision of the Torrejonian is named the *Anisonchus dracus* zone, which I believe is based on the Dragon fauna whose age would remain somewhat ambiguous on the basis of fauna alone. The *Anisonchus dracus* zone probably represents the same time interval as the *Periptychus-Loxolophus* zone, but the basis for defining the *Anisonchus dracus* zone has not been stated. Thus, I propose the *Periptychus-Loxolophus* zone as defined above. The *Periptychus-Loxolophus* zone is probably equivalent to the *A. dracus* zone of Sloan, but should supersede it because of the more rigorous definition of the *Periptychus-Loxolophus* zone.

The Torrejonian, then, is the first land mammal "age" to be subdivided into chronozones (chronostratigraphic units). Torrejonian chronozones are, in ascending order, the *Periptychus-Loxolophus* chronozone in anomaly 27 and the base of the overlying reversed magnetozone, the *Deltatherium* chronozone in the reversed interval (except its base) between anomalies 27 and 26, and the *Pantolambda* chronozone in anomaly 26. The *Periptychus-Loxolophus* chronozone is bounded below by the Puercan Land Mammal "Age" (which is a biochronologic unit but also bears a chronozone concept since it has an independent chronologic control, e.g., magnetostratigraphy), and the *Deltatherium* chronozone is bounded above by the *Pantolambda* zone. The *Pantolambda* zone, however, cannot be considered as a complete chronozone at this time. Neither the *Plesiadapis praecursor* zone (a range zone; Gingerich, 1976) nor the Tiffanian Land Mammal "Age" (a biochronologic unit) are recorded stratigraphically superjacent to faunas of the *Pantolambda* zone, nor are those units presently tied to an independent chronologic ordering (e.g. magnetostratigraphy). Work is presently underway to relate the *Pantolambda* zone to Tiffanian range zones in the Polecat Bench Formation of Wyoming by magnetostratigraphy, which would unequivocally substantiate the *Pantolambda* zone as a chronozone.

SUMMARY AND CONCLUSION

Magnetostratigraphic study in the North Horn Formation identified the stratigraphic interval of the San Juan Basin sequence where "Dragonian" fossils should be found. Fossils discovered from this interval have been identified as cf. *Paromomys* sp., *Protictis* cf. *P. haydenianus*, cf. *Loxolophus kimbetovius*, *Triisodon* cf. *T. antiquus*, cf. *Mioclaenus turgidus*, *Periptychus carinidens*, and *Anisonchus sectorius*.

Correlation of this fauna and the Dragon fauna with magnetostratigraphic chronologic control demonstrates that the "Dragonian" Land Mammal "Age" is best treated as the earliest subdivision of the Torrejonian because of its magnetic polarity sequence and the dominance of Torrejonian genera along with the lack of its own characteristic, widespread genera. The status of the "Dragonian" as a distinct land mammal "age" should be abandoned. The "Dragonian" time interval is distinct from the *Deltatherium* zone of the Torrejonian as evidenced by its magnetic polarity sequence, by the presence of *Loxolophus* and *Oxyclaenus*, and by the absence of *Deltatherium* and *Tetraclaenodon*. The "Dragonian" time interval is here named the *Periptychus-Loxolophus* zone.

The *Pantolambda* and *Deltatherium* zones of the Torrejonian Land Mammal "Age" were originally established as biostratigraphic units, but with an independent chronologic ordering (correlation with the magnetic polarity time scale), these zones, as well as the *Periptychus-Loxolophus* zone, can be recognized as chronozones (chronostratigraphic units). Thus, the Torrejonian Land Mammal "Age" is subdivided into three chronozones: the *Periptychus-Loxolophus* in anomaly 27 and the base of the overlying reversed magnetozone, the *Deltatherium* in the reversed interval (except its base) between anomalies 27 and 26, and the *Pantolambda* in anomaly 26. The *Pantolambda* chronozone is not yet fully documented because its upper boundary is presently defined (and recognized) on the basis of a biochronologic unit (Tiffanian Land Mammal "Age") rather than on biostratigraphic and magnetostratigraphic ordering.

ACKNOWLEDGMENTS

I am grateful for the guidance and constructive criticism given by Everett H. Lindsay throughout the course of this study. The manuscript was read by George G. Simpson and Louis L. Jacobs. Lawrence J. Flynn and Louis H. Taylor provided much valuable discussion on the concept of chronostratigraphy. Assistance in the field was given by Everett H. Lindsay, Robert F. Butler, Louis L. Jacobs, Lawrence J. Flynn, Louis H. Taylor, Barbara R. Standhardt, Richard A. Haskin, and Robert D. McCord. Richard H. Tedford provided an opportunity to compare specimens at the American Museum of Natural History, and Earl Manning gave me helpful suggestions on the identification of a few specimens. George S. Marcek prepared photographs. To all the above I express my sincere thanks for their help.

Financial support for this study was contributed by the National Science Foundation (NSF Grant EAR 78-03326) and the State of Arizona.

REFERENCES

Bown, T. M. and K. D. Rose. 1976. New early Tertiary primates and a reappraisal of some Plesiadapiformes. Folia Primatol. 26:109-138.

Evernden, J. F., D. E. Savage, G. H. Curtis, and G. T. James. 1964. Potassium-argon dates and the Cenozoic mammalian chronology of North America. Am. J. Sci. 262:145-198.

Gazin, C. L. 1938. A Paleocene mammalian fauna from central Utah. J. Washington Acad. Sci. 28:271-277.

———. 1939. A further contribution to the Dragon Paleocene fauna of central Utah. J. Washington Acad. Sci. 29:273-286.

———. 1941. The mammalian faunas of the Paleocene of central Utah, with notes on the geology. Proc. U.S. Natl. Mus. 91:1-53.

Gingerich, P. D. 1976. Cranial anatomy and evolution of early Tertiary Plesiadapidae (Mammalia, Primates). Univ. Michigan Pap. Paleontol. 15:1-141.

LaBrecque, J. L., D. V. Kent, and S. C. Cande. 1977. Revised magnetic polarity time scale for Late Cretaceous and Cenozoic time. Geology 5:330-335.

Lindsay, E. H., R. F. Butler, and N. M. Johnson. 1981. Magnetic polarity zonation and biostratigraphy of Late Cretaceous and Paleocene continental deposits, San Juan Basin, New Mexico. Am. J. Sci. 281:390-435.

Lindsay, E. H., L. L. Jacobs, and R. F. Butler. 1978. Biostratigraphy and magnetostratigraphy of Paleocene terrestrial deposits, San Juan Basin, New Mexico. Geology 6:425-429.

MacIntyre, G. T. 1966. The Miacidae (Mammalia, Carnivora) part 1. The systematics of *Ictidopappus* and *Protictis*. Bull. Am. Mus. Nat. Hist. 131:115-209.

Matthew, W. D. 1937. Paleocene faunas of the San Juan Basin, New Mexico. Am. Phil. Soc. Trans., n.s. 30:1-510.

Osborn, H. F. 1929. The titanotheres of ancient Wyoming, Dakota, and Nebraska, U.S. Geol. Surv. Monograph 55:1–701.

Osborn, H. F. and W. D. Matthew. 1909. Cenozoic mammal horizons of western North America. U.S. Geol. Surv. Bull. 361:1–138.

Russell, D. E. 1967. Le Paleocene continental d'Amerique du Nord. Mus. Natl. Hist. Nat., Mem., Ser. C. 16:37–99.

Simpson, G. G. 1959. Fossil mammals from the type area of the Puerco and Nacimiento strata, Paleocene of New Mexico. Am. Mus. Novitates 1957:1–22.

Sinclair, W. J. and W. Granger, 1914. Paleocene deposits of the San Juan Basin, New Mexico. Bull. Am. Mus. Nat. Hist. 33:297–316.

Sloan, R. E. 1970. Cretaceous and Paleocene terrestrial communities of western North America. North Am. Paleont. Convention, Chicago, 1969, p. 427–453.

Taylor, L. H. and R. F. Butler. 1980. Magnetic-polarity stratigraphy of Torrejonian sediments, Nacimiento Formation, San Juan Basin, New Mexico. Am. J. Sci. 280:97–115.

Tedford, R. H. 1970. Principles and practices of mammalian geochronology in North America. North Am. Paleont. Convention, Chicago, 1969, p. 666–703.

Tomida, Y. and R. F. Butler. 1980. Dragonian mammals and Paleocene magnetic polarity stratigraphy, North Horn Formation, central Utah. Am. J. Sci. 280:787–811.

Van Valen, L. 1978. The beginning of the age of mammals. Evol. Theory 4:45–80.

Wilson, R. W. 1951. Preliminary survey of a Paleocene faunule from the Angels Peak area, New Mexico. Univ. Kansas Publ. Mus. Nat. Hist. 5(1):1–11.

Wood, H. E., 2nd, R. W. Chaney, J. Clark, E. H. Colbert, G. L. Jepsen, J. B. Reeside, Jr., and C. Stock. 1941. Nomenclature and correlation of the North American Continental Tertiary. Geol. Soc. Am. Bull. 52:1–48.

11 The Kutz Canyon Local Fauna, Torrejonian (Middle Paleocene) of the San Juan Basin, New Mexico

Louis H. Taylor
Department of Geosciences, University of Arizona, Tucson, Arizona

INTRODUCTION

Kutz Canyon (Coots Cañon of Granger, 1917 and Coots Canyon of Matthew, 1937) is a steep-walled, badland-containing valley eroded into the Paleocene Nacimiento Formation by an intermittent tributary of the San Juan River. This northwesterly draining canyon covers a fan-shaped area about 23 km long and is about 15 km wide at its greatest width. The head of Kutz Canyon is located about 24 km south of Bloomfield, New Mexico. The most notable landmark in Kutz Canyon is Angel Peak, an erosional remnant near the east side of the head of the canyon (Text-fig. 11.1).

It has long been known that the strata exposed in Kutz Canyon contain vertebrate fossils. David Baldwin, a collector who was, at one time or another, employed by O. C. Marsh and E. D. Cope, collected near Kutz Canyon (Simons, 1963). It seems unlikely that Baldwin would have ignored such a large area of badlands in his search for what he thought were Cretaceous mammals, but there is no record of his having collected from Kutz Canyon. Walter Granger (1917) published the first account of fossil collecting from Kutz Canyon. His 1916 American Museum of Natural History expedition recovered mammalian fossils "on various horizons from near the base of the beds in the center of the basin to within 150 feet of the top along the western wall" (Granger, 1917, p. 822). His localities included "1 mile west of Angel Peak," "2 miles west of Angel Peak," "Sec. 26, T27N, R11W," and "Sec. 3, T27N, R11W" (Text-fig. 11.1).

Granger's (1917) faunal list from Kutz Canyon was composed of the following Torrejonian taxa:

Tetraclaenodon	*Protogonodon* sp. nov
Mioclaenus turgidus	*Tricentes*
Periptychus rhabdodon	*Deltatherium*
Anisonchus sectorius	*Psittacotherium*

Periptychus rhabdodon has been synonymized with *P. carinidens* by Simpson (1959) and *Tricentes* from Kutz Canyon has been referred to *Mimotricentes* by Van Valen (1978). *Triisodon antiquus* was reported by Matthew (1937) as belonging to Granger's (1917) collection, and Van Valen (1978) reported *Deuterogonodon noletil* from the same collection.

R. W. Wilson led a collecting expedition to the San Juan Basin in 1948 (Wilson, 1951). His University of Kansas party discovered an extremely productive vertebrate fossil locality in Kutz Canyon. This locality (Angels Peak "pocket" in Wilson, 1951; KUVP Loc. 13 in Wilson, 1956a and 1956b: and Locality NM 13 in Wilson & Szalay, 1972), situated about 49 m below the rim on the western wall of Kutz Canyon in Sec. 14, T27N, R11W, has come to be known informally as "Big Pocket." Specimens from Big Pocket greatly expanded the faunal list from Kutz Canyon; Granger's (1917) 10 taxa were joined by 17 others (Wilson, 1956a, 1956b; Wilson & Szalay, 1972; Kay & Cartmill, 1977). The Big Pocket collection, as identified by Wilson (1956a, 1956b), Wilson & Szalay (1972), and Kay & Cartmill (1977), is included in the biostratigraphical analysis and faunal composition comparisons in this paper.

Since the summer of 1972 workers from The University of Arizona have been studying the fossiliferous Late Cretaceous through Eocene strata of the San Juan Basin. The first University of Arizona Laboratory of Paleontology (UALP) vertebrate fossil locality in the interior of Kutz Canyon was discovered near the end of the 1975 field season. Over 30 additional UALP localities were discovered in Kutz Canyon during the following three field seasons. One UALP locality near Big Pocket was discovered in 1974. These new UALP localities, the specimens they yielded, their correlation with the magnetic-polarity time scale, and their comparison with other Torrejonian faunas are the subject of this paper. Vertebrates other than mammals are not considered in this study. At present it appears that lower vertebrate fossils recovered from Kutz Canyon are not distinct from those collected elsewhere in the San Juan Basin Paleocene. Their biostratigraphy is being studied by other workers.

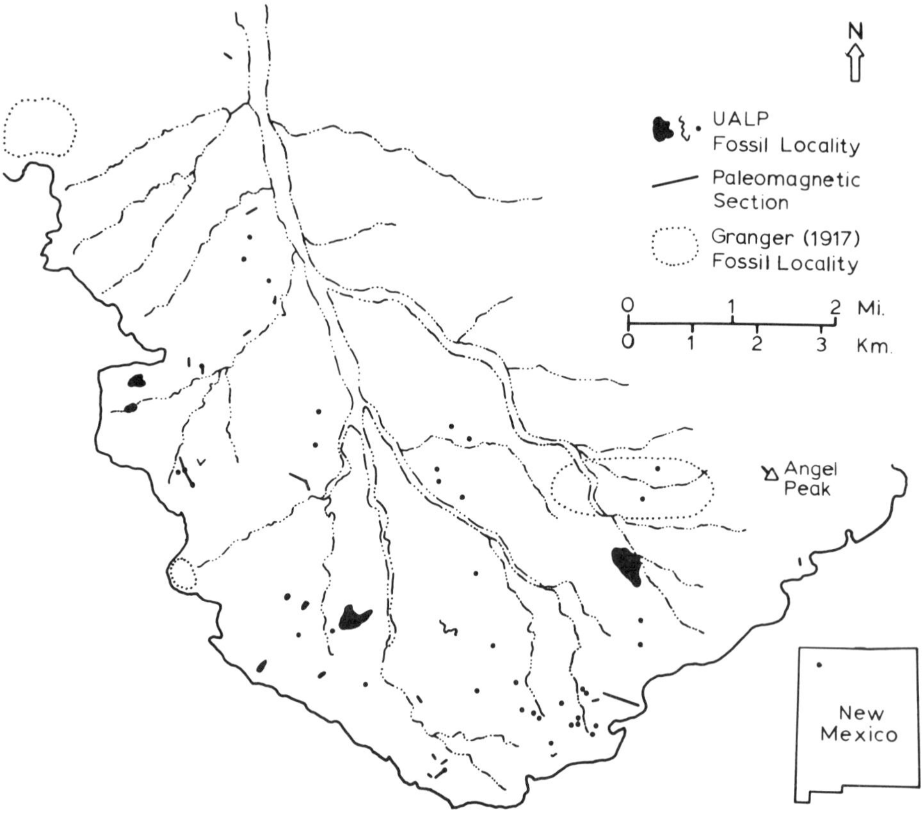

TEXT-FIG. *11.1*—Map of Kutz Canyon, San Juan Basin, New Mexico. Inset map with position of Kutz Canyon within New Mexico.

MAGNETIC-POLARITY STRATIGRAPHY

The collection of paleomagnetic sites from Kutz Canyon was undertaken during the 1975–1977 field seasons. Over 280 sites were collected with three samples per site. An approximately three meter stratigraphic interval was maintained throughout the over 400 m long stratigraphic section. Only the finest-grained lithologies were collected. Details of collecting methods, laboratory techniques, results of magnetic-polarity determinations, Curie point studies, magnetic separate studies, reliability of the magnetic data, and correlation of the Kutz Canyon magnetic-polarity column with the time scale have been published elsewhere (Butler & Taylor, 1978; Taylor & Butler, 1980; Lindsay et al., 1981).

The Kutz Canyon magnetic-polarity column used in Text-figures 11.2 and 11.3 was correlated with that portion of the LaBrecque et al. (1977) magnetic-polarity time scale containing anomalies 25, 26, and 27 (Taylor & Butler, 1980). This correlation is in agreement with the assignment of a middle Paleocene age to Kutz Canyon strata by earlier workers.

THE UALP FOSSIL LOCALITIES

Four hundred seventeen identifiable fossil mammal specimens have been recovered from 33 UALP vertebrate fossil localities in Kutz Canyon (Text-fig. 11.1). The 27 localities that yielded at least one fossil mammal specimen and could be stratigraphically correlated with the Kutz Canyon magnetic-polarity column were used in interpreting the biostratigraphy.

Several of the UALP localities were of the "pocket" type described by Wilson (1951). Others extended along the same stratum for perhaps a kilometer or more. At all localities only one or two strata yielded fossils. These strata varied from red or gray to purple or black siltstones and claystones and, at many localities, contained an abundance of ferruginous concretions. No mammalian fossils were recovered from sandstone.

The most productive UALP vertebrate fossil localities were UALP 75106 (Taylor Mound), UALP 7640 (Larry's Beaucoup), and UALP 7671 (Bab's Basin). UALP 75106 and UALP 7640 were extensive localities along cliff faces in the interior of Kutz Canyon, whereas UALP 7671 was a large pocket near Angel Peak. All of the Kutz Canyon localities are within the region included on the U.S. Geological Survey 15 Minute Bloomfield Quadrangle Map. (Precise geographical position of these localities is on file in the Geosciences Department, The University of Arizona.)

Specimen preservation was generally good, although a large number of specimens were encased within the concretions so common to the San Juan Basin; few specimens were unidentifiable. Most specimens were found on the surface without benefit of quarrying or underwater screening. A few specimens from UALP 7640, UALP 7650, and UALP 7671 were collected by underwater screening, but no quarrying was undertaken. Many localities were collected by close hands and knees scrutiny in such a way that individual teeth of minute size were recovered. The specimens were usually fragments of upper or lower jaws, although some isolated teeth and three skulls were recovered. Postcranial material was relatively rare.

BIOSTRATIGRAPHY

Text-figure 11.2 indicates the stratigraphic position of 27 vertebrate fossil localities from Kutz Canyon. The total length of the useful Kutz Canyon magnetic-polarity section is about 425 m. The mammalian fauna was found within about 165 m of the section; only those rocks found from 93 to 258 m stratigraphic elevations were fossiliferous. A large number of localities were clustered in the 157 to 164 m stratigraphic interval. This interval includes Wilson's (1951) Big Pocket

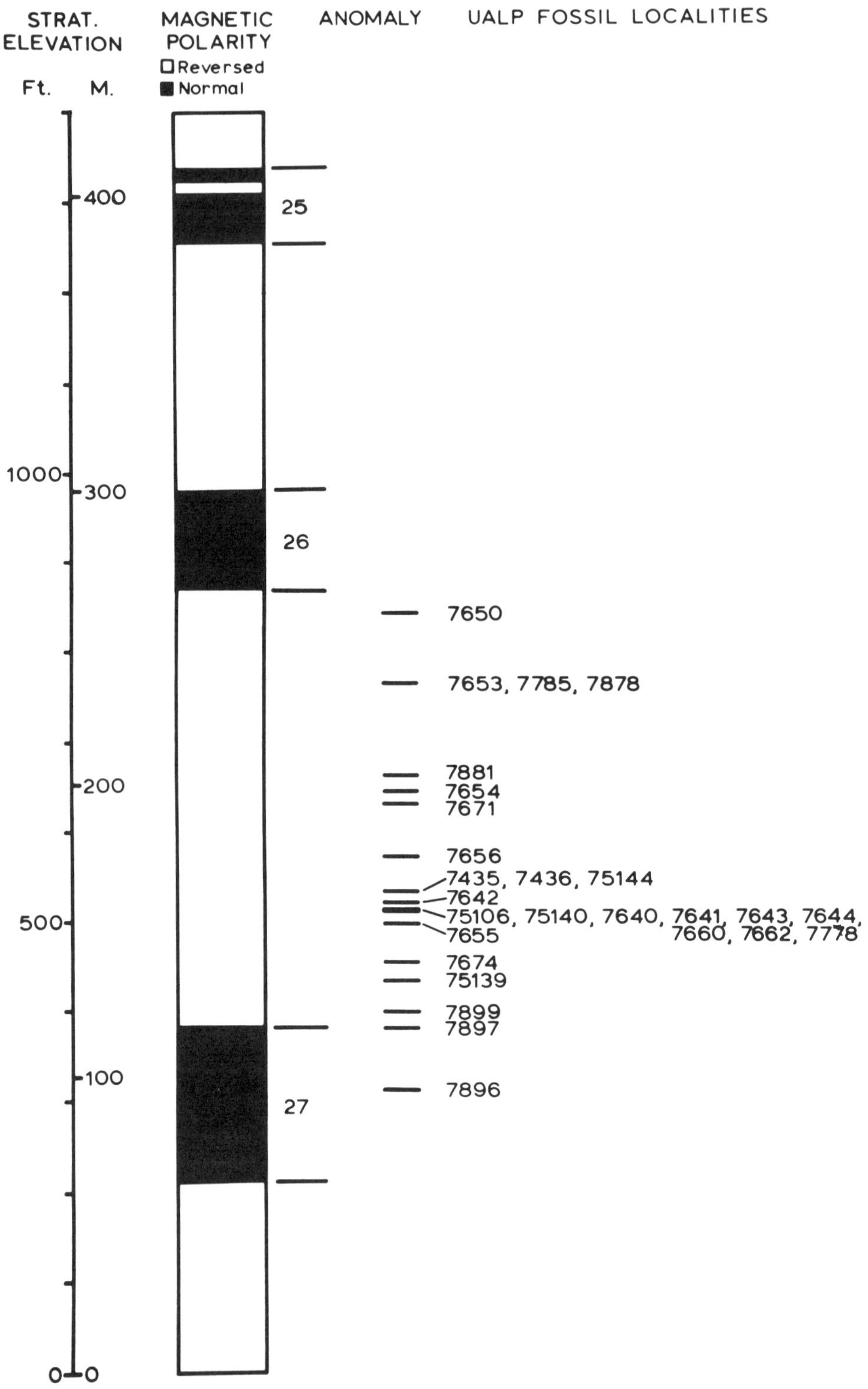

TEXT-FIG. *11.2*—Correlation of UALP vertebrate fossil localities from Kutz Canyon with the magnetic-polarity column from Kutz Canyon. Black represents normal polarity, and white represents reversed polarity.

locality. Only one large UALP locality falls outside of this interval; UALP 7671 is situated at the 192 m level.

Early workers in the San Juan Basin recognized two distinct mammalian collecting levels ("zones") within Torrejonian sediments. The lower of these was named the "*Deltatherium* zone" by Osborn (1929), and the upper, "*Pantolambda* zone", was named by Osborn & Matthew (1909). Wilson (1956a) referred his "Angels Peak" Faunule to the *Deltatherium* zone on the basis of its fauna. Later workers (Taylor & Butler, 1980) agreed with this referral and established a paleomagnetic description of the San Juan Basin Torrejonian. The *Pantolambda* zone is present within the normal magnetozone correlated with magnetic anomaly 26, and the *Deltatherium* zone is present within the directly underlying reversed magnetozone. No mammalian fossils from Kutz Canyon were recovered from the *Pantolambda* zone (Text-fig. 11.2). The Kutz Canyon local fauna is, therefore, of early, but not earliest, Torrejonian age (see below).

The paleomagnetic assignment of the Kutz Canyon local fauna to the *Deltatherium* zone is in agreement with the paleontological definition of that zone (Wilson, 1956a; Taylor & Butler, 1980). No *Pantolambda*, *Claenodon*, or *Mixodectes pungens*, characteristic *Pantolambda* zone fossils, were recovered from Kutz Canyon.

The three stratigraphically lowest localities, UALP 7896, UALP 7897, and UALP 7899, are considered to be of Torrejonian age but may be stratigraphically lower than the *Deltatherium* zone. These localities lie within, or slightly above, a normal magnetozone correlated with anomaly 27. Sediments correlated with this anomaly have been found in Dragon Canyon, Utah (Tomida & Butler, 1980). They represent the "Dragonian Land Mammal Age." Tomida (this volume) considers the "Dragonian" to be a subdivision of the Torrejonian, without land mammal "age" status. The Kutz Canyon localities do not, however contain any distinctive Dragonian (*sensu* Wood et al., 1941) taxa; all are common to the Torrejonian. "Dragonian" specimens are discussed in detail by Tomida (this volume).

The stratigraphic positions of all presently recognized genera from Kutz Canyon are shown in Text-figure 11.3. All positions above or below the 164 m level are extensions of previously recognized taxonomic ranges in Kutz Canyon. Many taxa are not recognized yet in Kutz Canyon beyond the Big Pocket level. These include the multituberculates *Ptilodus* nr. *P. trovessartianus* and *Ectypodus*? sp., the insectivores *Palaeoryctes* cf. *P. puercensis* and *Prodiacodon* n. sp., and the primates *Palaechthon nacimienti* and *Talpohenach torrejonius*.

The stratigraphic ranges of the taxa listed below are here extended downward (earlier) within Kutz Canyon:

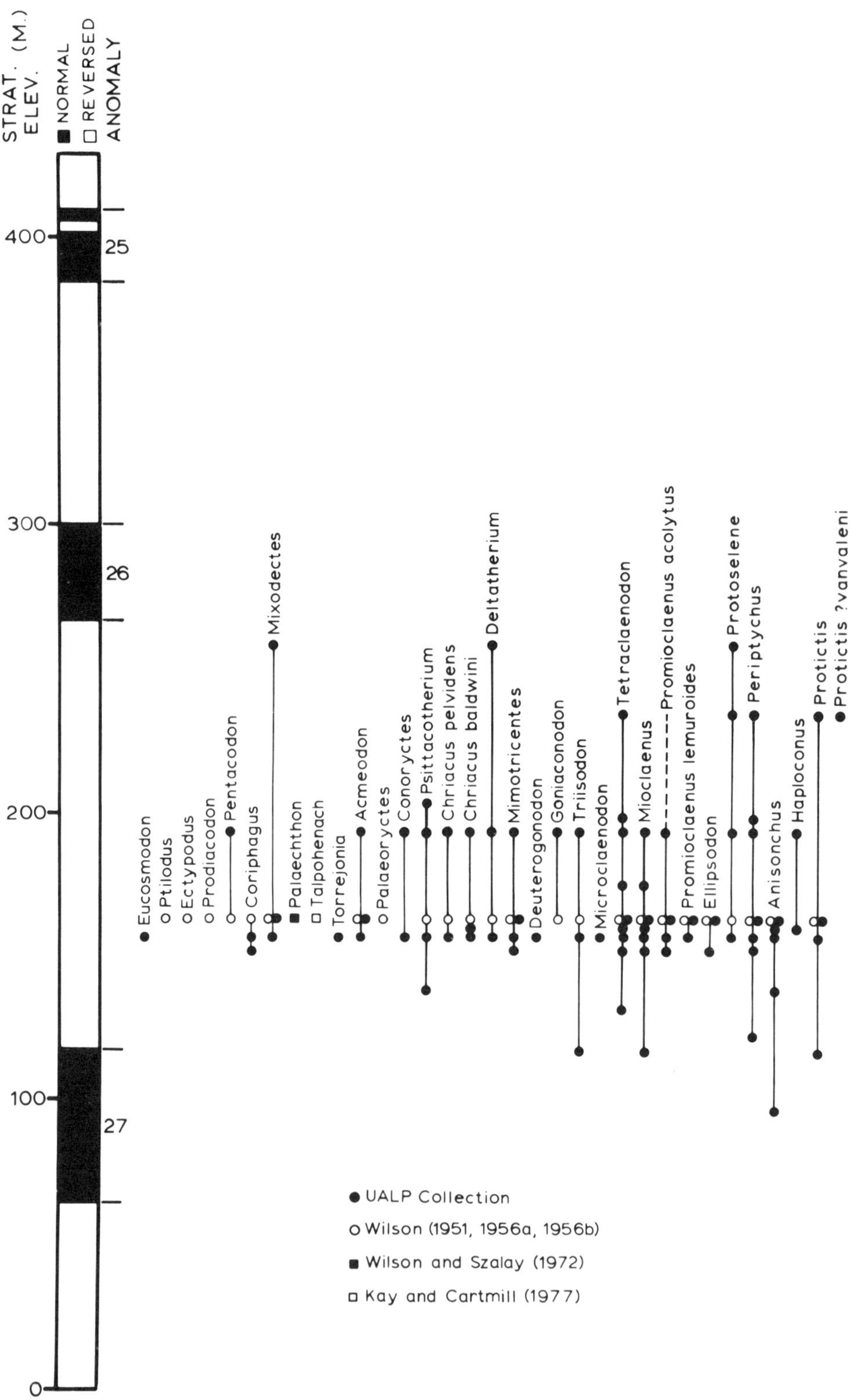

TEXT-FIG. *11.3*—Biostratigraphy of the Kutz Canyon local fauna. The magnetic-polarity colors are the same as in Text-figure 11.2.

Multituberculata
Eucosmodon
Insectivora
Mixodectes malaris
Coriphagus encinensis
Taeniodonta
Psittacotherium multifragum
Carnivora
Protictis
Condylarthra
Mimotricentes subtrigonus
Chriacus baldwini
Triisodon
Mioclaenus
Ellipsodon
Promioclaenus
Protoselene
Tetraclaenodon puercensis
Periptychus carinidens
Anisonchus sectorius

The stratigraphic ranges of the following taxa are here extended upward (later) within Kutz Canyon:

Insectivora
Mixodectes malaris
Pentacodon
Deltatheridia
Acmeodon secans
Taeniodonta
Psittacotherium multifragum
Carnivora
Protictis
Condylarthra
Mimotricentes subtrigonus
Chriacus baldwini
Deltatherium fundaminis
Triisodon
Mioclaenus turgidus
Goniacodon levisanus
Promioclaenus
Protoselene
Tetraclaenodon puercensis
Periptychus carinidens
Anisonchus sectorius

The comparison of the Kutz Canyon stratigraphic ranges with the San Juan Basin local range zones of some taxa is somewhat problematical. Wilson (1956a) reported a vertebrate fossil locality from the head of Kimbetoh Arroyo, some 40 km south of Kutz Canyon. This was University of Kansas (KUVP) Locality 9. Wilson (1956a) referred the fauna from KUVP 9 to the *Deltatherium* zone. Paleomagnetic samples collected during the University of Arizona 1978 field season indicate that the entire section at locality KUVP 9 exhibits reversed polarity. Thus, it is correlative with the *Deltatherium* zone in Kutz Canyon. However, the stratigraphic position of KUVP 9 within the *Deltatherium* zone is indeterminate. Taxa from Kutz Canyon that also were collected from locality KUVP 9 cannot be shown to have their range zones extended downward, as the possibility exists that locality KUVP 9 is stratigraphically lower than the Kutz Canyon *Deltatherium* zone fossils. These taxa include *Mixodectes* cf. *M. malaris*, *Coriphagus encinensis*, *Mimotricentes subtrigonus*, *Chriacus baldwini*, *Promioclaenus*, *Protoselene opisthacus*, *Tetraclaenodon puercensis*, *Haploconus angustus*, and *Periptychus carinidens*. The lowest stratigraphic level for *Psittacotherium multi-*

fragum in Kutz Canyon is, however, the lower limit of its local range zone. This is also true for the lowest Kutz Canyon occurrence of *Protictis, Triisodon*, cf. *Mioclaenus* and *Anisonchus sectorius*. The lowest stratigraphic occurrence of these taxa falls within the magnetozone correlative with anomaly 27. They first appeared in Kutz Canyon during "Dragonian" time, earliest Torrejonian (Tomida, this volume).

A similar problem exists in the interpretation of the highest stratigraphic levels of taxa present in locality KUVP 9. The unknown stratigraphic placement of the KUVP magnetic section does not permit the highest Kutz Canyon stratigraphic levels to be interpreted as the upper limit of the local range zones for *Mixodectes* cf. *M. malaris, Deltatherium fundaminis, Triisodon*, or *Haploconus angustus*. Other taxa from Kutz Canyon with stratigraphic ranges higher than the 164 m level (Text-fig. 11.3) have been reported from the *Pantolambda* zone in other San Juan Basin localities, and, therefore, have local range zones extending far above the stratigraphic level of the Kutz Canyon local fauna.

Taxa not previously reported from Kutz Canyon are:

Eucosmodon sp.	*Conoryctes comma*
Torrejonia cf. *T. wilsoni*	*Microclaenodon assurgens*
Protictis ?*vanvaleni*	*Haploconus angustus*
	Chriacus pelvidens

All have been reported from elsewhere in the San Juan Basin. Kutz Canyon represents the lowest San Juan Basin occurrence of *Conoryctes comma, Microclaenodon assurgens*, and possibly *Torrejonia*. *Eucosmodon*? sp. and *Haploconus angustus* were reported from locality KUVP 9 by Wilson (1956a). The stratigraphic level of the *Deuterogonodon noletil* specimen collected by Granger and reported by Van Valen (1978) is unknown. The stratigraphic level shown for *Deuterogonodon* in Text-figure 11.3 was determined from UALP specimens. Taxa new to Kutz Canyon are discussed in more detail below.

TAXA NEW TO KUTZ CANYON

In addition to the taxa new to Kutz Canyon listed above, a possible new species of *Protictis* and a new genus of hyopsodontid are reported here, but not formally described.

The multituberculate *Eucosmodon* is represented by specimen UALP 8955 (Pl. 11.1, fig. 1) from locality UALP 75106. This anterior portion of a right P^4 compares favorably with the illustration of *Eucosmodon* sp. (American Museum of Natural History specimen 16534) in Granger and Simpson (1929, p. 636) and of *Eucosmodon americanus* (KUVP specimen 9401) in Wilson (1956c, p. 120). Specimen UALP 8955, however, is larger than either of the other specimens mentioned.

Specimen UALP 8952 (Pl. 11.1, fig. 8), also from locality UALP 75106, is identified as *Torrejonia* cf. *T. wilsoni*. This specimen is a partial dentary with broken P_4 and M_1. It is encased in concretionary material. The P_4 and M_1 are slightly longer than those of USNM (U.S. National Museum) specimen 25255 from the *Pantolambda* zone near the head of Torreon Arroyo, described by Gazin (1968). The paraconid of the M_1 is somewhat lower, relative to the metaconid and protoconid, in the UALP specimen. However, the widths of the trigonid and talonid of the UALP specimen are identical to that of USNM 25255. The P_4 of the UALP specimen exhibits the elongated talonid of *T. wilsoni*. Only a few specimens of the lower dentition of *T. wilsoni* are known. UALP 8952 could fall within the expected variation of that species.

Specimen UALP 10727 (Pl. 11.1, figs. 2 & 3) from locality UALP 7653 is tentatively referred to *Protictis vanvaleni*, previously reported and described only from the *Pantolambda* zone near the head of Torreon Arroyo (MacIntyre, 1966). The size of UALP 10727 is within the range of *P. vanvaleni*, but it is a partial right dentary with broken M_1 and M_2 and cannot be positively identified. The trigonid cusps are missing from both teeth.

Conoryctes comma, previously reported only from the *Pantolambda* zone, is represented in the Kutz Canyon local fauna by three specimens from locality UALP 7671 and one specimen from UALP 7662. Specimen UALP 11661, a partial right maxilla with M^1 or M^2, from locality UALP 7662, is shown in Plate 11.1, figure 4.

Two specimens of *Microclaenodon assurgens* have been recovered from Kutz Canyon locality UALP 75106. Specimen UALP 8946 (Pl. 11.1, fig. 5) is a partial right maxilla with P^4–M^1, and specimen UALP 10858 (Pl. 11.1, figs. 6 & 7) is a partial right dentary with M_{1-2}. This taxon has been reported previously only from the southern part of the San Juan Basin and presumably only from the *Pantolambda* zone.

Haploconus is distinct from *Anisonchus* in the absence of a paraconid and more widely separated hypoconid and hypoconulid on the lower molars and in reduction of the P^3 relative to the P^4. The P_4 of *Haploconus* lacks a paraconid and metaconid and possesses a large, rectangular-shaped talonid with entoconid, hypoconid, and hypoconulid as distinct cusps. The talonid of *Anisonchus* is smaller relative to the trigonid, triangular in outline, and without distinct cusps. Two specimens from locality UALP 7642 and eight specimens from locality 7671 are referred to *Haploconus angustus*. Specimens UALP 10777, a partial left dentary with M_{1-3}, and UALP 10778, a partial left maxilla with P^3–M^2, are figured in Plate 11.2, figures 3 and 1, respectively. Also shown on Plate

11.2 are two specimens of *Anisonchus sectorius*: UALP 10865 (Pl. 11.2, fig. 4) and UALP 10945 (Pl. 11.2, fig. 2).

Chriacus pelvidens is represented in the Kutz Canyon local fauna by three specimens. UALP 10570 is a partial dentary with right P_3, broken P_4, and M_1, from locality UALP 75106. UALP 10639, a partial left dentary with broken P_3 and P_4–M_2, and UALP 10648, a partial left maxilla with P^4–M^2, are both from locality UALP 7640.

Specimen UALP 10612 (Pl. 11.2, fig. 5) from locality UALP 7640, a broken palate with left P^2–M^3 and right M^{1-3}, is similar to *Deuterogonodon noletil* as described and illustrated by Van Valen (1978, p. 76). Minor differences are present in the molars; the cingulum of the UALP specimen does not completely encircle the protocone on the M^1, and the parastyle of M^3 is somewhat larger and projects more labially. The P^4 of AMNH 17078 (the holotype of *D. noletil*) as figured by Van Valen (1978) has a large posterior-labial cusp (?metacone) that is absent in UALP 10612. The molars from both specimens are very close in size, but the P^4 of AMNH 17078 appears to be somewhat larger than that of specimen UALP 10612. It is possible that one of the specimens has deciduous dentition. The less worn condition of the AMNH 17078 molars and the larger size of its P^4 suggest that the P^4 of that specimen is actually DP^4. Thus UALP 10612 is referred to *Deuterogonodon noletil*.

Protictis n. sp. exhibits an M_1 that is slightly larger, an M_2 that is about 12% larger, and a P_4 that is about 15% larger than in *Protictis haydenianus* as reported by MacIntyre (1966) from the San Juan Basin. Specimens of *P. haydenianus* from the Crazy Mountain Field are slightly smaller than those from the San Juan Basin (Simpson, 1937; MacIntrye, 1966). *Protictis* n. sp. is represented by specimens UALP 10563, a partial dentary with left M_{1-2}, from locality UALP 75106; and by UALP 10638, a partial dentary with left P_4–M_2, from locality UALP 7640. The UALP specimens have not been compared with *Protictis* n. sp. A. and *Protictis* n. sp. B of Wilson (1956a). Until such comparison is made, formal description will be deferred.

A new hyopsodontid genus is represented by specimens UALP 10599 and UALP 10704 from localities UALP 7640 and UALP 7643, respectively. Both specimens of this taxon are partial right dentaries with M_{2-3}. Formal description of these specimens is forthcoming.

FAUNAL COMPOSITION AND COMPARISON

About 87% of the mammals in the Kutz Canyon local fauna were recovered from three stratigraphic levels. These levels are, in ascending order, the Taylor Mound, Big Pocket, and Bab's Basin levels. The Taylor Mound level comprises 9 UALP vertebrate fossil localities and has yielded 225 identifiable specimens. It is situated at a stratigraphic eleva-

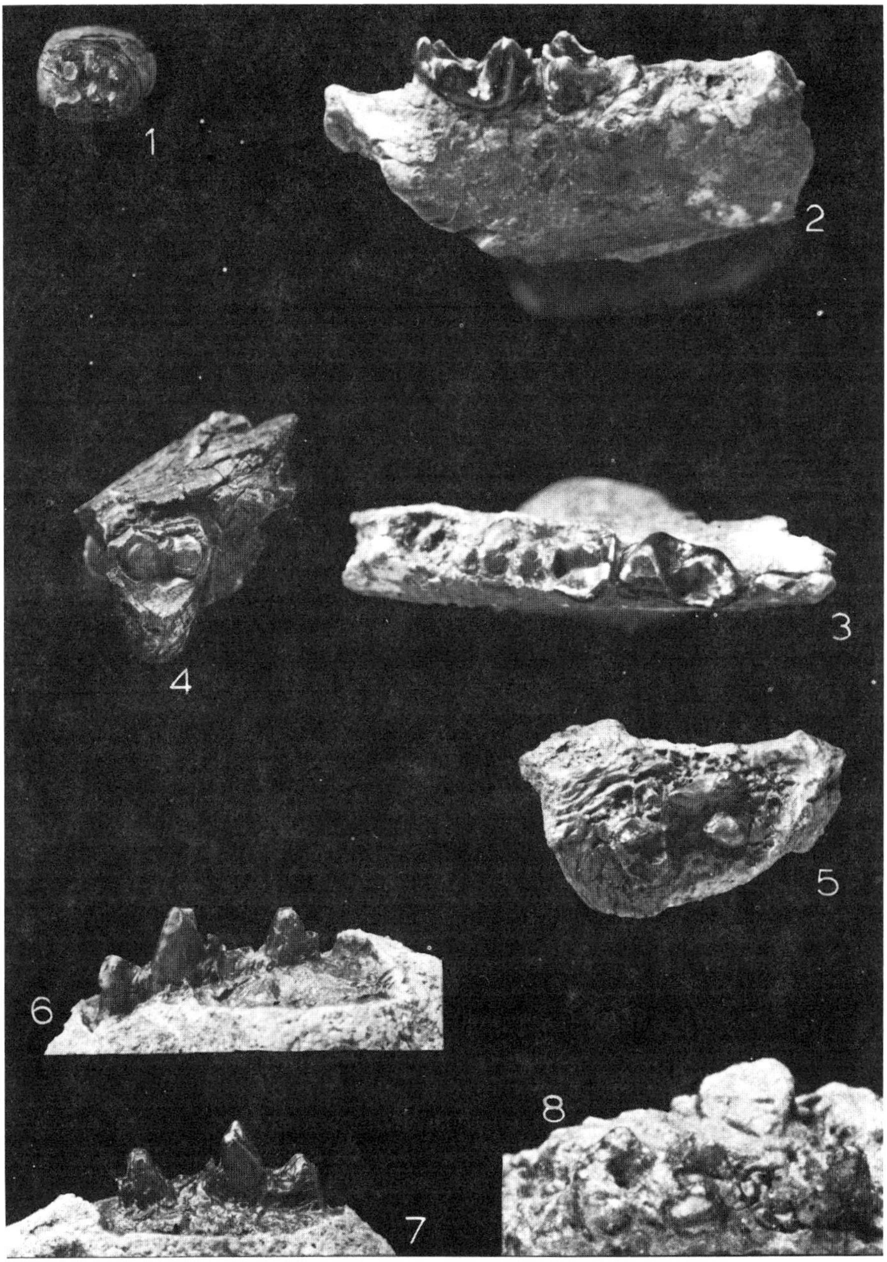

EXPLANATION OF PLATE 11.1

FIG. *1*—UALP 8955, *Eucosmodon* sp., broken RP^4, occlusal view, × 2.7.

2–3—UALP 10727, *Protictis* ?*vanvaleni*, RM_{1-2}. *2*, labial view, × 4.5; *3*, occlusal view, × 4.5.

4—UALP 11661, *Conoryctes comma*, RM_1 or RM_2, occlusal view, × 1.4.

5—UALP 8946, *Microclaenodon assurgens*, RP^4–M^1, occlusal view, × 1.8 (P^4 is lying on its side and is embedded in the concretion).

6–7—UALP 10858, *Microclaenodon assurgens*, RM_{1-2}. *6*, lingual view, × 1.8; *7*, labial view, × 1.8.

8—UALP 8952, *Torrejonia* cf. *T. wilsoni*, LP_4–M_1, occlusal view, × 5.4.

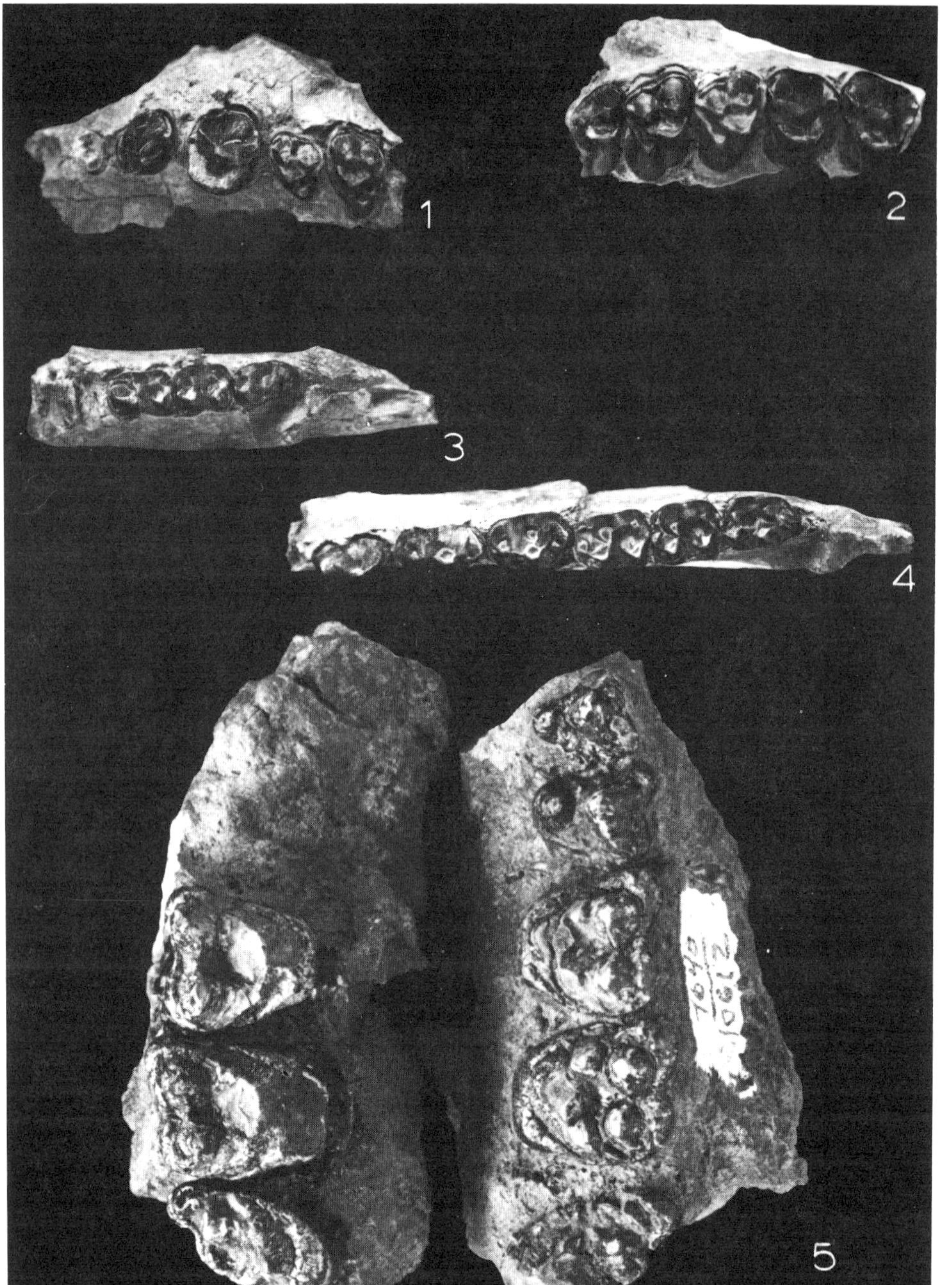

EXPLANATION OF PLATE 11.2

FIG. *1*—UALP 10778, *Haploconus angustus*, LP^{3}–M^{2}, occlusal view, ×2.
2—UALP 10945, *Anisonchus sectorius*, RP^{3}–M^{3}, occlusal view, ×2.
3—UALP 10777, *Haploconus angustus*, LM$_{1-3}$, occlusal view, ×2.
4—UALP 10865, *Anisonchus sectorius*, LP$_{2}$–M$_{3}$, occlusal view, ×2.
5—UALP 10612, *Deuterogonodon noletil*, RM$^{1-2}$, LP^{3}–M^{3}, occlusal view, ×2.

tion of 156 m (Text-fig. 11.2). The Big Pocket level, at the 163 m stratigraphic level (Text-fig. 11.2), includes three localities, including Big Pocket. Some 312 identifiable specimens have been recovered from the Big Pocket level. The Bab's Basin level is situated at the 192 m stratigraphic level (Text-fig. 11.2) and is represented by locality UALP 7671, which yielded 77 identifiable specimens.

The composition of the three levels mentioned above, as well as that of the total mammalian aspect of the Kutz Canyon local fauna and of the Gidley and Silberling Quarries from the Crazy Mountain Field, Montana (Simpson, 1937), is shown in Table 11.1 and Text-figure 11.4.

The faunal compositions from the Taylor Mound and Bab's Basin stratigraphic levels are greatly dominated by condylarths with a low percentage of small mammals (multituberculates, insectivores, and primates). Wilson (1951) accounted for this small percentage as possibly due to the bias of surface collecting. Collecting bias may have caused the lower percentages of small mammals from the Taylor Mound level because this level was not collected as intensively as were the other levels from Kutz Canyon. However, even the UALP localities that were collected on hands and knees, as Big Pocket was collected, produced results similar to Wilson's—few small mammals. Some sites in Kutz Canyon were collected by the technique of underwater screening. A few multituberculates and insectivores were recovered in this manner, but the process was generally unproductive. Specimens recovered by screening were excluded from this study. The low numbers of minute mammals appear to be real, not due to the bias of collecting techniques.

The low representation of taeniodonts and carnivores also seems to be real. They are large enough that, if abundant, they should have been recovered easily by surface collecting. These orders are poorly represented in other Paleocene collections. The mesonychid condylarths, represented only by the genus *Microclaenodon*, are also rare, with only two specimens found in all of Kutz Canyon. Mesonychids, generally not abundant in Paleocene faunas, are unknown from faunas older than Torrejonian.

Wilson (1951, 1956a) commented on the absence of *Haploconus* from Kutz Canyon as being an enigma, in light of the widespread abundance of the closely related *Anisonchus*. He thought that perhaps this resulted from some local peculiarity of the environment. However, two different UALP localities from two different stratigraphic levels produced both *Anisonchus* and *Haploconus*. The enigma is removed.

The most common taxa from the Kutz Canyon local fauna are, in decreasing abundance: *Tetraclaenodon puercensis*, *Periptychus carinidens*, *Anisonchus sectorius*, *Mimotricentes subtrigonus*, *Promioclaenus acolytus*, *Mioclaenus turgidus*, *Chriacus baldwini*, *Protoselene opisthacus*, and *Mixodectes malaris*. None of these taxa are of minute

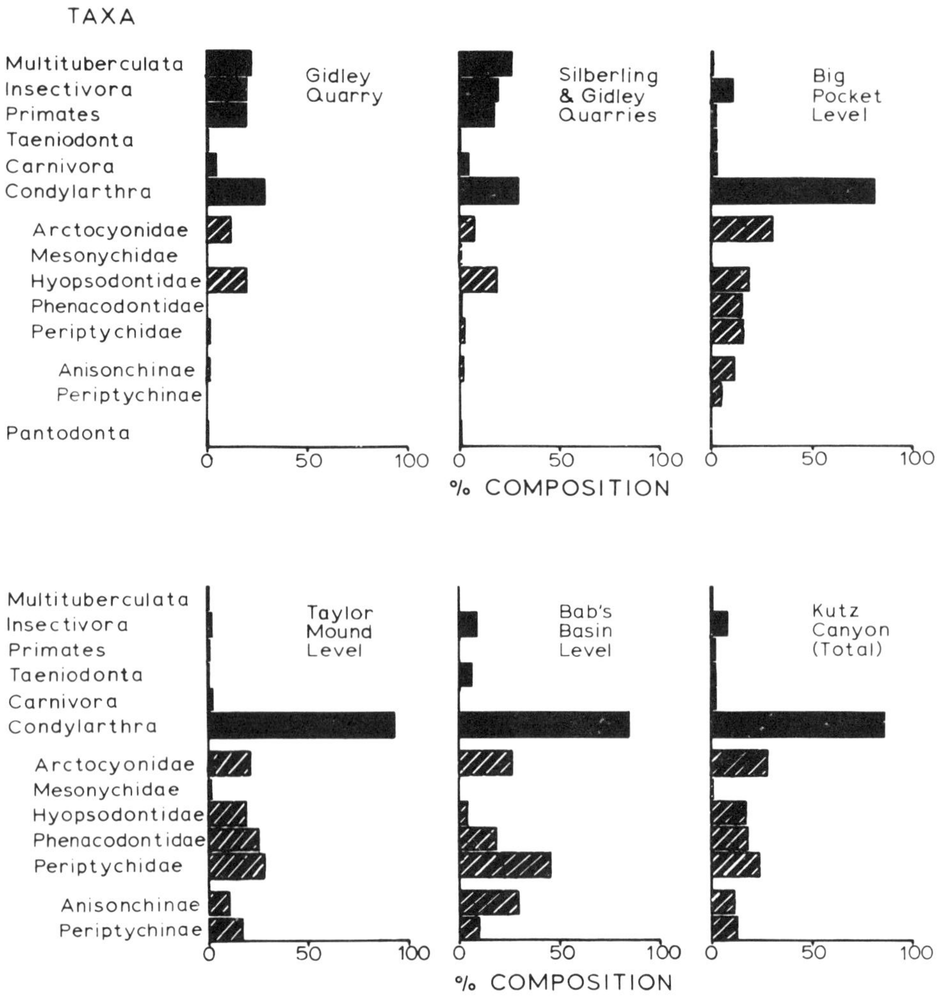

TEXT-FIG. *11.4*—Composition of UALP vertebrate fossil localities from Kutz Canyon and the Gidley and Silberling Quarries from the Crazy Mountain Field, Montana. Solid bars represent mammalian orders, and black and white bars represent families and subfamilies of Condylarthra. (Crazy Mountain Field data are from Simpson, 1937, and part of the Big Pocket data are from Wilson, 1956a. Some of these data also appear in Table 11.1.)

size. A complete Kutz Canyon local fauna mammalian faunal list is included in Table 11.2.

The composition of the Kutz Canyon local fauna is quite different from that of the Gidley and Silberling Quarries from the Crazy Mountain Field, Montana (Table 11.1; Text-fig. 11.4). The most common taxa

Table 11.1—Faunal composition of vertebrate fossil localities from the Crazy Mountain Field, Montana and Kutz Canyon, New Mexico. The Crazy Mountain Field and Kutz Canyon totals include all Torrejonian localities from those areas, both surface and quarry localities. The vertical totals include the order subtotals. (Crazy Mountain Field data from Simpson, 1937 and part of "Big Pocket" data from Wilson, 1956a.)

Taxon	Gidley Quarry Level		Silberling Quarry Level		Crazy Mountain Field Total		Big Pocket Level		Taylor Mound Level		Bab's Basin Level		Kutz Canyon Total	
	No.	%	No.	%	No.	%	No.	%	No.	%	No.	%	No.	%
Multituberculata	91	23.8	25	43.1	123	23.7	3	1.0	1	0.4			5	0.7
Insectivora	76	19.9	9	15.4	87	16.8	33	10.5	6	2.6	7	9.1	51	7.3
Primates	78	20.4	2	3.4	80	15.4	8	2.6	2	0.9			11	1.6
Taeniodonta	1	0.3	2	3.4	5	1.0	8	2.6	2	0.9	5	6.5	19	2.7
Carnivora														
Miacidae	20	5.2	3	5.2	27	5.2	8	2.6	5	2.2			15	2.1
Condylarthra	113	29.6	17	29.4	189	36.4	252	80.7	209	92.9	65	84.4	600	85.6
Arctocyonidae	27	7.1	7	12.2	59	11.4	95	30.4	47	20.9	21	27.2	189	27.0
Mesonychidae	1	0.3	1	1.7	2	0.4			2	0.9			2	0.9
Hyopsodontidae	76	19.9	8	13.8	88	17.0	59	18.9	41	18.2	3	3.9	112	16.0
Phenacodontidae	2	0.5	1	1.7	21	4.0	48	15.4	56	24.9	14	18.2	129	18.4
Periptychidae	7	1.8			19	3.6	50	16.0	63	28.0	27	35.1	168	23.9
Anisonchinae	7	1.8			19	3.6	35	11.2	23	10.2	16	20.8	82	11.7
Periptychinae							15	4.8	40	17.8	11	14.3	86	12.2
Pantodonta	3	0.8			8	1.5								
TOTAL	382		58		519		312		225		77		701	

from these quarries were *Ptilodus montanus*, *Parectypodus sinclairi*, *McKennatherium ladae*, *Aphronorus fraudator*, *Paromomys maturus*, *Palaechthon alticuspis*, *Metachriacus punitor* (= *Chriacus orthogonius*), *Protictis microlestes*, and *Ellipsodon aquilonius* (Simpson, 1937). These taxa are, as are most taxa from the quarries, of relatively small size. Indeed, even the most abundant *Protictis* and periptychid species from the Crazy Mountain Field are smaller than most common in the San Juan Basin; *Protictis microlestes* is considerably smaller than *Protictis haydenianus*. The only periptychid present in the Crazy Mountain Field is *Anisonchus*, whereas the larger genus *Periptychus* is quite common in Kutz Canyon. One possible reason for the greater abundance of smaller taxa in the Crazy Mountain Field could be the method of collecting; quarrying should be less biased with respect to size. However, because the techniques of underwater screening and hands and knees inspection were used in Kutz Canyon with no appreciable increase in small specimens, the size difference between the faunas from Kutz Canyon and the Crazy Mountain Field is thought to be due to some other factor or factors.

Table 11.2—Mammalian Faunal List for the Kutz Canyon Local Fauna

Class Mammalia Linnaeus, 1758
- Subclass Allotheria Marsh, 1880
  - Order Multituberculata Cope, 1884
    - Suborder Taeniolabidoidea (Granger and Simpson, 1929)
      - Family Eucosmodontidae (Jepsen, 1940)
        - *Eucosmodon* sp.
    - Suborder Ptilodontoidea (Simpson, 1927)
      - Family Ptilodontidae Gregory and Simpson, 1926
        - *Ptilodus* nr. *P. trovessartianus* (Cope, 1882)
      - Family Neoplagiaulacidae Ameghino, 1890
        - *Ectypodus*? sp. Wilson, 1956a
- Subclass Theria Parker and Haswell, 1897
  - Infraclass Eutheria Gill, 1872
    - Order Insectivora Bowditch, 1821
      - Family Leptictidae Gill, 1872
        - *Prodiacodon* n. sp. Wilson, 1956a
      - Family Palaeoryctidae (Winge, 1917)
        - *Palaeoryctes* cf. *P. puercensis* Matthew, 1913
        - *Acmeodon secans* Matthew and Granger, 1921
        - *Acmeodon* cf. *A. secans* Matthew and Granger, 1921
      - Family Pentacodontidae (Simpson, 1937)
        - *Pentacodon inversus* (Cope, 1888)
        - *Pentacodon* n. sp. Wilson, 1956a
        - *Coriphagus encinensis* (Matthew and Granger, 1921)
      - Family Mixodectidae Cope, 1883
        - *Mixodectes malaris* (Cope, 1883)

Order Primates Linnaeus, 1758
  Family Paromomyidae Simpson, 1940
    Palaechthon nacimienti Wilson and Szalay, 1972
    Talpohenach torrejonius Kay and Cartmill, 1977
    Torrejonia cf. *T. wilsoni* Gazin, 1968
Order Taeniodonta (Cope, 1876)
  Family Stylinodontidae Marsh, 1875
    Conoryctes comma Cope, 1881
    n. gen. and n. sp. Wilson, 1956a
    Psittacotherium multifragum Cope, 1882
Order Condylarthra Cope, 1881
  Family Arctocyonidae Giebel, 1855
    Chriacus pelvidens (Cope, 1881)
    Chriacus baldwini (Cope, 1882)
    Deltatherium fundaminis Cope, 1881
    Mimotricentes subtrigonus (Cope, 1881)
    Mimotricentes sp. indet. Wilson, 1956a
    Deuterogonodon noletil Van Valen, 1978
    Goniacodon levisanus (Cope, 1888)
    Triisodon antiquus (Cope, 1882)
    T. sp. Wilson, 1956a
    T. sp. Tomida, this volume
  Family Mesonychidae Cope, 1875
    Microclaenodon assurgens Scott, 1892
  Family Phenacodontidae Cope, 1881
    Tetraclaenodon puercensis (Cope, 1881)
    n. gen. and n. sp. Wilson, 1956a
  Family Hyopsodontidae Lydekker, 1889
    Mioclaenus turgidus Cope, 1881
    Mioclaenus? n. sp. Wilson, 1956a
    Promioclaenus acolytus (Cope, 1882)
    Promioclaenus lemuroides (Matthew, 1897)
    Ellipsodon grangeri Wilson, 1956b
    Protoselene opisthacus (Cope, 1882)
    n. gen. and n. sp.
  Family Periptychidae Cope, 1882
    Periptychus carinidens Cope, 1881
    Anisonchus sectorius (Cope, 1881)
    Haploconus angustus (Cope, 1882)
Order Carnivora Vicq d'Azyr, 1792
  Family Miacidae Cope, 1880
    Protictis haydenianus (Cope, 1882)
    Protictis vanvaleni? MacIntyre, 1966
    Protictis n. sp. A Wilson, 1956a
    Protictis n. sp. B Wilson, 1956a
    Protictis n. sp.

The major difference between the Kutz Canyon local fauna and that from the Gidley and Silberling Quarries is perhaps more than specimen size. The most abundant Gidley and Silberling taxa are not only small, but they also represent orders not well represented in Kutz Canyon. The orders Multituberculata, Insectivora, and Primates are much more diverse and abundant in the Gidley and Silberling Quarries than in Kutz Canyon. Simpson (1937) suggested that many members of these orders were arboreal in habitus. He also suggested that perhaps some of the smaller hyopsodontids may have been arboreal. It is possible that the difference between the Crazy Mountain Field and Kutz Canyon faunas could be one of environment rather than due to collecting methods or time differences. Indeed, the comparison of the Kutz Canyon local fauna with the faunas of the Gidley and Silberling Quarries seems to duplicate the comparison that Simpson (1937) made between these two quarries and nearby surface localities. He discovered that surface localities contained about 90% condylarths and contained larger members than did the quarry faunas, even though surface localities lacked many of the very large Paleocene taxa. He considered the surface localities to be the result of normal floodplain facies, whereas the quarries resulted from a swampy, heavily forested facies. Paucity of arboreal species from the Kutz Canyon local fauna is suggestive of the absence of a heavily forested environment in that area. Wilson (1951) considered the locality KUVP 13 fauna as not completely a floodplain facies because of the absence of *Claenodon ferox*, large *Chriacus*, *Triisodon*, and mesonychids. This argument is no longer tenable as *Chriacus pelvidens*, *Triisodon*, and *Microclaenodon* are now known from Kutz Canyon, and the absence of *Claenodon ferox* seems to be one of stratigraphy. I therefore conclude that the Kutz Canyon local fauna is one dominantly from floodplain deposits and that the fauna represents an open, shrub-inhabiting community. Some areas of black organic-rich claystone were observed in Kutz Canyon, but these paludal facies have produced only reptilian fossils.

Another difference between the faunas of the Crazy Mountain Field and Kutz Canyon is the presence of *Pantolambda* and *Claenodon* in the former. These taxa are characteristic of the San Juan Basin *Pantolambda* zone. Also, the characteristic *Deltatherium* zone taxa, *Deltatherium*, *Triisodon*, and *Haploconus*, are represented in the Crazy Mountain Field Torrejonian fauna by a single *Deltatherium* tooth of questionable provenance (Van Valen, 1978). This leaves open the possibility that the Kutz Canyon local fauna is older than Torrejonian faunas from Montana. Of course, an independent study of the Crazy Mountain Field paleomagnetic stratigraphy or radiometric dating is required for resolution of its age.

The Crazy Mountain Field fauna does not appear to contain higher percentages of orders of small mammals than does the San Juan Basin *Pantolambda* zone. Screening has, however, been more productive in the *Pantolambda* zone from near the head of Torreon Arroyo than in Kutz

Canyon and should provide a broader perspective of small mammals. Further comparison of the Crazy Mountain Field and San Juan Basin faunas will then be possible.

EVOLUTION OF *PERIPTYCHUS* AND *TETRACLAENODON*

Some 58 specimens of *Periptychus* and 68 specimens of *Tetraclaenodon* from the Kutz Canyon local fauna were measured and subjected to stratophenetic analysis as used by Gingerich (1976). The log of M_1 length times width was plotted against stratigraphic elevation for both genera. The results indicate that no morphometric trends were present in either genus in the *Deltatherium* zone in Kutz Canyon. The reversed magnetozone containing most of the Kutz Canyon local fauna is probably less than two million years in duration (LaBrecque et al., 1977). Further analyses of this type will be conducted on various taxa through the complete San Juan Basin Torrejonian section.

The observed ranges for dentitions of *Tetraclaenodon* were well within that given by West (1976) for the single species *Tetraclaenodon puercensis*. Bivariate analysis of molar length and width reconfirmed West's synonymies.

The observed range of *Periptychus* was slightly above and below that given in Simpson (1959) for San Juan Basin specimens. However, bivariate plots of molar length versus width indicated that only one species is represented. This is *Periptychus carinidens*, as suggested by Simpson (1959).

SUMMARY

Over 400 mammalian fossil specimens were collected from 33 University of Arizona vertebrate fossil localities within a stratigraphic interval of 165 m. Three UALP localities were situated in strata deposited during magnetic anomaly 27. The others were from within the reversed magnetozone overlying anomaly 27. Sediments deposited during anomaly 27 time are of "Dragonian age" (see Tomida, this volume) and those correlated with the reversed magnetozone overlying anomaly 27 are from the Torrejonian *Deltatherium* zone.

The earliest San Juan Basin occurrence of *Psittacotherium multifragum* is from the *Deltatherium* zone of Kutz Canyon. *Protictis*, *Triisodon*, cf. *Mioclaenus*, and *Anisonchus sectorius* first appeared in the San Juan Basin during "Dragonian time" in Kutz Canyon. Several other taxa may have made their first San Juan Basin appearance in Kutz Canyon, but the unknown stratigraphic position of a Torrejonian *Deltatherium* zone locality from Kimbetoh Arroyo could represent their first

San Juan Basin occurrences. The Kimbetoh Arroyo locality also precludes similar interpretation of the highest stratigraphic occurrence of mammals in Kutz Canyon.

Eucosmodon sp., *Torrejonia* cf. *T. wilsoni*, *Protictis ?vanvaleni*, *Conoryctes comma*, *Microclaenodon assurgens*, and *Haploconus angustus* are reported from Kutz Canyon for the first time.

The Kutz Canyon local fauna is dominated by the presence of condylarths with few taxa of small mammals. This differs greatly from the composition of the Torrejonian fauna from the Gidley and Silberling Quarries in Montana, where multituberculates, insectivores, and primates are nearly co-dominant with condylarths. The difference between these faunas may be due to environmental factors, temporal factors, or both. The small, arboreal species characteristic of the Montana faunas are poorly represented in the Kutz Canyon local fauna due to the probable absence of a heavily forested environment in that area.

No morphometric trends were recognized in *Periptychus carinidens* or *Tetraclaenodon puercensis* from the *Deltatherium* zone of Kutz Canyon.

ACKNOWLEDGMENTS

I thank Dr. Everett H. Lindsay, Dr. George Gaylord Simpson, Dr. Louis L. Jacobs, and Yukimitsu Tomida for valuable suggestions concerning the writing of this paper; Dr. Everett H. Lindsay, Dr. Robert F. Butler, Dr. Louis L. Jacobs, James G. Honey, Lawrence J. Flynn, Barbara R. Standhardt, Yukimitsu Tomida, Richard A. Haskin, and Robert D. McCord for valuable assistance in the field; Kevin B. Moodie for specimen preparation; George S. Marcek for photographs and photographic advice; and Earl Manning for help in specimen identification.

Financial support for this project was provided by The State of Arizona and by the National Science Foundation, Grants DES-75-13616 and EAR-78-03326.

REFERENCES

Butler, R. F. and L. H. Taylor. 1978. A middle Paleocene paleomagnetic pole from the Nacimiento Formation, San Juan Basin, New Mexico. Geology 6:495–498.

Gazin, C. L. 1968. A new primate from the Torrejonian middle Paleocene of the San Juan Basin, New Mexico. Proc. Biol. Soc. Wash. 81:629–634.

Gingerich, P. D. 1976. Cranial anatomy and evolution of early Tertiary Plesiadapidae (Mammalia, Primates). Papers on Paleont., Mus. Paleontol., Univ. Michigan 15:1–141.

Granger, W. 1917. Notes on Paleocene and lower Eocene mammal horizons of northern New Mexico and southern Colorado. Bull. Am. Mus. Nat. Hist. 37:821–830.

Granger, W. and G. G. Simpson. 1929. A revision of the Tertiary Multituberculata. Bull. Am. Mus. Nat. Hist. 61:601–676.

Kay, R. F. and M. Cartmill. 1977. Cranial morphology and adaptations of *Palaechthon nacimienti* and other Paromomyidae (Plesiadapoidea, ?Primates), with a description of a new genus and species. J. Human Evol. 6:19–53.
LaBrecque, J. L., D. V. Kent, and S. C. Cande. 1977. Revised magnetic polarity time scale for Late Cretaceous and Cenozoic time. Geology 5:330–335.
Lindsay, E. H., R. F. Butler, and N. M. Johnson. 1981. Magnetic polarity zonation and biostratigraphy of Late Cretaceous and Paleocene continental deposits, San Juan Basin, New Mexico. Am. J. Sci. 281:390–435.
Matthew, W. D. 1937. Paleocene faunas of the San Juan Basin, New Mexico. Am .Phil. Soc. Trans., n.s. 30:1–510.
MacIntyre, G. T. 1966. The Miacidae (Mammalia, Carnivora), Part 1. The systematics of *Ictidopappus* and *Protictis*. Bull. Am. Mus. Nat. Hist. 133:115–210.
Osborn, H. F. 1929. The titanotheres of ancient Wyoming, Dakota, and Nebraska. U.S. Geol. Survey Monograph 55:1–701.
Osborn, H. F. and W. D. Matthew. 1909. Cenozoic mammal horizons of western North America. U.S. Geol. Survey Bull. 361:1–138.
Simons, E. L. 1963. David Baldwin, O. C. Marsh and the discovery of the first continental Paleocene faunas of the New World. Postilla 75:1–11.
Simpson, G. G. 1937. The Fort Union of the Crazy Mountain Field, Montana and its mammalian faunas. U.S. Natl. Mus. Bull. 169:1–287.
———. 1959. Fossil mammals from the type area of the Puerco and Nacimiento strata, Paleocene of New Mexico. Am. Mus. Novitates 1957:1–22.
Taylor, L. H. and R. F. Butler. 1980. Magnetic-polarity stratigraphy of Torrejonian sediments, Nacimiento Formation, San Juan Basin, New Mexico. Am. J. Sci. 280: 97–115.
Tomida, Y. and R. F. Butler. 1980. Dragonian mammals and Paleocene magnetic polarity stratigraphy, North Horn Formation, central Utah. Am. J. Sci. 280:787–811.
Van Valen, L. 1978. The beginning of the age of mammals. Evol. Theory 4:45–80.
West, R. M. 1976. The North American Phenacodontidae (Mammalia, Condylarthra). Contr. Biol. Geol. Milwaukee Publ. Mus. 6:1–78.
Wilson, R. W. 1951. Preliminary survey of a Paleocene faunule from the Angels Peak area, New Mexico. Univ. Kansas Publ., Mus. Nat. Hist. 5:1–11.
———. 1956a. A new multituberculate from the Paleocene Torrejonian fauna of New Mexico. Kansas Acad. Sci. Trans. 59:76–84.
———. 1956b. The condylarth genus *Ellipsodon*. Univ. Kansas Publ., Mus. Nat. Hist. 9:105–116.
———. 1956c. Additional remains of the multituberculate *Eucosmodon*. Univ. Kansas Publ., Mus. Nat. Hist. 9:117–123.
Wilson, R. W. and F. S. Szalay. 1972. New paromomyid primate from middle Paleocene beds, Kutz Canyon area, San Juan Basin, New Mexico. Am. Mus. Novitates 2499:1–18.
Wood, H. E., 2nd, R. W. Chaney, J. Clark, E. H. Colbert, G. L. Jepsen, J. B. Reeside, Jr., and C. Stock. 1941. Nomenclature and correlation of the North American continental Tertiary. Geol. Soc. Am. Bull. 52:1–48.

12 Mammalian biostratigraphy of the Middle Paleocene (Torrejonian) strata of the San Juan Basin: Notes on Torreon Wash and the status of the *Pantolambda* and *Deltatherium* faunal "zones"

Costas Tsentas
Department of Anthropology, New York University
New York, New York

INTRODUCTION

The San Juan Basin, occupying the entire northwest quarter of the state of New Mexico and the southwest corner of Colorado, has some of the best middle Paleocene exposures in the world. The type area of the North American middle Paleocene (Torrejonian) Land Mammal "Age" (Wood et al., 1941; Evernden et al., 1964) is found here, in beds of the upper part of the Nacimiento Formation at the head of Torreon Wash. Sinclair & Granger (1914), working in this area of the wash, discovered what they believed to be two distinct faunal horizons in these beds. Up until that time the Torrejon "phase" of the "Basal Eocene" was known as the "*Pantolambda* zone" (Osborn & Matthew, 1909) because the species of *Pantolambda* were restricted to that particular horizon. However, from their studies in the San Juan Basin in 1912 and 1913, Sinclair & Granger (1914) determined that there was a second faunal horizon in the Torrejon, located about 30 m below the original horizon and characterized by the presence of *Deltatherium*. From that time to the present there have been questions as to the true nature of the Torrejonian "zones." The problem of whether the *Pantolambda* (upper) and *Deltatherium* (lower) faunal "zones" have a temporal relationship, or instead reflect facies differences, or perhaps are even a product of collecting biases has not been satisfactorily resolved. The answer is of importance to paleontology. A temporal relationship would mean possible evolutionary relationships between the taxa of the two "zones." A facies relationship would mean environmental differences and, therefore, adaptive differences between the faunas of the two "zones." If collecting bias is at work, then, of course, there are no zones. The nature of the "zones" must be understood if we are to comprehend the past history of life in the basin. The problem is a difficult one to resolve, in part because of the failure of most early workers to keep detailed notes of the localities, horizons and strata from which fossils were obtained. Also, the Torrejonian fauna has been collected mainly in only five areas: the Animas River valley, Kutz Canyon, Kimbeto Arroyo, the headlands of Escavada Wash southeast of Kimbeto, and Torreon Wash (Text-fig. 12.1). The *Pantolambda* "zone" is limited to Torreon Wash, the area southeast of Kimbeto, and a small area just east of Cedar Hill in the upper Animas River valley (Granger, 1913, 1916, 1917); the remainder of the collecting areas are considered as part of the *Deltatherium* "zone," including all of Kutz Canyon, the lower Animas River valley, Kimbeto Arroyo, and the lower fossiliferous horizon in Torreon Wash. The two "zones" are superimposed only in the area of Torreon Wash (Sinclair & Granger, 1914; Lindsay et al., 1978). Resolution of the problem is thus hampered by a lack of data on the precise provenance of the fossils, so that faunal and stratigraphic analyses in terms of fossils are not easily done, and by the fact that the "zones" occur together at only a single location.

The problem is reviewed here by presenting a historical background of the development of our knowledge of the geology and paleontology of the San Juan Basin middle Paleocene followed by a discussion of two different concepts of the "zones." A new screening site in Torreon Wash that adds to our knowledge of the "zones" is reported, along with detailed measured geologic sections of strata of the Nacimiento Formation on the east and west flanks of Torreon Wash; an updated faunal list of the fossil mammals found in the Torrejonian of the San Juan Basin is also provided.

Abbreviations used in the text are:

AMNH American Museum of Natural History, New York.

BUNM Bureau of Land Management and University of New Mexico survey locality.

UNM University of New Mexico, Albuquerque, Department of Geology vertebrate fossil collection.

HISTORICAL BACKGROUND

The historical development of our knowledge of the paleontology and geology of the San Juan Basin middle Paleocene is given in detail by Simpson (1948, 1959, this volume) and Taylor (1977). A brief historical

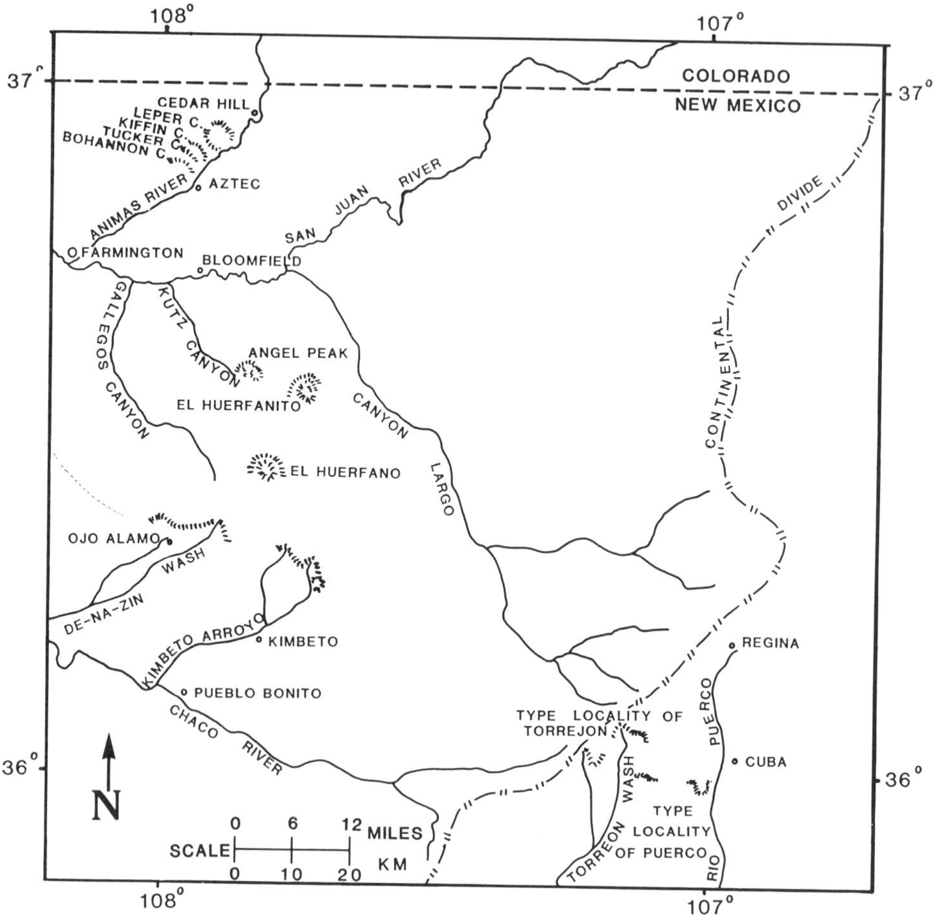

TEXT-FIG. *12.1*—Map of the east-central part of the San Juan Basin showing locations discussed in the text (modified from Matthew, 1937).

review is given here to set forth a framework for discussing the problem of the "zones."

The Tertiary deposits of the San Juan Basin first became known to modern geologists when E. D. Cope in 1874 explored a part of the Basin which he called the "Eocene Plateau" (Cope, 1875a, b, c). Cope believed the deposits to be Eocene because the "variegated marls" were similar to the Wyoming "Wahsatch" that was previously known as Eocene and because of the relative stratigraphic position of the beds and the discovery of a molar of *"Bathmodon"* (= *Coryphodon*), an Eocene pantodont. Cope named the deposits underlying the "variegated marls" in the southeast part of the Basin near Cuba (Text-fig. 12.1) "Puerco" (Spanish for muddy) because of the soapy, muddy consistency of the "marls" when wet. These "Puerco Marls," on the south side of Cuba Mesa, came to be regarded as the type section of the Puerco, which at that time was considered to be a single unit of basal Eocene or Late Cretaceous age (Simpson, 1948, 1959; Baltz, 1967).

Cope described the first fossil mammals from beds equivalent to the Puerco Marls in 1881: *Periptychus carinidens* and *Deltatherium fundaminis* (Cope, 1881a). These had been found by David Baldwin of Farmington, New Mexico, a collector hired by Cope in 1880 (Matthew, 1897; Gardner, 1910). Baldwin collected for Cope through the 1880's, and during this time Cope published a series of papers, some of them describing what later were determined to be Paleocene mammals (e.g., Cope, 1881a, b, 1882a, b, c, 1883a, b, 1884, 1885, 1886, 1887, 1888). In 1892, and again in 1896, the American Museum of Natural History sent expeditions to the San Juan Basin led by J. L. Wortman (Osborn & Earle, 1895; Matthew, 1897). Osborn & Earle studied the Puerco fossils and concluded that there were actually two distinct horizons containing widely different taxa within the Puerco. Matthew (1897) determined that the fossils came from separate upper and lower beds, each having distinct species. He credited J. L. Wortman with proposing the name "Torrejon" for the upper beds and with the retention of the name Puerco for the lower beds. The Torrejon "formation" was named for "Torrejon Arroyo," the area where the upper bed fossils were discovered (Reeside, 1924). The name Torrejon, however, as Simpson (1959) pointed out, is not quite correct since the wash is known locally as Torreon. Further, since it is mapped as Torreon Wash (U.S.G.S. Deer Mesa, New Mexico 7.5′ Quadrangle, 1966) it will be referred to as such, although many early workers called the wash Torrejon Arroyo or Arroyo Torrejon.

In 1909, the Osborn and Matthew study of North American mammal horizons designated the Torrejon as the "second phase" of the "Basal Eocene," or the "*Pantolambda* zone," distinguishing it from the Puerco or "first phase" of the "Basal Eocene" or "*Polymastodon* (= *Taeniolabis*) zone." Gardner (1910), in his study of the geology of the San Juan Basin completed several years earlier, proposed that the Torrejon and Puerco be named as parts of the Nacimiento Group, under the incorrect assumption that the two were geologic formations. Simpson (1948), however, correctly pointed out that the Torrejon is actually a fauna and not a formation. He proposed that the Nacimiento Group of Gardner (1910) be considered the Nacimiento Formation containing the Puerco and Torrejon faunas, as the Puerco and Torrejon deposits are not distinct lithologically and therefore by definition are not formations (American Commission on Stratigraphic Nomenclature, 1970).

Sinclair & Granger (1914) in 1912 and 1913 did extensive studies of the Paleocene areas of the Basin; they concluded that there were two separate horizons within the Torrejon, with *Pantolambda* restricted to the upper zone and with *Deltatherium* confined to the lower. They found that the upper horizon was fossiliferous eastward and nearly barren westward, while the lower horizon was rich in fossils in the western but not in the eastern part of the Basin. Granger (1916, 1917) returned to the San Juan Basin in 1916 and collected fossils in "Coots" (= Kutz) Canyon and the Animas River valley. He determined that the two distinct fossil horizons of the type locality in Torreon Wash were not present in either of these two areas. The fauna was described as resembling that of the "*Deltatherium* zone," except near Cedar Hill, further up the Animas River valley, where *Pantolambda* was found. Granger concluded that *Deltatherium* occurs only in the lower horizon, where it is common, while *Pantolambda* occurs only in the upper horizon. He considered that there was no really great distinction between the two horizons since they were separated by only 100 ft (about 30 m) of continuous sedimentation, the "zones" differing mainly in the relative abundance of a number of species they held in common. Osborn (1929) agreed that there probably was continuous sedimentation without a break between the two Torrejonian "zones," but he held that the faunas of the two horizons were sharply distinct. He validated naming the upper Torrejonian level the "*Pantolambda* zone" and formally named the lower the "*Deltatherium* zone." The "*Pantolambda* zone" was characterized by the presence of *Pantolambda cavirictus*, "*Claenodon*" (=*Arctocyon*) *ferox* (see Table 12.1) and *Tetraclaenodon puercensis*; the "*Deltatherium* zone" was characterized by the presence of *Deltatherium fundaminis*, *Mioclaenus turgidus*, and *Haploconus lineatus*. However, Matthew (1937), who did a comprehensive study of the Paleocene fossils of the San Juan Basin, concluded that there were no valid zonal distinctions. He contended that there was no evidence that the upper level fauna was of a significantly later age and stated that certain species occurred in both levels without discernible morphological difference; any differences in fauna were thus due to either collecting bias or facies differences.

In 1941 the Torrejonian was formalized as a provincial time term by the Wood Committee (Wood et al., 1941) based on the Torrejon "formation" in the San Juan Basin, with the heads of Torreon Wash (Pl. 12.1) as the type locality. The Torrejonian was recognized by the index fossils *Conoryctes*, *Deltatherium*, *Mioclaenus*, *Mixodectes*, and *Triisodon*; by the first appearance of *Chriacus*, "*Claenodon*" (=*Arctocyon*), "*Didymictis*" (=*Protictis*) (Table 12.1), *Dissacus*, *Pantolambda*, *Psittacotherium*, and *Tetraclaenodon*; by the last appearance of *Ellipsodon*, *Eucosmodon*, and *Haploconus*; and by the characteristic fossils *Anisonchus*, *Periptychus*, and *Ptilodus*. The currently accepted concept of the Torrejonian was set forth in 1964 when the Torrejonian was designated

EXPLANATION OF PLATE 12.1

The type area of the Torrejonian Land Mammal "Age" (Wood et al., 1941) on the east and west flanks of Torreon Wash.

FIG. *1*—View of east flank of Torreon Wash, near head of the Wash, looking southeastward across the flats separating the two flanks; approximately located in the SE¼ Sec. 32, T21N, R4W, Deer Mesa Quadrangle.

2—West flank of Torreon Wash (in distance) as seen from the east flank (foreground) looking northwestward from Ceja Pelon Mesa, SE¼ Sec. 32 and SW¼, Sec. 33, T21N, R4W, Deer Mesa (in part) and Ojo Encino Mesa (in part) Quadrangles.

as a North American Land Mammal "Age" and was dated indirectly by the potassium/argon radioactive decay method as being about 59.5 to 62 million years old, placing it in the middle Paleocene (Evernden et al., 1964).

Table 12.1—List of fossil mammals found in the Torrejonian of the San Juan Basin, New Mexico. Sources are a: Russell, 1967; b: Taylor, 1977; c: Rigby, 1976; d: Van Valen, 1978; e: UNM locality forms; f: Matthew & Granger, 1921; g: locality BUNM-77-184; h: Wood et al., 1979; i: Kay & Cartmill, 1977; j: Tomida, this volume; k: Sloan, this volume; l: Taylor, this volume.

Taxon	Source
Class Mammalia	
Order Multituberculata	
Taeniolabididae	
Catopsalis fissidens Cope, 1884	a
Eucosmodontidae	
Stygimys teilhardi (Granger & Simpson, 1929)	a
Eucosmodon molestus (Cope, 1886)	a
Eucosmodon? sp. (Wilson, 1956c)	a
Ptilodontidae	
Ptilodus mediaevus Cope, 1881	a
Ectypodontidae	
Parectypodus trovessartianus (Cope, 1882)[1]	a
Parectypodus clemensi Sloan, 1981	k
Neoplagiaulax macrotomeus (Wilson, 1956)	a
Ectypodus szalayi Sloan, 1981	k
Mimetodon krausei Sloan, 1981	k
Cimolodontidae	
Anconodon gidleyi Sloan, 1981	k
Order Marsupialia	
Didelphidae	b
Order Carnivora	
Miacidae	
Protictis haydenianus (Cope, 1882)	c
Protictis n. sp. *a* (Wilson, 1956a)	a
Protictis n. sp. *b* (Wilson, 1956a)	a
Protictis n. sp. (Taylor, this volume)	l
Bryanictis vanvaleni (MacIntyre, 1966)	c
Order Condylarthra	
Arctocyonidae	
Chriacus baldwini (Cope, 1882)[2]	a
Chriacus pelvidens (Cope, 1881)	a
Mimotricentes subtrigonus (Cope, 1881)	a
Deltatherium fundaminis Cope, 1881	a
Arctocyon ferox (Cope, 1883)[3]	a
"*Neoclaenodon*" *procyonoides* Matthew, 1937[4]	a
Goniacodon levisanus (Cope, 1883)	a
Triisodon quivirensis Cope, 1881	a
Triisodon antiquus (Cope, 1882)	b

Triisodon crassicuspis (Cope, 1882)	a
Prothryptacodon ambiguus (Van Valen, 1967)[5]	d
Deuterogonodon noletil Van Valen, 1978	d
Loxolophus sp. Cope, 1885	e
cf. *Loxolophus kimbetovius* (Matthew, 1937)	j
Hyopsodontidae	
Mioclaenus turgidus Cope, 1881[6]	a
Mioclaenus? n. sp.? (Wilson, 1956a)	a
Ellipsodon inaequidens (Cope, 1884)	a
Ellipsodon grangeri Wilson, 1956b	a
Ellipsodon yotankae Van Valen, 1978	d
Promioclaenus acolytus (Cope, 1882)	a
Promioclaenus aequidens (Matthew, 1937)	a
Promioclaenus lemuroides (Matthew, 1897)	a
Protoselene opisthacus (Cope, 1882)	a
Litomylus osceolae Van Valen, 1978	d
n. gen. et sp. (Taylor, this volume)	l
Phenacodontidae	
Tetraclaenodon puercensis (Cope, 1881)	a
n. gen. et sp. (Wilson, 1956a)	a
Periptychidae	
Anisonchus sectorius (Cope, 1881)	a
Haploconus angustus (Cope, 1881)	a
Haploconus corniculatus Cope, 1888	a
Periptychus carinidens Cope, 1881	a
Oxyacodon tecumsae Van Valen, 1978	d
Mesonychidae	
Dissacus navajovius (Cope, 1881)	a
Ankalagon saurognathus (Wortman, in Matthew, 1897) Van Valen, 1980	a
Microclaenodon assurgens (Cope, 1884)	a
Order Deltatheridia	
Palaeoryctidae	
Acmeodon secans Matthew & Granger, 1921	f
Palaeoryctes puercensis Matthew, 1913	a
cf. *Gelastops* sp. Simpson, 1935	e
Order Insectivora	
Leptictidae	
Prodiacodon puercensis (Matthew, 1918)	a
Prodiacodon n. sp.? (Wilson, 1956a)	a
Pentacodontidae	
Pentacodon inversus (Cope, 1888)	a
Pentacodon occultus Matthew, 1937	a
Pentacodon n. sp. (Wilson, 1956a)	a
Coriphagus encinensis (Matthew & Granger, 1921)	a
Mixodectidae	
Mixodectes pungens Cope, 1883[7]	a
Mixodectes malaris (Cope, 1883)	a
Order Pantodonta	
Pantolambdidae	
Pantolambda bathmodon Cope, 1882	a
Pantolambda cavirictus Cope, 1883	a

Order Primates	
Paromomyidae	
Palaechthon nacimienti Wilson & Szalay, 1972	b
Palaechthon woodi Gazin, 1971	h
Palaechthon problematicus (Jepsen, 1930) Rigby, 1980[8]	b
Talpohenach torrejonius Kay & Cartmill, 1977	i
cf. *Paromomys* sp. (Tomida, this volume)	j
Order Taeniodonta	
Stylinodontidae	
Conoryctes comma Cope, 1881	a
Psittacotherium multifragum Cope, 1882[9]	a
n. gen. et sp. (Wilson, 1956a)	a

— NOTES —

1. = *Mimetodon trovessartianus* (Van Valen & Sloan, 1966).
2. "*Tricentes*" *truncatus* in Taylor (1977) should have been *Chriacus truncatus* (Van Valen & Sloan, 1965; Wilson, 1956a) as *Tricentes* is a junior synonym of *Chriacus* (Van Valen, 1978). *Tricentes crassicollidens* is a junior synonym of *Chriacus truncatus* (Van Valen & Sloan, 1965) which in turn is a junior synonym of *C. baldwini* (Van Valen, 1978). *C. baldwini* had been sunk into *C. pelvidens* by Rigby (1976) but was resurrected by Van Valen (1978).
3. According to Russell (1964) *Claenodon* Scott, 1892 is a junior synonym of *Arctocyon* Blaineville, 1841, but according to Van Valen (1978), *Claenodon* is a junior synonym of *Arctocyonides* Lemoine, 1891. Rigby (1980) agrees with Russell (1964) that *Claenodon* is a junior synonym of *Arctocyon*. The latest revision is followed here.
4. = "*Colpoclaenus*" *procyonoides* (Van Valen, 1978).
5. *Pantinomia ambigua* is a synonym (Van Valen, 1978).
6. Includes *Mioclaenus lydekkerianus* (Van Valen, 1978).
7. *Mixodectes crassiusculus* is a synonym (Szalay, 1969).
8. *Torrejonia wilsoni* is a junior synonym of *Palaechthon problematicus* according to Rigby (1980). *Torrejonia* was considered to be a junior synonym of *Plesiolestes* by Szalay (1973) but both are junior synonyms of *Palaechthon* according to Rigby (1980).
9. *Psittacotherium aspasiae* is a junior synonym (Schoch, this volume).

THE *DELTATHERIUM* AND *PANTOLAMBDA* FAUNAL "ZONES"

The most comprehensive studies of the Torrejonian "zones" were done by Wilson (1951, 1956a). He reported finding a Torrejonian fauna in Kutz Canyon at a locality he called Angels Peak pocket (Wilson, 1951). Among the taxa found in this pocket were *Mixodectes malaris*, *Mimotricentes subtrigonus*, *Chriacus* "*truncatus*" (= *C. baldwini*), *Mioclaenus turgidus*, *Promioclaenus acolytus*, *Protictis*, *Protoselene opisthacus*, *Tetraclaenodon puercensis*, *Deltatherium fundaminis*, and the first two primates ever found in the middle Paleocene of the San Juan Basin, later identified as *Palaechthon nacimienti* (Wilson & Szalay, 1972) and *Talpohenach torrejonius* (Kay & Cartmill, 1977). Wilson considered this fauna close to the "*Deltatherium* zone" fauna and as having come from a probable floodplain facies. He attributed the complete absence of *Haploconus*, a genus supposedly diagnostic of the

"*Deltatherium* zone" (Osborn, 1929), to some facies or environmental factors.[1] In Kimbeto Arroyo, however, *Haploconus angustus* specimens constituted 27% of the fossils that were found in association with a new multituberculate described by Wilson (1956a). Apparently Wilson felt that there can be a wide range of variation between localities within a "zone."

A study of the Torrejonian collections at the American Museum of Natural History, the U.S. National Museum, and the University of Kansas (Wilson, 1956a) determined that *Mimotricentes subtrigonus*, *Mioclaenus turgidus*, and *Tetraclaenodon puercensis* seem to be present throughout the Torrejonian in the San Juan Basin; *Deltatherium*, *Triisodon*, and *Haploconus* are restricted to the lower "zone," and *Pantolambda* and *"Claenodon"* (=*Arctocyon*)[2] are restricted to the upper "zone." The larger form of *Promioclaenus*, *P. lemuroides*, was common in the *Pantolambda* "zone" but rare in the *Deltatherium* "zone" (Wilson, 1956a). The "zones" were thus interpreted by Wilson as facial in nature, with the "*Deltatherium* zone" fauna representing a forest border environment, while the "*Pantolambda* zone" fauna represented a more riparian setting. These interpretations may be valid. However, when attempting to determine the facial or ecological context of Paleocene fossils, one must question whether definitive statements as to habitat and ecology of these early mammals can be made, as the incomplete anatomical evidence of most forms hinders firm functional-ecological conclusions. We do not even fully understand the relationship between tooth morphology and diet (e.g., Van Valen & Sloan, 1966). Simons (1960), in his study of the Pantodonta, presented good evidence for making an analogy between *Pantolambda* and the capybara, *Hydrochoerus*, which is a semi-aquatic inhabitant of dense forests near bodies of water (Walker, 1975). It is interesting to note that a reconstruc-

---

[1]But see Taylor (this volume) for reports of *Haploconus* from Kutz Canyon.

[2]Wilson (1956a) considered *"Claenodon"* (=*Arctocyon*) along with *Pantolambda* as restricted to the upper "zone." Granger (1917, p. 824), in the report of his collecting along the lower Animas River valley in 1916, described an instance where *Arctocyon* apparently was found below *Deltatherium* at a draw south of the Aztec station. *Arctocyon* was obtained from the "lower element" of this section while *Deltatherium fundaminis* was found at the head of this same draw at a "somewhat higher level." Wilson (1956a, p. 83) stated that certain identifications of *"Claenodon"* (=*Arctocyon*) by Granger were questionable, and that he (Wilson) had not yet seen a specimen positively identified as *"Claenodon"* (=*Arctocyon*) and positively coming from the lower "zone." A check of specimens at the American Museum of Natural History verified that AMNH 17037 is correctly identified as *Deltatherium* from "Along Animas River . . . 2 mi. S-E of Aztec, upper level" in Granger's (1916) field notebook. However, no fossils that are listed in Granger's field notebook as coming from the same area have been subsequently identified as *Arctocyon*. This corroborates Wilson's contentions regarding the identification of specimens and the relative positions of *Arctocyon* and *Deltatherium*.

tion of *Pantolambda cavirictus* (Simons, 1960, p. 64) closely resembles the capybara in size and appearance (Walker, 1975, p. 1021). Simons thus tended to agree with Wilson (1956a) that *Pantolambda* may have been riparian, pointing out several similarities between the feet of *Hydrochoerus* and pantolambdids, especially the similar shape of the unguals, the relative length of the phalanges and the sub-equal size of the digits. All possibly indicate similar habits of the two animals; the feet of both groups have small hooves and relatively separable digits which could splay out and provide support on marshy ground as an adaptation to a riparian habitat. These apparent adaptive parallels are of interest but cannot help solve the problem of the "zones" without similar data for other animals in each of the "zones" for comparison; unfortunately very few such data exist.

Finally, Wilson (1956a) in faunal analyses of the "zones" using "available material," determined that about one out of three species present in the lower level were absent from the upper level; he interpreted this as evidence that there was no significant span of time separating the "zones." Thus it is apparent that Wilson (1951, 1956a, personal communication cited in Taylor & Butler, 1980) considered the "zones" to be real but concluded from his studies that a facies difference is the primary factor determining the faunal differences and not a time difference. This essentially agrees with the conclusions of Granger (1917) that no great length of time separates the "zones," and with Matthew (1937) and Russell (1967), who held that a facies difference between the "zones" is the most likely explanation for the faunal differences.

A different conclusion is reached by Taylor (1977, this volume), Taylor & Butler (1980) and Tomida (this volume) in their paleomagnetic studies of the Ojo Encino area (head of Torreon Wash), the Big Pocket area of Kutz Canyon, and the upper reaches of Kimbeto Arroyo. They argue against a facies relationship between the "zones" by pointing out that there are no discernible lithological differences between the two horizons. This argument is not unreasonable since it is difficult to postulate past environmental differences without any correlative lithological differences. The paleomagnetic studies (e.g., Taylor & Butler, 1980) recognize two levels in the Torrejonian: an upper *Pantolambda* "zone" characterized by *Pantolambda* and "*Claenodon*" (=*Arctocyon*) and a lower *Deltatherium* "zone" defined by the presence of *Deltatherium*, *Triisodon*, *Haploconus* and *Mixodectes malaris*. The studies of the three areas essentially show that in one instance the *Pantolambda* "zone" is within a normal polarity magnetozone, while the *Deltatherium* "zone" in three separate areas is within a reversed polarity magnetozone. The *Pantolambda* "zone" on the west flank of Torreon Wash (Locality 10 of Sinclair & Granger, 1914) was found to be in a normal polarity magnetozone directly overlying a reversed polarity magnetozone which contained the *Deltatherium* "zone." It is shown that the *Pantolambda* "zone" on the east flank of Torreon Wash (Locality 11 of Sinclair & Granger, 1914) is also within a normal polarity magnetozone, and it is assumed that this is the same magnetozone as the one containing the *Pantolambda* "zone" on the west flank. This is reasonable, judging from the lithologic and faunal correlations between the two flanks (Taylor & Butler, 1980). The *Deltatherium* "zone" in the Big Pocket area of Kutz Canyon and in Kimbeto Arroyo, like the *Deltatherium* "zone" on the west flank[3] of Torreon Wash, is in a reversed polarity magnetozone. It is concluded that since the *Pantolambda* and *Deltatherium* "zones" are in different magnetozones in Torreon Wash, they are separated chronologically throughout the basin, as a magnetozone boundary is a time line.

Viewed objectively, the evidence shows only that one *Pantolambda* producing horizon (it is shown that it is the same *Pantolambda* producing horizon on both sides of the Wash) is in a normal polarity magnetozone, while three separated *Deltatherium* producing horizons are in reversed polarity magnetozones. It is assumed that because one *Pantolambda* producing horizon is in a given magnetozone, the other occurrences of the *Pantolambda* producing horizon are in the same magnetozone; it is thus assumed that the *Pantolambda* "zone" is a single horizon occurring in the same stratigraphic position throughout the Basin. Taylor (1977, this volume), Taylor & Butler (1980) and Tomida (this volume) appear to accept this premise as fact; there is no evidence that the *Pantolambda* "zone" is a single horizon at a single stratigraphic level everywhere in the San Juan Basin. There may well be *Pantolambda* "zone" occurrences that are in different normal or reversed magnetozones; since only one *Pantolambda* "zone" was tested, it is difficult to see how it is concluded that the two faunal "zones" can "now be defined by their magnetic polarity stratigraphy" (Taylor, 1977, p. 63). The evidence could be interpreted differently: it could be that the faunal "zones" indicate facies changes which occurred intermittently throughout the Torrejonian, with some facies changes occurring later than others, therefore appearing higher in the section. Finding a *Pantolambda* "zone" occurrence in a magnetozone above a magnetozone containing a *Deltatherium* "zone" occurrence could be evidence of a facies

[3]The *Deltatherium* "zone" on the east flank of Torreon Wash (locality 11 of Sinclair & Granger, 1914) was not measured for the paleomagnetic study because, according to Taylor (1977, p. 45) and Taylor and Butler (1980, p. 107), it is not present at locality 11. The *Deltatherium* "zone" *is* well represented on the east flank of Torreon Wash at locality 11 according to Sinclair and Granger (1914, p. 316) and according to Granger's (1913) record of specimens, which lists 22 specimens, including *Deltatherium* (AMNH 16610), from the "East fork Torrejon Arroyo Lower Level." Locality 11 is one of only two localities where both the *Pantolambda* and *Deltatherium* "zones" are present.

change that took place later in time. It must be shown that a *Pantolambda* "zone" at least in more than one locality exists in a subsequent magnetozone above a *Deltatherium* "zone," and the magnetozones must be correlated between localities and shown to be the same reversed and normal magnetozones from locality to locality. The paleomagnetic evidence would be more acceptable if a larger number of localities were measured, including the *Pantolambda* producing horizons southeast of Kimbeto (Escavada Wash) and the area east of Cedar Hill (Granger, 1913, 1916, 1917), and if it were thereby repeatedly shown that the *Pantolambda* "zone" was in a later magnetozone than the *Deltatherium* "zone."

It should be noted that Taylor (1977) and Taylor & Butler (1980), apparently following Szalay (1969), considered *Mixodectes malaris* as one of the taxa characterizing the *Deltatherium* "zone." This is shown here to be incorrect by the findings from a new site located within the *Pantolambda* "zone" in Torreon Wash.

A NEW SCREENING SITE IN TORREON WASH

Fauna.—UNM field locality BUNM-77-184 (in previous works referred to as BUNM-77-199a) is currently being screened by New York University and Brown University. The site (Pl. 12.2) is located in the SE¼ NW¼, Sec. 27, T21N, R5W, U.S.G.S. Deer Mesa, New Mexico 7.5′ Quadrangle (1966). Screening is being undertaken by the author and Dr. G. C. Conroy of Brown University under Antiquities Permit No. 79-NM-066.

The locality, discovered in 1977 by Spencer Lucas and Eric Devor, is located on the west flank of Torreon Wash within the "*Pantolambda* zone" (Text-fig. 12.2), about 1.7 km west of the head of Toledo Arroyo (Text-fig. 12.3). Screening began in the summer of 1978 and has shown the site to be highly fossiliferous; the equivalent of only one ton of matrix has been picked, but already several hundred mammal teeth have been found. The most common taxon represented is *Mixodectes*, including both *M. malaris* and *M. pungens*. A preliminary faunal list of the mammals found includes: *Ptilodus mediaevus*, *Ectypodus*? sp., Eucosmodont?, *Mimotricentes subtrigonus*, *Arctocyon ferox*, *Goniacodon levisanus*, *Chriacus baldwini*?, *Tetraclaenodon puercensis*, *Promioclaenus lemuroides*, *P. acolytus*?, *Protoselene opisthacus*, *Periptychus* sp., *Acmeodon* sp., cf. *Gelastops* sp., *Mixodectes malaris*, *M. pungens*, *Pantolambda cavirictus*?, *Palaechthon woodi*, "*Torrejonia wilsoni*" (= *Palaechthon problematicus*) (Table 12.1), and Taeniodonta?. Note that *Mixodectes malaris*, *M. pungens*, *Arctocyon ferox*, and *Pantolambda* are present, evidence that *Mixodectes malaris* is not a good in-

1

2

EXPLANATION OF PLATE 12.2

FIG. *1*—Screening locality BUNM-77-184, seen looking southward as a flattened area (indicated by arrow) after several cm of matrix had been removed. The gray-white clayey sandstones covered with brown calcitic concretions are typical of the *Pantolambda* "zone" in Torreon Wash.

2—Screening locality BUNM-77-184, looking northward. Site is flat area (indicated by arrow) immediately beyond 1.8 m tall human provided for scale. Note the massive sandstones of the Cuba Mesa Member of the San Jose Formation overlying the Nacimiento Formation in the background.

dicator of the *Deltatherium* "zone," *contra* Taylor (1977) and Taylor & Butler (1980). It had previously been thought that *Mixodectes malaris* and *M. pungens* do not occur in association (Szalay, 1969), but this is not the case, as both *M. malaris* (UNM NP-339, B-926) and *M. pungens* (UNM NP-337, NP-338) are found at the BUNM-77-184 screening site. The two species are also found associated at locality BUNM-77-183 in the NW¼ NE³¼, Sec. 5, T20N, R4W, Ojo Encino Mesa, New Mexico 7.5′ Quadrangle (1961), *M. malaris* specimen numbers (UNM) B-1354B, B-199G; *M. pungens* number B-199E, according to UNM locality forms.[4]

A relatively large number of *Mixodectes* specimens have been recovered, and no doubt more will be found. A large sample of this genus may allow the determination of whether *M. malaris* and *M. pungens* are valid species. The teeth of the two groups are very similar morphologically (Szalay, 1969), being differentiated mainly on the basis of size (*M. pungens* is larger); a large sample should show if there is overlap of the size ranges of the two groups or if there is a bimodal size distribution, indicating that there are two species represented.

New primates.—The new site (BUNM-77-184) has produced two primate taxa, the first middle Paleocene primates reported from the Basin in the last twelve years (Wood et al., 1979), "*Torrejonia wilsoni*" (Gazin, 1968) and *Palaechthon woodi* (Gazin, 1971). These two taxa with *Palaechthon nacimienti* (Wilson & Szalay, 1972), cf. *Paromomys* sp. (Tomida, this volume), and *Talpohenach torrejonius* (Kay & Cartmill, 1977) constitute the known primate fauna from the Torrejonian of the San Juan Basin (Conroy, this volume). "*Torrejonia wilsoni*" and *Palaechthon woodi* are found in the *Pantolambda* "zone," while *Palaechthon nacimienti* and *Talpohenach torrejonius* are found in the *Deltatherium* "zone" (Kay & Cartmill, 1977). "*Torrejonia*" (Taylor, this volume) and cf. *Paromomys* sp. (Tomida, this volume) apparently also come from the *Deltatherium* "zone." *Palaechthon woodi* was previously unknown from the middle Paleocene of the San Juan Basin; the only other occurrence of this taxon is in the Fort Union Formation of Wyoming. The association of *P. woodi* with "*Torrejonia wilsoni*" in the type area of the Torrejonian in Torreon Wash is of interest in terms of biostratigraphic correlation; the two taxa are otherwise found associated only in the Shotgun local fauna of the Fort Union Formation of Wyoming (Wood et al., 1979).

Geologic section.—Detailed stratigraphic sections (Text-fig. 12.2;

---

[4]UNM locality forms give detailed information on all UNM localities, including precise map coordinates, all specimens found, specimen numbers and identifications, formation names and stratigraphic position. They are available at the Vertebrate Paleontology Laboratory, Department of Geology, University of New Mexico, Albuquerque.

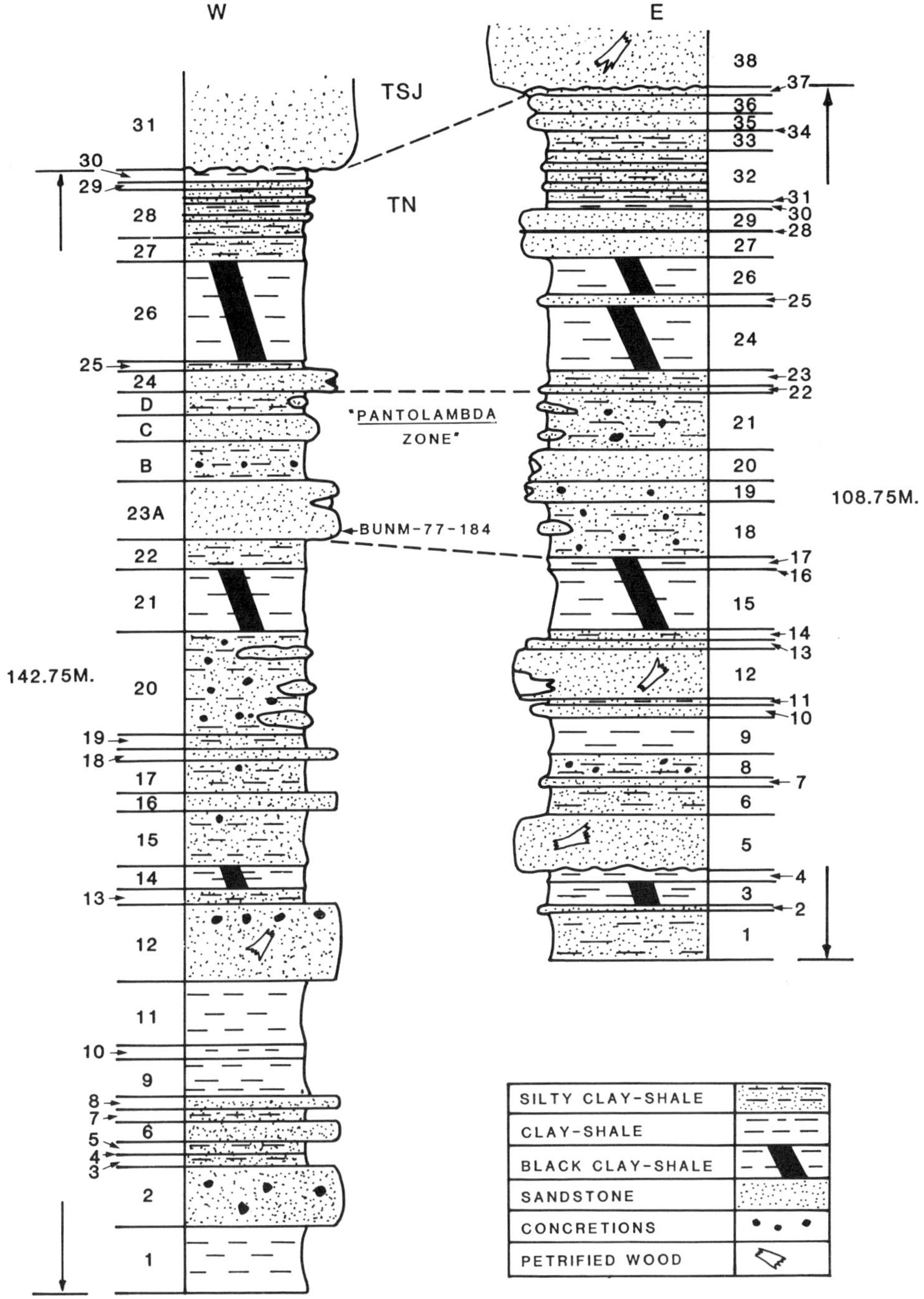

TEXT-FIG. *12.2*—Measured stratigraphic sections of the Nacimiento Formation and San Jose Formation on the east and west flanks of Torreon Wash. Numbers adjacent to each unit correlate with unit numbers assigned in the description of the lithology in Appendix 12.1. See Text-fig. 12.3 for precise location of sections. W: west flank; E: east flank; TSJ: San Jose Fm.; TN: Nacimiento Fm.

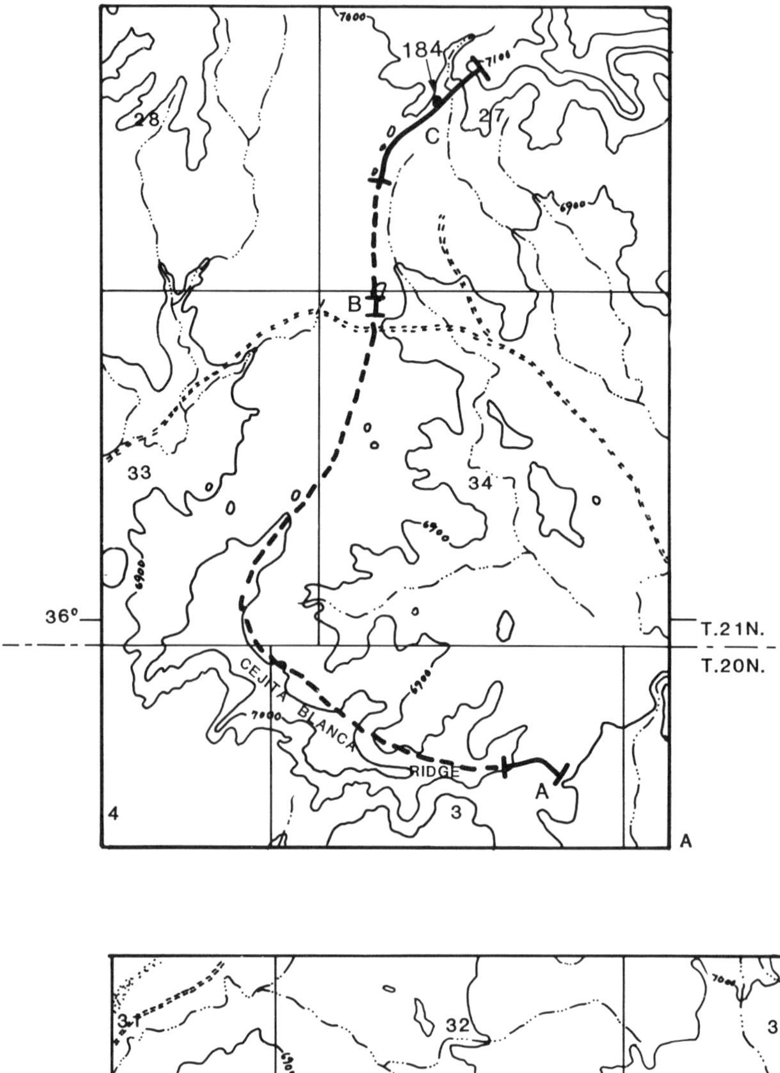

7000
184
7100
28
27
C
6900
B
33
34
6900
6900
36°
T.21N.
T.20N.
CEJITA BLANCA
7000
6900
RIDGE
A
4
3
A

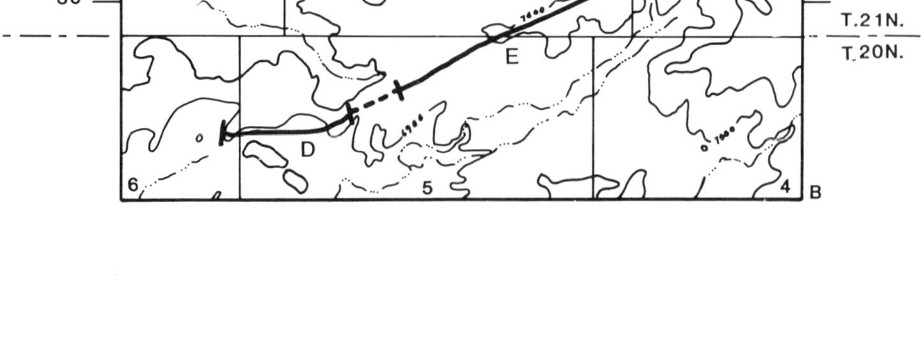

7000
31
32
33
6900
6900
36°
7000
T.21N.
T.20N.
E
6900
7000
D
6
5
4
B

Appendix 12.1) were measured on both the west and east flanks of Torreon Wash (Pl. 12.1; Text-fig. 12.3) in order to place the screening site within the section and to provide information on the stratigraphy of the type area of the Torrejonian.

BUNM-77-184 (Pl. 12.2, fig. 1) is located in unit 23a of the west flank section in a gray-white sandstone (Pl. 12.3, fig. 1) that is typical of the lithology of the *Pantolambda* "zone" in this area: red silty clay-shales interbedded with gray-white sandstones containing reddish brown calcitic concretions (Pl. 12.3, fig. 2; Pl. 12.4, fig. 3). The site is located on the top of a small ridge, but the tops of adjacent ridges are also fossiliferous and were formerly part of the same horizontal bed before erosion took place to form the ridges (Pl. 12.3, fig. 1). The actual horizontal excavation area of the main ridge being worked is about 4 m^2 (Pl. 12.2, fig. 2).

The suggested preliminary correlation of the sections (Text-fig. 12.2) links together the three black (montmorillonitic?) clay-shales. These clay-shales are carbonaceous, possibly from the deposition of large amounts of organic matter, such as occurs in a swamp (Pl. 12.3, fig. 2; Pl. 12.4, fig. 1). The black clay-shales are easily distinguished by their tendency to form a "popcorn"-like surface when dry. The brown sandstones seen in the sections are indistinguishable and variable in thickness and cannot be used for correlation. They sometimes contain large rust-colored ("cannonball") concretions (Pl. 12.4, fig. 3).

There is an apparent erosional unconformity at the top of the sections at the base of the Cuba Mesa Member of the San Jose Formation (Baltz, 1967). This unconformity is obvious when the geology is seen firsthand, at least in the area of Torreon Wash. This erosional unconformity is seen

TEXT-FIG. *12.3*—Topographic maps showing precise location of measured sections on east and west flanks of Torreon Wash. A, west flank, measured section begins in the SE¼ NE¼, Sec. 3, T20N, R5W, Ojo Encino Mesa Quadrangle (1961) and proceeds west and north to NW¼ NW¼, Sec. 34, T21N, R5W, Deer Mesa Quadrangle (1966) then to the SW¼ Sec. 27, ending in the NW¼ Sec. 27 of the Deer Mesa Quadrangle. Solid line is area of actual measurement; dashed line indicates an offset. Units 1–12 measured at point A, units 13–21 at point B, units 22–San Jose Fm. at point C. Arrow indicates location of locality BUNM-77-184. B, east flank, measured section begins in the NE¼ Sec. 6, and runs east to the NW¼ Sec. 5, T20N, R4W, Ojo Encino Mesa Quadrangle, then northeast to end in the SE¼ Sec. 32, T21N, R4W, Deer Mesa Quadrangle. Units 1–23 measured at point D, units 24–San Jose Fm. at point E. (Contour interval is 100 feet; topography from U.S.G.S. Deer Mesa and Ojo Encino Mesa Quadrangles.)

1

2

quite well in the measured sections (Text-fig. 12.2), where there is 10 m less sediment between the top of the highest black clay-shale and the Cuba Mesa Member sandstone on the west flank than on the east flank, presumably due to erosion during deposition of the sandstone (Pl. 12.2, fig. 2). Evidence of a high energy depositional environment is seen in the basal San Jose sandstone, where large prone petrified tree trunks occur (Pl. 12.4, fig. 2).

CONCLUDING REMARKS

Although certain general statements can be made concerning the fauna of the supposed *Pantolambda* and *Deltatherium* "zones" (Wilson, 1951, 1956a), no definitive answer can be given as to the nature of the "zones." It has not been shown for certain that the "zones" as such even exist. Hopefully, fossil collectors in the future will carefully note the localities, horizons, and strata where specimens were found, thus allowing faunal and stratigraphic analyses in terms of the actual fossils. If adequate evidence were available, assuming the "zones" are real, a facies relationship could be recognized if the two faunas were found in the same horizon or if the faunas were intermittently absent and present through time. If there was a temporal relationship, one would expect to find the faunas segregated stratigraphically and in the same relative position throughout the Basin. Without detailed information as to the provenience of the fossils, it is difficult to determine either facies or temporal relationships. Since there are no recognizable lithological differences between the "zones," this leaves only a functional-ecological analysis of the animals as an attempt to determine facies relationships. As adaptive differences between Paleocene mammals can only be deduced in general terms, it may not be possible to make definitive com-

EXPLANATION OF PLATE 12.3

FIG. *1*—West side of small arroyo where locality BUNM-77-184 is located (indicated by arrow). Site is within a gray-white sandstone identified as unit 23a in the measured section of the west flank of Torreon Wash (Text-fig. 12.2; Appendix 12.1); unit 22 is the dark silty clay-shale immediately below the light colored sandstone unit 23a.

2—An example of the geology of the measured section on the west flank of Torreon Wash: the lower light gray bed is the equivalent of the sandstone in which locality BUNM-77-184 is found; the arrow indicates the uppermost black clay-shale on the west flank (unit 26 in Text-fig. 12.2 and Appendix 12.1). All the units can be seen, from the light gray sandstone at the bottom (unit 23a) up to the basal Cuba Mesa Member sandstone of the San Jose Formation which caps the ridge (see Text-fig. 12.2 and Appendix 12.1).

1

2

3

parisons between faunal "zones" in terms of ecological adaptations. Phylogenetic relationships between the taxa of the two "zones" are also vague. Even if direct phylogenetic relationships were known, it would still not necessarily mean that the "zones" were temporal, since it is possible to have ancestor and descendant taxa existing at the same time but in different ecological niches. The "zones" still could be only facies.

Paleomagnetic studies hold some promise in helping to resolve the problem, but at present insufficient data exist, as well as the problem of correlating paleomagnetozones between widely separated areas. Also, the paleomagnetic studies (Taylor, 1977, this volume; Taylor & Butler, 1980; Tomida, this volume) assume as fact the very premise that should be investigated. Recognizing or defining the two "zones" in terms of magnetozone polarity stratigraphy based on a single *Pantolambda* "zone" occurrence assumes that the two "zones" are in the same relative stratigraphic position throughout the Basin—the premise that is in question. Nor has adequate evidence been presented to show why the *Pantolambda* and *Deltatherium* "zones" are accepted to be real entities. In addition, one of the taxa (*Mixodectes malaris*) used as diagnostic of the *Deltatherium* "zone" has been shown to occur in association with *Pantolambda* (BUNM-77-184 locality). The way the problem may have to be resolved is by faunal and stratigraphic analysis. Such analysis may reveal whether the localities segregate into zones or whether there was intergradation of localities.

As a matter of interest a list of taxa established in the literature as being common to both "zones" has been compiled (Table 12.2). It is less reliable to consider which taxa are not found in common, since this does not necessarily mean that the taxa do not occur together but may reflect

EXPLANATION OF PLATE 12.4

FIG. *1*—A view down-section (south) from the base of the San Jose Formation in the area of the measured section on the west flank of Torreon Wash. The dark appearing bed at left center is the middle black clay- shale (unit 21 of Text-fig. 12.2 and Appendix 12.1); the light colored bed at the lower right is the sandstone unit 23a.

2—A prone petrified log within the basal Cuba Mesa Member of the San Jose Formation capping the Nacimiento Formation in unit 38 of the section measured on the east flank of Torreon Wash (SE¼ Sec. 32, T21N, R4W). Rock hammer is 28 cm long.

3—An example of the "cannonball" concretions found within some of the sandstones of the Nacimiento Formation, in unit 12 of the section measured on the west flank of Torreon Wash (NE¼ Sec. 3, T20N, R5W). Rock hammer is 28 cm long.

collecting bias. The list is tentative and will no doubt become larger as further collecting takes place and more specimens are identified.

Table 12.2—List of mammal taxa common to both the *Pantolambda* and *Deltatherium* "zones." Sources: Granger, 1912, 1913, 1916; Matthew & Granger, 1921; Osborn, 1929; Matthew, 1937; Wilson, 1951, 1956a, b, c; MacIntyre, 1966; Van Valen & Sloan, 1966; Van Valen, 1967, 1978; Szalay, 1969; Rigby, 1976, 1980; locality BUNM-77-184.

Mioclaenus turgidus	*Chriacus baldwini*?
Tetraclaenodon puercensis	*Anisonchus*
Goniacodon levisanus	*Pentacodon*
Protictis haydenianus	*Periptychus*
Protoselene opisthacus	*Dissacus*
Mimotricentes subtrigonus	*Promioclaenus acolytus*?
Promioclaenus lemuroides	Eucosmodontidae?
Mixodectes malaris	

There is no evidence that most of the taxa commonly held to be restricted to each "zone" are anywhere found in association (*Pantolambda cavirictus*, *P. bathmodon*, and *Arctocyon ferox* of the upper "zone" and *Haploconus*, *Deltatherium*, and *Triisodon* of the lower "zone" [Osborn, 1929; Wilson, 1956a; Taylor, 1977; Taylor & Butler, 1980]). However, as noted earlier, *Mixodectes malaris* must be eliminated as a taxon diagnostic of the *Deltatherium* "zone," as also must *Mioclaenus turgidus* (Osborn, 1929), which was shown by Wilson (1956a) to not be restricted to either "zone." *Tetraclaenodon puercensis* is also present in both "zones" (Wilson, 1956a) and is eliminated from the list of taxa characterizing the *Pantolambda* "zone" (Osborn, 1929).

Finally, the separation of the *Deltatherium* "zone" based on the presence of *Deltatherium* and the absence of *Pantolambda* may not be a good distinction. *Pantolambda* and *Deltatherium* are relatively rare in the fossil record (Matthew, 1937; Simons, 1960), and it may not be wise to characterize zones by the presence or absence of rare forms. The absence of *Pantolambda* in the *Deltatherium* "zone" may be due in part to the fact that *Pantolambda* in general is rarely found. It is possible that it is only by chance of collecting that *Pantolambda* has not yet been found in association with *Deltatherium*.

ACKNOWLEDGMENTS

I wish to express my grateful thanks to Spencer Lucas for many informative discussions, and for reviewing the manuscript, and to Dr. J. Keith Rigby, Jr. and the Bureau of Land Management, Albuquerque for their logistical support. Thanks also to the Department of Vertebrate Paleontology of the American Museum of Natural History and the Department of Geology of the University of New Mexico for their assistance and for permitting me to examine the specimens in their care, and to Jeanne Tsentas for encouragement and help in so many ways.

REFERENCES

American Commission on Stratigraphic Nomenclature. 1970. Code of Stratigraphic Nomenclature. Am. Assoc. Petrol. Geol., Tulsa, 22 p.

Baltz, E. H. 1967. Stratigraphy and regional tectonic implications of part of Upper Cretaceous and Tertiary rocks east-central San Juan Basin, New Mexico. U.S. Geol. Surv. Prof. Pap. 552:1–98.

Cope, E. D. 1875a. Report on the geology of that part of northwestern New Mexico examined during the field season of 1874, p. 981–1036. *In*: G. M. Wheeler, Ann. Rept. Geographic Explorations Surv. West 100th Meridian . . . , Appendix LL, Ann. Rept. Chief Engineers, Wash., D. C.

————. 1875b. The geology of New Mexico. Proc. Acad. Nat. Sci. Phila. 27:263–268.

————. 1875c. The geology of New Mexico. Am. J. Sci. 10:152–153.

————. 1881a. Mammalia of the lower Eocene beds. Am. Nat. 15:337–338.

————. 1881b. Mammalia of the lowest Eocene. Am. Nat. 15:829–831.

————. 1882a. A new genus of Tillodonta. Am. Nat. 16:156–157.

————. 1882b. Two new genera of the Puerco Eocene. Am. Nat. 16:417–418.

————. 1882c. New marsupials from the Puerco Eocene. Am. Nat. 16:684–686.

————. 1883a. New Mammalia from the Puerco Eocene. Am. Nat. 17:191.

————. 1883b. First addition to the fauna of the Puerco Eocene. Proc. Am. Philos. Soc. 20:545–563.

————. 1884. Second addition to the knowledge of the Puerco epoch. Proc. Am. Philos. Soc. 21:309–324.

————. 1885. Marsupials from the lower Eocene of New Mexico. Am. Nat. 19:493–494.

————. 1886. The Plagiaulacidae of the Puerco epoch. Am. Nat. 20:451.

————. 1887. Some new Taeniodonta of the Puerco. Am. Nat. 21:469.

————. 1888. Synopsis of the vertebrate fauna of the Puerco series. Trans. Am. Philos. Soc. 16:298–361.

Evernden, J. F., D. E. Savage, G. H. Curtis and G. T. James. 1964. Potassium-argon dates and the Cenozoic mammalian chronology of North America. Am. J. Sci. 262:145–198.

Gardner, J. H. 1910. The Puerco and Torrejon Formations of the Nacimiento Group. J. Geol. 18:702–741.

Gazin, C. L. 1968. A new primate from the Torrejon middle Paleocene of the San Juan Basin, New Mexico. Proc. Biol. Soc. Wash. 81:629–634.

————. 1971. Paleocene primates from the Shotgun Member of the Fort Union Formation in the Wind River Basin, Wyoming. Proc. Biol. Soc. Wash. 84:13–38.

Granger, W. 1912. Eocene Exp. 1912 New Mexico Wyoming. Unpublished field notebook in the Archives of the Dept. of Vertebrate Paleontology, Am. Mus. Nat. Hist., New York.

————. 1913. Record of Specimens New Mexico 1913. Unpublished field notebook in the Archives of the Dept. of Vertebrate Paleontology, Am. Mus. Nat. Hist., New York.

————. 1916. Eocene New Mexico Colorado 1916. Unpublished field notebook in the Archives of the Dept. of Vertebrate Paleontology, Am. Mus. Nat. Hist., New York.

————. 1917. Notes on Paleocene and lower Eocene mammal horizons of northern New Mexico and southern Colorado. Bull. Am. Mus. Nat. Hist. 37:821–830.

Kay, R. F. and M. Cartmill. 1977. Cranial morphology and adaptations of *Palaechthon nacimienti* and other Paromomyidae (Plesiadapoidea, ?Primates), with a description of a new genus and species. J. Hum. Evol. 6:19-53.

Lindsay, E. H., L. L. Jacobs and R. F. Butler. 1978. Biostratigraphy and magnetostratigraphy of Paleocene terrestrial deposits, San Juan Basin, New Mexico. Geology 6:425-429.

MacIntyre, G. T. 1966. The Miacidae (Mammalia, Carnivora). Bull. Am. Mus. Nat. Hist. 131:115-210.

Matthew, W. D. 1897. A revision of the Puerco fauna. Bull. Am. Mus. Nat. Hist. 9:259-323.

———. 1937. Paleocene faunas of the San Juan Basin, New Mexico. Trans. Am. Philos. Soc. n.s. 30:1-510.

Matthew, W. D. and W. Granger. 1921. New genera of Paleocene mammals. Am. Mus. Novitates 13:1-7.

Osborn, H. F. 1929. The titanotheres of ancient Wyoming, Dakota, and Nebraska, Vol. I. U.S. Geol. Surv. Monogr. 55:1-701.

Osborn, H. F. and C. Earle. 1895. Fossil mammals of the Puerco beds. Collection of 1892. Bull. Am. Mus. Nat. Hist. 7:1-70.

Osborn, H. F. and W. D. Matthew. 1909. Cenozoic mammal horizons of western North America with faunal lists of the Tertiary Mammalia of the West. U.S. Geol. Surv. Bull. 361:1-138.

Reeside, J. B., Jr. 1924. Upper Cretaceous and Tertiary formations of the western part of the San Juan Basin Colorado and New Mexico. U.S. Geol. Surv. Prof. Pap. 134:1-70.

Rigby, J. K., Jr. 1976. Geologic setting and mammalian fauna of Swain Quarry, middle Paleocene (Torrejonian), Carbon County, Wyoming. Unpublished Ph.D. dissert., Columbia Univ., 296 p.

———. 1980. Swain Quarry of the Fort Union Formation, middle Paleocene (Torrejonian), Carbon County, Wyoming: geologic setting and mammalian fauna. Evol. Monogr. 3:1-179.

Russell, D. E. 1964. Les Mammiferes Paleocenes d'Europe. Mem. Mus. Natl. Hist. Nat., Ser. C 13:133-139.

———. 1967. Le Paleocene continental d'Amerique du Nord. Mem. Mus. Natl. Hist. Nat., Ser. C 16:1-99.

Simons, E. L. 1960. The Paleocene Pantodonta. Trans. Am. Philos. Soc. 50:1-99.

Simpson, G. G. 1948. The Eocene of the San Juan Basin, New Mexico. Part 1. Am. J. Sci. 246:257-282.

———. 1959. Fossil mammals from the type area of the Puerco and Nacimiento strata, Paleocene of New Mexico. Am. Mus. Novitates 1957:1-22.

Sinclair, W. J. and W. Granger. 1914. Paleocene deposits of the San Juan Basin, New Mexico. Bull. Am. Mus. Nat. Hist. 33:297-316.

Szalay, F. S. 1969. Mixodectidae, Microsyopidae, and the insectivore-primate transition. Bull. Am. Mus. Nat. Hist. 140:193-330.

———. 1973. New Paleocene primates and a diagnosis of the new suborder Paromomyiformes. Folia Primat. 19:73-87.

Taylor, L. H. 1977. Geochronology of Torrejonian sediments, Nacimiento Formation, San Juan Basin, New Mexico. Unpublished MS thesis, Univ. of Arizona, 93 p.

Taylor, L. H. and R. F. Butler. 1980. Magnetic-polarity stratigraphy of Torrejonian sediments, Nacimiento Formation, San Juan Basin, New Mexico. Am. J. Sci. 280:97-115.

Van Valen, L. 1967. New Paleocene insectivores and insectivore classification. Bull. Am. Mus. Nat. Hist. 135:217-284.

———. 1978. The beginning of the Age of Mammals. Evol. Theory 4:45-80.

Van Valen, L. and R. E. Sloan. 1965. The earliest primates. Science 150:743–745.
———. 1966. The extinction of the multituberculates. Syst. Zool. 15:261–278.
Walker, E. P. 1975. Mammals of the World. John Hopkins Univ. Press, Baltimore, 1500 p.
Wilson, R. W. 1951. Preliminary survey of a Paleocene faunule from the Angels Peak area, New Mexico. Univ. Kans. Pub. Mus. Nat. Hist. 5:1–11.
———. 1956a. A new multituberculate from the Paleocene Torrejon fauna of New Mexico. Trans. Kans. Acad. Sci. 59:76–84.
———. 1956b. The condylarth genus *Ellipsodon.* Univ. Kans. Pub. Mus. Nat. Hist. 9:105–116.
———. 1956c. Additional remains of the multituberculate genus *Eucosmodon.* Univ. Kans. Pub. Mus. Nat. Hist. 9:117–123.
Wilson, R. W. and F. S. Szalay. 1972. New paromomyid primate from middle Paleocene beds, Kutz Canyon area, San Juan Basin, New Mexico. Am. Mus. Novitates 2499:1–18.
Wood, C. B., G. C. Conroy and S. G. Lucas. 1979. New discoveries of fossil primates from the type Torrejonian (middle Paleocene) of New Mexico. Folia Primat. 32:1–7.
Wood, H. E., 2nd, R. W. Chaney, J. Clark, E. H. Colbert, G. L. Jepsen, J. B. Reeside, Jr. and C. Stock. 1941. Nomenclature and correlation of the North American continental Tertiary. Bull. Geol. Soc. Am. 52:1–48.

Appendix 12.1—Measured stratigraphic sections.

Section W

Section measured on the west flank of Torreon Wash. The base of the section is on dune sand and stabilized soil covering a plain north of Ojo Encino School. Measured section begins in the SE¼ NE¼ Sec. 3, T20N, R5W, Ojo Encino Mesa Quadrangle (1961), offsets along unit 12 to the NW¼ NW¼ Sec. 34, T21N, R5W, Deer Mesa Quadrangle (1966), then offsets again along unit 21 to the W½ Sec. 27, T21N, R5W, Deer Mesa Quadrangle. Measured 8 August 1979 by C. Tsentas and S. Lucas.

Unit Number		Thickness (Meters)
San Jose Formation, Cuba Mesa Member:		
31.	Sandstone, coarse-grained and conglomeratic; brown, yellow and white, cross-bedded, clay ball rip-up clasts, cliff former, irregular base that channels down in places. Thickness not measured.	
Unconformity (erosional)		
Nacimiento Formation:		
30.	Silty clay-shale, green purple.	1.50
29.	Siltstone, purplish gray, weathers brown and blocky; well indurated, ledge former.	.75
28.	Interbeds of soft, fine-grained yellow sandstone and green and gray silty clay-shale. The two sandstone beds are each approximately .5 m thick.	6.25
27.	Silty clay-shale, gray and green, sandstone debris on surface.	3.00
26.	Clay-shale, black, carbonaceous (montmorillonitic?), with some green tints and gypsum veins.	13.00
25.	Silty clay-shale, red purple with green mottling.	1.00
24.	Sandstone, light green and gray, fine-grained, friable, some clayey bands.	2.50
23d.	Silty clay-shale (red) and soft pink gray sandstone, similar to sandstones of 23a and 23c.	3.00

23c.	Sandstone, same lithology as 23a.	3.25
23b.	Silty clay-shale, red, many red brown calcitic concretions, unionid shell hash; holds up bench.	5.00
23a.	Sandstone, gray white, medium-grained, quartzose, in places lenticular, giving way to 23b lithology. BUNM-77-184 is about 3 m above the base of this unit.	7.50
22.	Silty clay-shale, green with some buff tints. At base of unit there is a 25 cm thick olive green medium-grained sandstone ledge.	3.75
(Offset)		
21.	Clay-shale, gray black (montmorillonitic?), carbonaceous, forms crest of ridge.	8.25
20.	Silty clay-shale and interbedded stringers of friable, white, fine-grained sandstone containing some cross-beds, clayey bands and medium-grained pockets. Clay-shales are green and pink red, many red brown calcitic concretions. Upper 1–2 meters before unit 21 is mottled red, purple, yellow, green and gray.	13.50
19.	Silty clay-shale, pale green.	1.50
18.	Sandstone, dark brown, fine to medium-grained; dark, platy ledges are set in a fine-grained gray to buff friable sandstone.	1.50
17.	Silty clay-shale, green and gray, calcitic concretions; a discontinuous thin silt ledge (gray) about half way up the unit.	4.00
16.	Sandstone, clayey, buff, fine-grained, friable.	2.50
15.	Silty clay-shale, green with some buff bands, gypsum veins (about 1–2 cm thick) and calcitic concretion debris on surface (some *in situ* in upper part of unit).	7.00
14.	Clay-shale, gray black with some purple tint, "popcorn" surface (montmorillonitic?); a .25 m ledge of gray green silt forms a discontinuous ledge at about the middle of the unit.	3.00
13.	Silty clay-shale, green with red, brown and purple mottling, petrified wood, gar scales.	2.00
(Offset)		
12.	Sandstone, brown, well indurated, medium-grained, slightly arkosic, cross-bedded; cliff former that holds up Cejita Blanca Ridge. This unit forms a prominent sheet that holds up ridges and cuestas in this area and can be traced for several km; cannonball concretions (Pl. 12.4, fig. 3) and much petrified wood; unit thickens and thins on strike, depending on the extent of channeling at its base or of weathering at its top.	± 10.00
11.	Clay-shale, buff.	8.25
10.	Clay-shale, prominent purple band.	1.50
9.	Clay-shale, variegated green, yellow, gray.	5.00
8.	Sandstone, gray white, quartzose, medium-grained, lenticular; mostly covered by weathered clay-shale, petrified wood fragments.	1.50
7.	Silty clay-shale, variegated green and yellow, petrified wood fragments.	1.75
6.	Sandstone, gray white, medium-grained, quartzose, some rusty brown Fe stains and petrified wood fragments.	2.50
5.	Silty clay-shale, yellow.	1.00
4.	Sandstone, platy, dark brown, fine-grained.	0.25
3.	Silty clay-shale and mostly clayey, soft, fine-grained sandstone; greenish gray with some rusty brown Fe stains.	1.50

2.	Sandstone, lower .5 m is gray white, quartzose, friable, medium-grained. Above that is copper brown, well indurated, medium-grained, slightly arkosic, cross-bedded sandstone that forms a cliff and holds up small ridges locally. Some evidence of soft sediment deformation and dark brown cannonball concretions up to .75 m in diameter. This unit thins and thickens on strike, and differential weathering produces dark brown platy concretions on top of the unit that are up to 2 m in diameter.	± 7.50
1.	Clay-shale, dark gray, slightly carbonaceous.	8.25

Section E

Section measured on the east flank of Torreon Wash. The base of the section is on a flat of dune sand and stabilized soil about 500 m north of a Navajo dwelling. Measured section begins in the NE¼ Sec. 6 and NW¼ Sec. 5, offsetting along unit 24 into the NE¼ NW¼ Sec. 5, T20N, R4W, Ojo Encino Mesa Quadrangle (1961), then through the NE¼ Sec. 5 into the SE¼ Sec. 32, T21N, R4W, Deer Mesa Quadrangle (1966). Measured 9 August 1979 by C. Tsentas and S. Lucas.

Unit Number		Thickness (Meters)
San Jose Formation, Cuba Mesa Member:		
38.	Sandstone, coarse-grained to conglomeratic, many gravelly lenses, large logs up to 2 m long (Pl. 12.4, fig. 2); erosional base cuts out unit 37 in places; brown, yellow. Note that Baltz (1967) did not map as San Jose Formation. Thickness not measured.	
Unconformity (erosional)		
Nacimiento Formation:		
37.	Siltstone, gray, weathers brown and blocky, forms prominent ledge traceable for over 1 km but cut out in places by channeling at base of San Jose Formation.	0.50
36.	Sandstone, clayey, fine-grained, friable; buff with rusty brown and gray mottling.	2.50
35.	Sandstone, clayey, gray with olive green plates in it in a few places.	1.75
34.	Same as unit 31.	0.25
33.	Silty clay-shale, gray and purple.	2.50
32.	Interbedded green silty clay-shale and gray white friable medium-grained sandstone. About 4.5 m above base is thin (about 4 cm) plate of sandstone.	6.00
31.	Silt, green gray, weathers brown and blocky, ledge former.	0.25
30.	Clay-shale, silty, purple and gray.	0.75
29.	Same lithology as unit 27 but no blocky weathering.	2.25
28.	Silt, purple and white, weathers blocky, ledge former.	0.25
27.	Sandstone, white, clayey, quartzose, friable, medium-grained; some blocky weathering.	3.25
26.	Clay-shale, black, carbonaceous (montmorillonitic?).	4.75
25.	Sandstone, white, medium-grained, quartzose, friable.	1.00
24.	Clay-shale, black, carbonaceous (montmorillonitic?), gypsum veins.	8.00
(Offset)		
23.	Silty clay-shale, mostly red with green and brown mottling.	2.00
22.	Sandstone, fine-grained, green gray, weathers blocky to rounded.	0.50

21.	Interbedded red silty clay-shale and gray white sandstone (same as unit 19); some calcitic concretions.	7.75
20.	Sandstone, medium-grained, gray white, some clayey lenses, quartzose, lenticular; gives way to red silty clay-shales like unit 19. Top of this unit is where a site yielding a *Pantolambda* skeleton is located (UNM locality NP-2).	3.75
19.	Sandstone, gray white, clayey in places; up to .5 m thick dark brown cannonball weathered sandstone and interbedded red silty clay-shales in places; much red brown calcitic concretions, black silt(?) blocks up to .5 m across on surface, forms bench, medium-grained.	2.50
18.	Silty clay-shale, red and green banded, red brown calcitic concretions, shell hash; laterally lenses of gray white sandstone cut out the shale.	7.00
17.	Same lithology as unit 15 but with reddish tints.	1.00
16.	Sandstone, fine-grained, olive green ledge that weathers into small pellets.	0.25
15.	Clay-shale, black, carbonaceous (montmorillonitic?); some green tints, prominent bench former.	7.75
14.	Silty clay-shale, buff and yellow with red mottling and rusty brown stains from small ironstone concretions.	1.50
13.	Sandstone, green and buff, clayey, fine-grained.	1.00
12.	Sandstone, buff and gray, petrified logs up to 1 m long, medium-grained, quartzose to very slightly arkosic; cliff former. This unit present locally and pinches out into green and gray silty clay-shales with interbedded soft gray sandstones.	6.50
11.	Silty clay-shale, gray and green, forms shoulder on hill.	1.25
10.	Sandstone, fine-grained, gray green, quartzose, friable.	1.50
9.	Clay-shale, green, grayer and silty near top.	4.50
8.	Silty clay-shale, mostly pinkish red with some green tints; calcitic concretions and roses (red brown) abundant.	3.00
7.	Siltstone, purple, weathers blocky, ledge former.	0.50
6.	Silty clay-shale, variegated green, gray and purple with some red bands and mottling; thin (± 0.25 m), platy, olive green fine-grained sandstone about 1.5 m above base of unit.	4.00
5.	Sandstone, brown, medium-grained; upper part weathers dark brown into spheroidal, platy blocks; slightly arkosic, unconformable base (thins and thickens on strike), some rusty brown ferric stains, petrified wood. This unit continuous for several km.	± 7.00
4.	Clay-shale, olive green and buff.	1.50
3.	Clay-shale, black, carbonaceous (montmorillonitic?), "popcorn" surface; unit holds up shoulder of a hill.	3.00
2.	Sandstone, fine-grained, purplish gray, blocky, quartzose; bench former.	0.50
1.	Silty clay-shale, mostly olive green but with some bands of buff and gray (particularly near the top of the unit) and an infrequent lens of olive green fine-grained sandstone.	6.50

13 First occurrence of *Homogalax* (Mammalia; Perissodactyla; Tapiroidea) in the Regina Member of the San Jose Formation, San Juan Basin, New Mexico

J. W. Froehlich and P. K. Reser
Department of Anthropology, University of New Mexico
Albuquerque, New Mexico

INTRODUCTION

In the fall of 1979, we began fossil prospecting in the Badland Hills region on the southeast corner of the Jicarilla Apache Reservation. The outcrops along the Continental Divide in this area are of the early Eocene (Wasatchian) San Jose Formation. They are contiguous with well known areas such as Almagre Arroyo to the northeast, where fossils have been collected from the San Jose for more than 100 years. It appears that the Jicarilla lands have never been systematically prospected (Simpson, 1951). From a preliminary survey and a review of stratigraphic work by Baltz (1967), we anticipated that prospecting in the Regina Member of the San Jose in this unprospected area would be similar to our previous collecting experiences in Almagre Arroyo.

Our preliminary impression, after only three days of field work, is that this new area is at least as fossiliferous as the Almagre Arroyo region. Moreover, in our initial collection of 13 catalogued specimens from two locaities, we have identified a typical example of *Homogalax protapirinus*, previously unrecorded from the San Jose Formation. Another, smaller specimen seems best referred to cf. *Homogalax* sp. In this paper, we describe this first occurrence of *Homogalax* in the San Jose and discuss the biostratigraphic and phylogenetic implications of this discovery.

Abbreviations used below are UNM for the University of New Mexico Department of Geology vertebrate fossil collection, SJE for the San Jose Formation, and J (specimen) or JIC (locality) for the Jicarilla Apache Reservation.

SYSTEMATIC PALEONTOLOGY

Order PERISSODACTYLA Owen, 1848
Superfamily TAPIROIDEA Gill, 1872
Family ISECTOLOPHIDAE Peterson, 1919
Genus HOMOGALAX Hay, 1899
HOMOGALAX PROTAPIRINUS (Wortman, 1896)
Text-fig. 13.1 A, B, C.

Referred Material.—UNM SJE-J20, a left M^3, missing the border anterior to the protoloph and the antero-lingual angle, but including the antero-labial angle and the parastyle.

Locality.—JIC-E8, located in the NW¼NE¼, Sec. 21, T22N, R2W, Jicarilla Apache Reservation, Sandoval County, New Mexico.

Horizon.—Upper half of the Regina Member, San Jose Formation.

Associated Fauna.—A few fragments of *Coryphodon* teeth were found nearby; otherwise the specimen was isolated.

Description.—In size and morphology, the specimen is readily referable to *Homogalax protapirinus* as defined by Radinksy (1963). It is trapezoidal in outline, due largely to the antero-labial projection of the parastyle. The bladelike parastyle is as high or higher than all other cusps except the paracone; it is widely separated from the latter, forming an acute antero-labial angle. The metaloph is prominent and continuous with the ectoloph, as apparently was the worn protoloph. There is no suggestion of a protoconule or metaconule. The conical paracone is relatively high and separated from the metacone by a notch in the ectoloph. A narrow cingulum is discontinuous around the base of the paracone. There is a relatively short posterior cingulum. Finally, there is a small cuspule on the cingulum at the antero-lingual corner of the hypocone, a feature possibly due to individual variation and thus of little taxonomic significance.

The specimen's dimensions (Table 13.1) are well within 0.5 standard deviations of the M^3 means for *Homogalax protapirinus* (Radinsky, 1963, Table 1). The length of this nearly complete tooth was estimated by extrapolating from a measurement of 9 mm between the protoloph and posterior border using Radinsky's illustrations for reference.

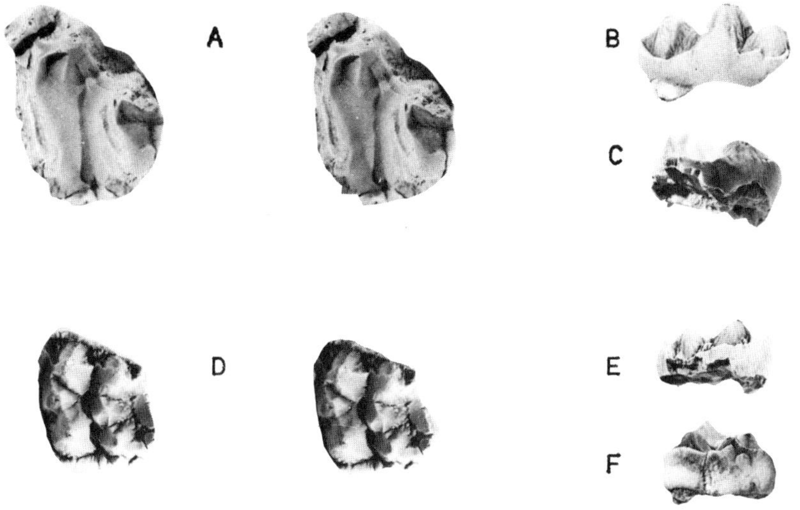

TEXT-FIG. *13.1*—San Jose Formation *Homogalax protapirinus* (UNM SJE-J20) compared to *Hyracotherium* sp. (UNM SJE-38a). *A*, stereo occlusal view of left M^3, *H. protapirinus*; *B*, labial view of left M^3, *H. protapirinus*; *C*, lingual view of left M^3, *H. protapirinus*; *D*, stereo occlusal view of left M^3, *Hyracotherium* sp.; *E*, labial view of left M^3, *Hyracotherium* sp.; *F*, lingual view of left M^3, *Hyracotherium* sp. All × 2.

Table 13.1—Tooth dimensions in mm for described specimens of *Homogalax protapirinus* and cf. *Homogalax* sp.

UNM-SJE No.	Tooth	Length	Width
J10	M^1	9.2	10.3
J10	M^2	10.2	12.1
J20	M^3	10.6[a]	13.8

[a]Estimate based on extrapolating from distance between protoloph and posterior border.

cf. *HOMOGALAX* sp.
Pl. 13.1, figs. 1, 3, 4.

Referred Material.—UNM SJE-J10, a right maxillary fragment with M^1, M^2, and one root of M^3.

Locality.—JIC-E6, located in the SE¼NE¼, Sec. 3, T22N, R2W, Jicarilla Apache Reservation, Sandoval County, New Mexico.

Horizon.—Upper half of the Regina Member, San Jose Formation, about 12 m below JIC-E8.

Associated Fauna.—The initial surface collection from this potential quarry site includes the mammalian genera *Coryphodon*, *Didymictis*, *Esthonyx*, *Hyopsodus*, *Hyracotherium*, *Microsyops*, *Oxyaena*, and incisor fragments referable to the Paramyidae. On the basis of its large size, a mandibular fragment with P_4–M_3(UNM SJE-J8, Pl. 13.2) seems referable to *Hyracotherium craspedotum* as defined by Kitts (1956).

Description.—This maxillary fragment approaches *Homogalax protapirinus* in morphology and size, but it is not as clearly referable to that taxon as UNM SJE-J20. The M^2 of UNM SJE-J10 has a subtrapezoidal outline. Both molars in the specimen have relatively prominent, blade-like parastyles. The metaloph is more weakly expressed than the protoloph, but both are continuous with the ectoloph. The protoconule and metaconule are barely distinct, especially on M^2. In both teeth, the conical paracone is relatively high and separated from the metacone by a moderately deep notch in the ectoloph. The cingula are relatively prominant, especially along the labial borders.

The dimensions of this specimen (Table 13.1) are well within two standard deviations of the means for *Homogalax protapirinus* from the Bighorn Basin of Wyoming (Radinsky, 1963, Table 1), except for M^1 breadth, which is close to three standard deviations. Moreover, M^2 length is within the first standard deviation, essentially indistinguishable from *Homogalax protapirinus.*

DISCUSSION

The two localities described above are clearly in the Regina Member of the San Jose Formation. They lie in a nearly straight and unfaulted line along the east side of the Continental Divide between two stratigraphic reference points described by Baltz (1967). A well log (Smith, 1-32 Jicarilla), about 4.2 km southwest of the *Homogalax protapirinus* locality, suggests that the Regina Member is roughly 178 m thick in this area. This view is corroborated by Baltz's (1967, Pl. 1, sec. BB) geologic cross-section 2.9 km northeast of the cf. *Homogalax* locality, and by his general description of the stratigraphy in this area. Baltz (1967, p. 50) stated that the Regina Member is 500 to 600 ft. (152–183 m) thick along the Continental Divide in most of the Penistaja Cuestas physiographic sector.

There are remnants of a thick sandstone on top of the mesas in the sections of both the above mentioned drill hole and the *Homogalax* locality, which Baltz (1967, p. 51) concluded are probably equivalent to the persistent medial sandstone tongue of the Llaves Member of the San Jose Formation. Since this sandstone is the stratigraphic cap above the Regina Member, and since the beds in this area have negligible dip ($<3°$), topographic contours can be used to estimate the distance of both localities from the top of the Regina Member. The *Homogalax* locality is estimated to be 55 m below the Llaves Member sandstone, and the cf.

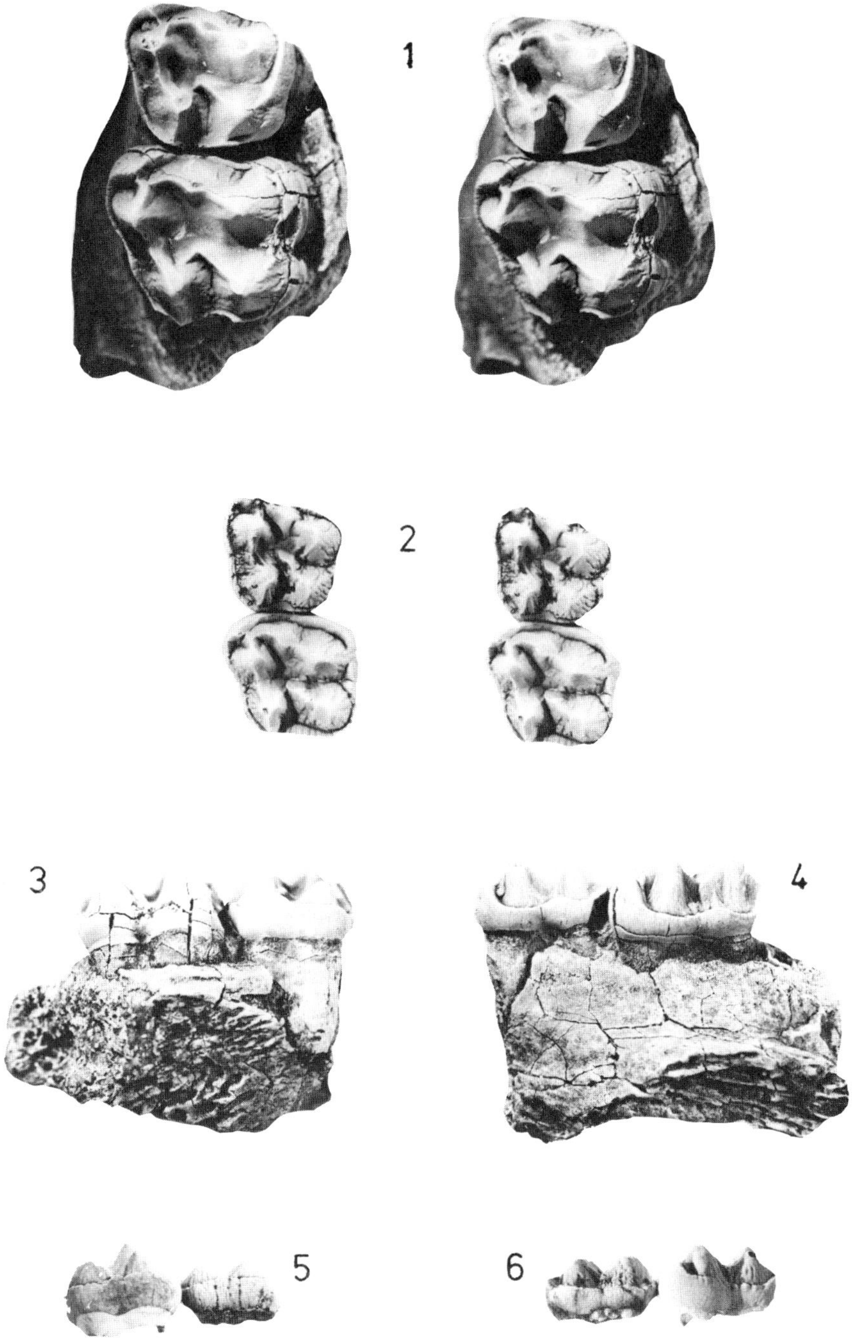

EXPLANATION OF PLATE 13.1

FIGS. *1–6*—San Jose Formation cf. *Homogalax* (UNM SJE-J10) compared to *Hyracotherium* sp. (M^1, UNM SJE-36; M^2, UNM SJE-38b). *1*, Stereo occlusal view of right maxillary fragment, cf. *Homogalax*; *2*, Stereo occlusal view of left M^1–M^2 (reversed image), *Hyracotherium* sp.; *3*, Lingual view of right maxillary fragment, cf. *Homogalax*; *4*, Labial view of right maxillary fragment, cf. *Homogalax*; *5*, Lingual view of left M^1–M^2 (reversed image), *Hyracotherium* sp.; *6*, Labial view of right M^1–M^2 (reversed image), *Hyracotherium* sp. All ×2.

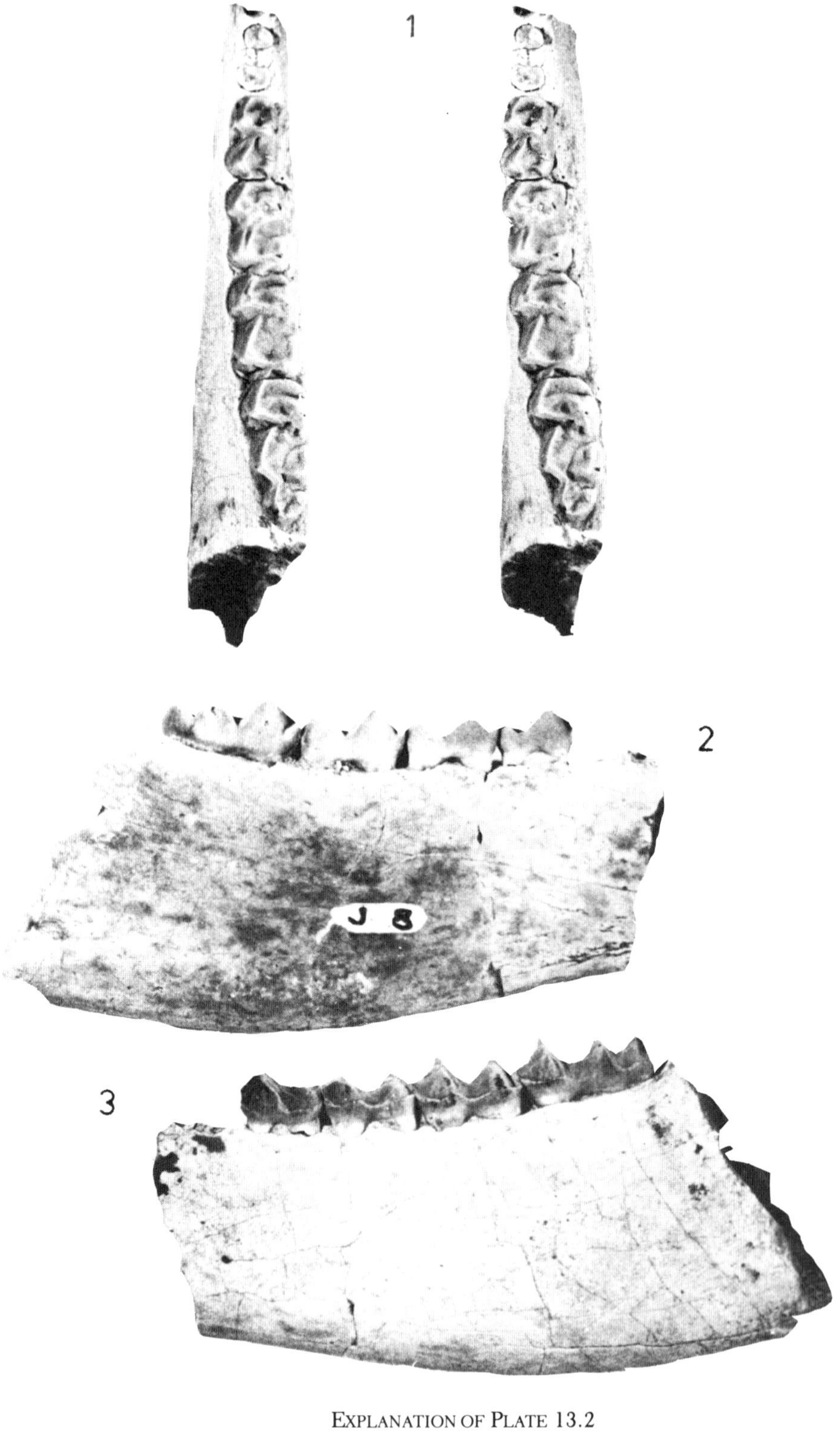

EXPLANATION OF PLATE 13.2

FIGS. *1–3*—Mandibular ramus of *Hyracotherium craspedotum* (UNM SJE-J8) from the San Jose Formation. *1*, Stereo occlusal view of left mandibular ramus; *2*, Lingual view of left mandibular ramus; *3*, Labial view of left mandible fragment. All × 1.5.

Homogalax locality is about 67 m below this sandstone. Thus, we are reasonably confident, until sections can be measured at the sites, that both localities are in the middle to the upper third of the Regina Member. This conclusion is supported by the location of our sites below prominent pale red to maroon shales with thick ledge-forming lenticular sandstones, described by Baltz (1967, p. 49) as common in the upper fourth of the Regina Member.

The specimen referred to cf. *Homogalax* was found in the same horizon as a *Hyracotherium craspedotum* jaw. This fact alone should foster caution in distinguishing between small *Homogalax* and large *Hyracotherium*. Indeed, we were initially tempted to consider both specimens as parts of the same individual. However, we have opted for the cf. *Homogalax* designation for the maxillary fragment because the two teeth are essentially the same as specimens previously referred to that taxon by McKenna (1960) and Gazin (1962). From a review of the literature, we are nevertheless impressed by the apparent intergradation between the two taxa in size and morphology. The bladelike parastyles, weak conules, high paracones, trapezoidal outlines, and uninterrupted transverse lophs of UNM SJE-J10 all suggest assignment to *Homogalax*, but the weakly expressed metalophs and the labial cingula are more characteristic of *Hyracotherium* (see Pl. 13.1). In overall morphology, UNM SJE-J10 is very close to specimens designated as *Homogalax protapirinus*, illustrated by Gazin (1962, Pl. 12) from the early Graybullian Bitter Creek locality in Wyoming, and by Guthrie (1971, Text-fig. 18a) from the Lost Cabin Member of the Wind River Formation in Wyoming. The dimensions of the M^2 in the UNM specimen are also very close to approximate measurements from the above cited illustrations.

As described above, the tooth dimensions of the cf. *Homogalax* specimen are in the lower part of the range for measurements of Graybullian *Homogalax protapirinus* described by Radinsky (1963, Table 1); only in the case of M^1 breadth do they fall outside the predicted 95% confidence limits. Moreover, in a comparison with samples of seven *Hyracotherium* sp. M^1's and M^2's in the UNM collection from Almagre Arroyo, the two molars in question are extremely large, falling well outside three standard deviations of the means. Considering that M^3 is typically larger than M^2 in *Homogalax*, the UNM cf. *Homogalax* M^2 is also much larger than third molars of cf. *Homogalax* "*primaevus*" from the Four Mile fauna of northwest Colorado described and illustrated by McKenna (1960, Table 8, Text-fig. 63). On the other hand, the tooth dimensions of the UNM specimen are within the range for two M^1 and three M^2 specimens of Lost Cabin *Hyracotherium craspedotum* measured by Guthrie (1971, Table 33). Despite the fact that this is a small sample, which is based on comparisons of lower molars from both formations (Kitts, 1956, Table 17), and although the Wind River Formation specimens are about 10% larger than those from the San Jose, we nevertheless conclude that there is nearly complete intergradation in size between the two taxa.

In sum, the UNM cf. *Homogalax* specimen has several morphological features which ally it to *Homogalax protapirinus*, but its size is smaller than Graybullian *Homogalax* and close to Lostcabinian *Hyracotherium craspedotum.* This general description also applies to all other previously described cf. *Homogalax* from non-Graybullian time. McKenna's (1960) specimens from the Four Mile fauna were relatively small, but he attributed them to a possibly ancestral species, older than *Homogalax protapirinus.* Both Gazin (1962) and Guthrie (1972) conceded that their specimens, referred to cf. *Homogalax* and *Homogalax* respectively, were within the size range of large *Hyracotherium* from the same Lostcabinian horizons. We are thus faced with the apparent dilemma that as *Hyracotherium* evolved to larger size in the Lostcabinian, *Homogalax* was apparently becoming smaller. This is the reverse of the general trends in isectolophid evolution toward increased size and a lengthened molar row (Radinsky, 1963).

As one possible explanation for this contradictory intergradation of the two taxa in Lostcabinian time, we offer the hypothesis that small *Homogalax* are not actually present. This is supported by our review of the literature; there may be no *bona fide Homogalax* from the Lost Cabin Member. Furthermore, we suggest that as *Hyracotherium* evolved to larger size, it converged on some *Homogalax* features, leading to the apparent confusion of mistaking particularly robust specimens for *Homogalax.* Such metric and morphological convergence in Wasatchian equid evolution might explain the eventual extinction of the isectolophids, due to competitive exclusion. However, further work is necessary to resolve this problem.

In contrast to the ambiguities of clearly referring the cf. *Homogalax* specimen to a known taxon, the UNM specimen here identified as *Homogalax protapirinus* offers no doubt as to its correct assignment. In virtually every feature listed above, it is typical of Graybullian *Homogalax protapirinus* as described and measured by Radinsky (1963, Table 1). Moreover, it contrasts markedly with *Hyracotherium* sp. (see Text-fig. 13.1). Metrically, the tooth is statistically indistinguishable from the means in Radinsky's sample. It is much larger than any specimens described or figured by McKenna (1960), Gazin (1962), or Guthrie (1971). For example, the UNM specimen is over seven standard deviations broader than the mean in McKenna's sample. Thus, the UNM SJE-J20 specimen is clearly *Homogalax protapirinus*; it is possibly the first *bona fide* example of this taxon from outside accepted Graybullian horizons.

Previously, *Homogalax protapirinus* has served as an index fossil for the Graybullian sub-"age" of the Wasatchian (Radinsky, 1963, p. 77). One explanation for its occurrence in New Mexico is that part of the Regina Member may be Graybullian. Such a viewpoint was suggested by Van Houten (1945) and Simpson (1948). Moreover, the presence of isectolophids in the Almagre "facies" of the Regina Member was previously reported by Lucas (1977), based on an undescribed specimen in the University of Arizona collection (S. Lucas, oral communication, 1980). The presence of cf. *Homogalax protapirinus* in New Mexico was also recently reported for an isolated upper molar from the Wasatchian part of the Galisteo Formation in north-central New Mexico, east of the San Juan Basin (Lucas & Kues, 1979). However, Lucas (1977) discounted the significance of *Homogalax* as a guide fossil and suggested a Lostcabinian age for the San Jose Formation. This biostratigraphic correlation is supported by the adapid primates (Gingerich & Simons, 1977), tillodonts (Gingerich & Gunnell, 1979), *Meniscotherium* (Gazin, 1965), and various other taxa (S. Lucas, oral communication, 1980). The correlation of our localities with Graybullian time because of the occurrence of *Homogalax* is also inconsistent with their apparently high stratigraphic position in the Regina Member and the occurrence of *Hyracotherium craspedotum* at the lower cf. *Homogalax* locality.

It would seem, therefore, that *Homogalax protapirinus* occurs in New Mexico at a time later than the Graybull sub-"age." This may be the first definitive example of such a temporal occurrence; certainly it extends the known geographical range for *Homogalax protapirinus* considerably to the south. There is no record of *Homogalax*, for example, from the Huerfano Formation of Colorado (Robinson, 1966). The New Mexico specimen also occurs in a faunal context which presently does not include other nonequid perissodactyls, such as *Heptodon* and *Lambdotherium*, typical of post-Graybullian time. Perhaps the southern survival of *Homogalax protapirinus* in New Mexico can be understood in terms of the absence of these other forms, which are common further north in Wyoming. If Radinsky's (1963) phylogeny is correct, *Homogalax protapirinus* or its derivative must have survived somewhere in order to give rise eventually to the larger, more specialized *Isectolophus* of middle Eocene (Bridgerian) time.

SUMMARY

From previously unprospected outcrops on the Jicarilla Apache Reservation, we report the first occurrence of *Homogalax protapirinus* in the upper part of the Regina Member of the San Jose Formation in the San Juan Basin. Another, smaller specimen is tentatively referred to cf. *Homogalax*, but its size is between large *Hyracotherium craspedotum* and typical *Homogalax protapirinus*. The apparent intergradation between these two perissodactyls, particularly in later Wasatchian time, may be due to large hyracotheres being mistaken for *Homogalax*. Possibly all previously suggested occurrences of *Homogalax* younger than Graybullian time should be discounted.

A left M^3 is unambiguously assigned to *Homogalax protapirinus*, a taxon previously used as a guide fossil for the Graybullian sub-"age." On the basis of stratigraphic position and the occurrence of *Hyracotherium craspedotum*, we conclude that the *Homogalax* locality is not Graybullian. Rather, it appears that *Homogalax protapirinus* may have survived later in New Mexico than elsewhere.

ACKNOWLEDGMENTS

We wish to thank President Leonard Atole and members of the Jicarilla Apache Tribe for their permission and hospitality during field work on the reservation. A special thanks goes to Mr. Joe Muniz, Tourism Director, for his logistical support and interest in the project. We are also grateful to Spencer Lucas for his suggestions and encouragement throughout the project, and for his comments on the text. Dr. Keith Rigby, Jr. helped with the preparation of illustrations. Any remaining oversights in the paper are our own. The research was partially supported by the University of New Mexico.

NOTE

A return trip to locality JIC-E8 in Fall, 1980 by Mr. Tom Logan, a UNM undergraduate, produced a second *Homogalax* specimen, possibly from the same individual. This complete tooth (UNM SJE-J25) is a right M^3, with a length of 11.3 mm and a width of 13.5 mm. Morphologically, it is identical to the broken specimen described above. In every respect, it is typical of *Homogalax protapirinus*, as described by Radinsky (1963).

REFERENCES

Baltz, E. H. 1967. Stratigraphy and regional tectonic implications of part of Upper Cretaceous and Tertiary rocks, east-central San Juan Basin, New Mexico. U. S. Geol. Survey Prof. Paper. 552, 101 p.

Gazin, C. L. 1962. A further study of the lower Eocene mammalian faunas of southwestern Wyoming. Smithsonian Misc. Coll. 144, 98 p.

———. 1965. A study of the early Tertiary condylarthran mammal *Meniscotherium*. Smithsonian Misc. Coll. 149, 98 p.

Gingerich, P. D. and E. L. Simons. 1977. Systematics, phylogeny and evolution of early Eocene Adapidae (Mammalia; Primates) in North America. Contr. Mus. Paleontol. Univ. Michigan. 24:245–279.

Gingerich, P. D. and F. Gunnell. 1979. Systematics and evolution of the genus *Esthonyx* (Mammalia; Tillodontia) in the early Eocene of North America. Contr. Mus. Paleontol. Univ. Michigan. 25:125–153.

Guthrie, D. A. 1971. The mammalian fauna of the Lost Cabin Member, Wind River Formation (lower Eocene) of Wyoming. Ann. Carnegie Mus. 43:34–47.

Kitts, D. B. 1956. American *Hyracotherium* (Perissodactyla; Equidae). Bull. Am. Mus. Nat. Hist. 110:1–60.

Lucas, S. G. 1977. Vertebrate paleontology of the San Jose Formation, east-central San Juan Basin, New Mexico. New Mexico Geol. Soc. Guidebook. 28:221–225.

Lucas, S. G. and B. S. Kues. 1979. Vertebrate biostratigraphy of the Eocene Galisteo Formation, north-central New Mexico. New Mexico Geol. Soc. Guidebook. 30:225–229.

McKenna, M. C. 1960. Fossil Mammalia from the early Wasatchian Four Mile Fauna, Eocene of northwest Colorado, Univ. Calif. Publ. Geol. Sci. 37:1–30.

Radinksy, L. B. 1963. Origin and early evolution of North American Tapiroidea. Bull. Peabody Mus. Nat. Hist. 17, 106 p.

Robinson, P. 1966. Fossil Mammalia of the Huerfano Formation, Eocene of Colorado. Bull. Peabody Mus. Nat. Hist. 21, 95 p.

Simpson, G. G. 1948. The Eocene of the San Juan Basin, New Mexico. Am. Jour. Sci. 246:257–282 and 363–385.

———. 1951. Hayden, Cope, and the Eocene of New Mexico. Proc. Acad. Nat. Sci. Philadelphia. 103:1–21.

Van Houten, F. B. 1945. Review of latest Paleocene and early Eocene mammalian faunas. J. Paleontol. 19:421–461.

Part 4
Paleobotany

14

Cretaceous and Tertiary floras of the San Juan Basin

William D. Tidwell
Department of Botany and Range Science,
Brigham Young University
Provo, Utah

Sidney R. Ash
Department of Geology, Weber State College
Ogden, Utah

and Lee R. Parker
Biological Science Department,
California Polytechnic State University
San Luis Obispo, California

INTRODUCTION

Fossil plants have been reported and described from the San Juan Basin of New Mexico for more than a century. These include petrified wood, leaves, reproductive structures and palynomorphs. Such plant remains clearly indicate that the area supported several kinds of terrestrial plant communities in a variety of environments throughout the late Mesozoic and early Cenozoic.

The purpose of this report is to summarize the paleobotanical work that has been undertaken in the San Juan Basin to this date. It is anticipated that this summary will not only demonstrate the variety and importance of the plants already examined, but will also reveal the significance of the area for future work. Excavations for a variety of reasons, particularly in coal mines, have already provided many otherwise unobtainable collections. However, in specific cases commercial interests are threatening areas which are unique and important. It is hoped that these areas will be preserved for further study.

In or near the San Juan Basin, fossil plants have been reported from the Cretaceous Dakota Sandstone, the Mesa Verde Group, and the Fruitland–Kirtland Formations; the Paleocene Ojo Alamo, Nacimiento, and Animas Formations; and the Eocene San Jose Formation. The Cretaceous rocks contain a nearly continuous fossil record of the later portion of that period, when the flowering plants became the dominant plant group on earth. Paleoecological and morphological studies of these plants continue to give clues as to their origin, distribution, and the environmental conditions under which they grew. Plants collected from Cretaceous rocks were among those which contributed to the coals currently being mined in the San Juan Basin. Careful examination of these fossils will be necessary when environments of coal deposition are considered during coal exploration. These same plants were probably heavily browsed by dinosaurs. This relationship should be examined further. Finally, the San Juan Basin floras present clues to the terrestrial environments during the Paleocene and Eocene Epochs, when broad floodplains existed near mountains. The changes in plant community structure from the Cretaceous swamp forests to the Tertiary flood plain and lacustrine communities present opportunities for additional study.

HISTORY OF INVESTIGATIONS

Cretaceous and Tertiary plant fossils were first discovered in the San Juan Basin by American Indians who utilized the petrified wood they found there for tools such as points and scrapers. Possibly some of the Indians in the region even realized that there was something special about petrified wood since the trunk of a cycadeoid, which probably came from the western San Juan Basin, was discovered in an Indian ruin in an adjacent area in eastern Arizona (Wieland, 1934). Although Spanish explorers and colonists were in the region for many years, they made no reference to the plant fossils in any of their known chronicles and narratives (Northrop, 1962). Also, none of the trappers and mountain men who frequented the Basin area recorded them, and it was not until 1852 that any were mentioned in print. In that year, Lieut. J. H. Simpson's journal of an 1849 "military reconnaissance from Santa Fe to the Navajo Country" was published. In it he reported the occurrence of what he called beautiful specimens of petrified wood near his 8th camp in the southeastern edge of the basin at the base of the Nacimiento Mountains. Newberry (1861) reported fossil wood along what is now the New Mexico–Arizona border and noted that the leaves associated with the coals in northwestern New Mexico were similar to the Cretaceous leaves in the Dakota Sandstone of Kansas and Nebraska.

Plant fossils from the basin were not seriously studied until the early part of this century. Some Cretaceous forms were mentioned in an article by Shaler (1907) that concerned the western part of the Durango–Gallup coal field of Colorado and New Mexico. These fossils were reportedly rare and fragmentary. A list of a few other Cretaceous plant fossils obtained in the Basin was given by F. H. Knowlton (in Lee, 1912). The earliest reasonably detailed discussion of fossil plants from this region was made by Knowlton (1916) when he described the Cretaceous flora of the Fruitland and Kirtland Formations. Later Knowlton (1924) described the Paleocene Animas flora in some detail. Wieland collected many cycadeoid trunks from the Cretaceous rocks in the western part of the basin intermittently between 1920 and 1940. He described some of these trunks and later Delevoryas (1959) revised his findings.

Others who have contributed to knowledge of the flora of the San Juan Basin include Gregory (1917), who reported on the occurrence of Cretaceous leaves in the western San Juan Basin near the Chuska Mountains; Anderson (1960, and in Baltz et al., 1966), who studied the palynomorphs of the Cretaceous and Tertiary rocks; and Brown (1962), who reported on a few Tertiary leaves from the basin. Ash and Read (1976) briefly reported on some petrified wood from the Cretaceous rocks in the northwestern part of the basin. In their reports on the paleontological resources of the San Juan Basin, Kues et al. (1977) and Ash et al. (1980) noted the occurrence of plant fossils in Cretaceous and Tertiary rocks at several localities in the basin.

FLORISTICS AND PALEOECOLOGY

In working with the Cretaceous and Tertiary plant fossils of the San Juan Basin, certain problems exist in their correct identification. The classification system of vascular plants is based upon reproductive features. Because few fossil reproductive structures are ever preserved or collected, the task of assigning most fossil plant material to modern families or genera is usually difficult or uncertain. Fossil ferns and gymnosperms are less troublesome than angiosperms because vegetative characters are often diagnostic. In addition, reproductive features such as fertile fern pinnules or woody conifer cones are more likely to be preserved with direct organic connection to vegetative parts (i.e., vegetative leaves on the pedicle of the cone). However, Cretaceous and early Tertiary angiosperms are usually placed with much less confidence into extant genera or families. Evidence indicates that 60% or more of the previously assigned Cretaceous and lower Tertiary leaves are incorrectly placed in living families and genera (Dilcher, 1971, 1974; Hickey & Wolfe, 1975; Doyle & Hickey, 1976; Roth & Dilcher, 1979). Most of these were assigned in the early part of this century. This discrepancy is due to four facts: 1) flowers are rare in the fossil record, 2) flowers are never attached to leaves or wood even in life, 3) very little is known about characteristic vegetative features of living taxa such as leaf venation and epidermal patterns, and 4) fossil angiosperm leaves with gross features similar to modern taxa might, upon detailed examination of microscopic structures, be found to belong to extinct taxa. Usually these examinations are very time consuming. They have never been done with all the angiosperm leaves in any large fossil flora. Early angiosperm wood and pollen are also difficult to place with modern genera, although they can often be placed in modern families with confidence.

Because of the summary nature of this report, we have purposely used common names, in addition to Latin binomials, for some of the fossil plants. We recognize that while this practice might be misleading with specific plants, most readers will find it useful. We have depended almost completely upon the accuracy of the workers we cite for correct identifications. In doing this we are confident that the ferns and conifers are related to extant groups. We are much less confident of the common names used for angiosperms, especially the dicotyledons. Common names for certain genera have been omitted because they have traditionally been used as catchall genera. These include: *Carpites*, *Cyperacites*, *Ficus*, *Laurus*, *Phyllites* and others.

CRETACEOUS

Dakota Sandstone.—Only a few fossil plants have been reported from the Dakota Sandstone of the San Juan Basin. These occur in the western part of the basin. Gregory (1917) listed a few plant fossils from the Chuska Mountains of New Mexico and adjacent areas in Arizona. Although some of the identifications are questionable, the flora appears to include a holly (*Ilex* sp.), a willow (*Salix* sp.), a bog rosemary (*Andromeda pfaffiana*), a walnut relative (cf. *Juglans crassipes*), a possible fig (*Ficus inequalis*), and a conifer, *Phyllocladus subintegrifolius*, which should be reassigned to *Protophyllocladus.*

Remains of the extinct tree fern *Tempskya* were reported from this formation north of Shiprock, N.M. (Ash and Read, 1976). The occurrence of these fossils suggests an Early Cretaceous age for this part of the Dakota Sandstone.

Mesaverde Group.—The Mesaverde Group consists of three formations throughout much of the San Juan Basin. These are the Point Lookout Sandstone at the bottom, the Menefee Formation in the middle, and the Cliff House Sandstone at the top. In the western part of the basin, several other formations are included in this group. Two of these are the Gallup Sandstone and the Crevasse Canyon Formation, forming the base of the Mesaverde Group.

Lee (1917) pointed out that fossil plants occur at certain horizons in the Mesaverde coalfields in central New Mexico, and that in some places plants are numerous and well preserved, but that they appear rare in the Mesaverde of northern New Mexico and southern Colorado. In correlating these strata with the Cretaceous beds of the Raton Basin, Knowlton (in Lee, 1917) listed 17 species of fossil plants from the Mesa Verde along the northeastern edge of the basin. He considered his identifications of these plants tentative. His list included a horsetail (*Equisetum* sp.), five coniferous species, the shield fern (*Dryopteris* sp.), and 10 species of angiosperms.

Wieland (1928, 1930, 1936) collected cycadeoid trunks, which supposedly came from the Mesaverde Group, from the western side of the San Juan Basin. Originally Wieland (1928) called these fossils *Navajoia*, but he later described them as three species of *Monanthesia*. After further study, Delevoryas (1959) reassigned them to one species, *Monanthesia magnifica*. There is some question about the source of these specimens of *Monanthesia*. Many of these fossils were traded to Wieland by the Navajo Indians; he apparently never saw their actual locality. Thus, it is possible that they were not derived from the Mesaverde. Wieland (1936) also mentioned collecting a specimen of *Tempskya* from the Mesaverde. Since this genus occurs in older rocks, such as the Dakota Sandstone, it is possible these specimens of *Monanthesia* may also be from older strata. *Monanthesia* and *Tempskya* are relatively abundant in the Lower Cretaceous Cedar Mountain Formation near Moab, Utah (Tidwell, 1975; Tidwell et al., 1976), which suggests that Wieland's specimens are older than Mesaverde.

Kues et al. (1977) reported fossil plants from the Point Lookout Sandstone, the Crevasse Canyon, and the Menefee formations. Only two species from the Point Lookout Sandstone and nine from the Crevasse Canyon were listed. Of this nine, three are conifers, and the remainder are angiosperms. One of the conifers, a relative of the star pine (*Araucaria* sp.), is fairly abundant in this formation.

The Menefee Formation contains numerous plant fossils at several localities in the San Juan Basin (O'Sullivan et al., 1972; Kues et al., 1977). Many of these localities are associated with coal beds, and most of the collected specimens are compressions of leaves. Two species of ferns represented in the flora are *Anemia hesperia* of the climbing fern family (Schizaeaceae) and a "spleenwort" (*Asplenium dicksonianum*) of the Polypodiaceae. The fern allies are represented by a species of horsetail (*Equisetum* sp.).

Five species of conifers occur in the Menefee Formation. The genus *Sequoia* is represented in the flora by three species (Text-fig. 14.1D), one of which may be the ancestor of the living California redwood (*Sequoia sempervirens*). Other conifers are *Araucaria* sp. (Text-fig. 14.2D) and *Protophyllocladus polymorpha*. The latter taxon is considered a relative of the living genus *Phyllocladus*, which presently grows on southern Pacific islands.

The Menefee flora is dominated by angiosperms (22 species), the Dicotyledonae being the most plentiful. One of the more common leaves in the flora belongs to the genus *Ficus*. There appears to be at least two species of this genus in the Menefee. A species of willow (*Salix* sp.), one of sycamore (*Platanus raynoldsi*), and a relative of the Katsura tree (*Cercidiphyllum arcticum*) are also present in the flora. The beech family (Fagaceae) is represented by an extinct genus and species of oaklike leaves (*Dryophyllum subfalcatum*). Other genera and families represented in the Menefee flora by one or more species include *Cinnamomum* and *Laurophyllum* of the Lauraceae, *Rhamnites* and *Rhamnus* of the buckthorn or coffeeberry family (Rhamnaceae), and *Cissus marginata* (Text-fig. 14.1I) and *Vitus stantoni* of the grape family (Vitaceae). *Cyperacites* and the palms *Sabalites* and *Phoenocites imperialis* are the only monocotyledons reported from this flora. Although it has been thought useful for its ecological connotations (Dorf, 1942), there is very little significance in the occurrence of *Cyperacites* except to record the presence of an additional monocotyledon (Dorf, 1938).

We observed two conifers, *Araucaria* sp. and *Sequoia cuneata* (Text-fig. 14.1D), and one leaf of the palm *Phoenocites imperalis* among the fossil plants in the roof of an abandoned coal mine entry in the Menefee Formation south of Cuba, N.M. Several small unidentified fragments of dicotyledon leaves were also present. In addition, we saw several three-toed dinosaur footprint casts, each almost 1 m in length. The same fossil coniferous forms, the palm, and the dinosaur footprint casts have been reported from time-equivalent, coal-bearing units in Utah and Wyoming (Parker & Balsley, 1979).

Six of the Menefee species possibly grew in bottomland environments, whereas nine were probably swamp plants. *Sequoia cuneata* and *Rhaminites eminens*, which occur in the Menefee flora, may have been codominant trees in this swamp environment (Parker, 1976). This is similar to the plant community in the swamps of the Mississippi River floodplain where baldcypress (*Taxodium distichum*), a conifer, and water tupelo (*Nyssa aquatica*), a dicotyledon, are the codominant trees. As pointed out by Parker (1976), these living forms are also similar morphologically to their fossil counterparts.

Fossil plants of the Blackhawk Formation (a time-equivalent of the Menefee) in Utah have been examined for the climatic factors they suggest. This included the preferred climates of living relatives of the fossil plants, leaf size and other leaf features, growth rings in petrified wood (Parker, 1976), and the presence of fusinite in all coal horizons (Balsley, personal communication). The climate suggested was subtropical with distinct dry seasons. Rainfall was probably about 1600 to 1900 mm annually (Parker, 1976). Since the Menefee Formation was deposited at the same time as the Blackhawk, in the same type of fluvial coastal plain environment, and with many of the same plant species, it is assumed that the climatic conditions for the Blackhawk and Menefee Formations were probably the same.

Fruitland–Kirtland Formations.—Plant fossils are numerous in both the Fruitland and Kirtland Formations. Because of the similarities of the flora of the Fruitland Formation to that of the overlying Kirtland Shale, the two are considered together.

The Fruitland–Kirtland floras as described by Knowlton (1916) and listed by Lee (1917), Reeside (1924), O'Sullivan et al. (1972), Kues et al. (1977), and this report, are composed of 73 species (Table 14.1; see also Text-figs. 14.1, 14.2, and Pl. 14.1–14.3). Of this number, nine are ferns, nine are conifers, 44 are dicotyledons, and 11 species are monocotyledons.

Ferns do not appear to be particularly abundant or varied in the Fruitland–Kirtland floras. This may be due to the lowland fluvial coastal plain environment in which these plants grew, because the largest variety and abundance of ferns in the modern tropics are those of upland or mountainous rainforests (Berry, 1924). Modern species of *Osmunda* (Text.-fig. 14.1A) and *Anemia* which occur in the floras are swamp-living forms. Extant species of *Cyathea* live predominately in humid forests. Some, however, are xerophytic and flourish in open sunny habitats (Eames, 1936). Certain species of *Cyathea* are commonly known as "tree ferns" because of their tall, slender palmlike growth habit. There is no evidence that the fossil *Cyathea pinnata* collected in the San Juan Basin attained such size. It could have been similar to the living *Cyathea glabra*, which has a small stem or is "stemless" and grows directly on the organic substrate of peat-forming swamps in Malaysia (Anderson, 1961). *Salvinia*, the water fern, although uncommon in fossil floras has been reported from other Cretaceous floras.

Living members of *Onoclea*, another member of the flora, are generally restricted to swamps and areas of acid soil. Knowlton (1916) considered *Onoclea neomexicana* from the Kirtland flora to be closely related to the living *Onoclea sensibilis*. The species *Onoclea sensibilis* is an important member of the herbaceous community in the baldcypress-tupelo swamps of Louisiana (Penfound, 1952; Parker, 1976). These ferns grow in environments similar to those postulated for their fossil counterparts.

Seven species of conifers occur in the Fruitland–Kirtland floras. Six are represented by leafy shoots and one by petrified wood. The species of petrified wood is assigned to *Cupressinoxylon* and is similar to *C. coloradense* from the Vermejo flora of the Raton Basin. It differs from the latter species by having longer rays (up to 35 cells high), uniseriate pitting, noncontiguous bordered pits and in lacking growth rings.

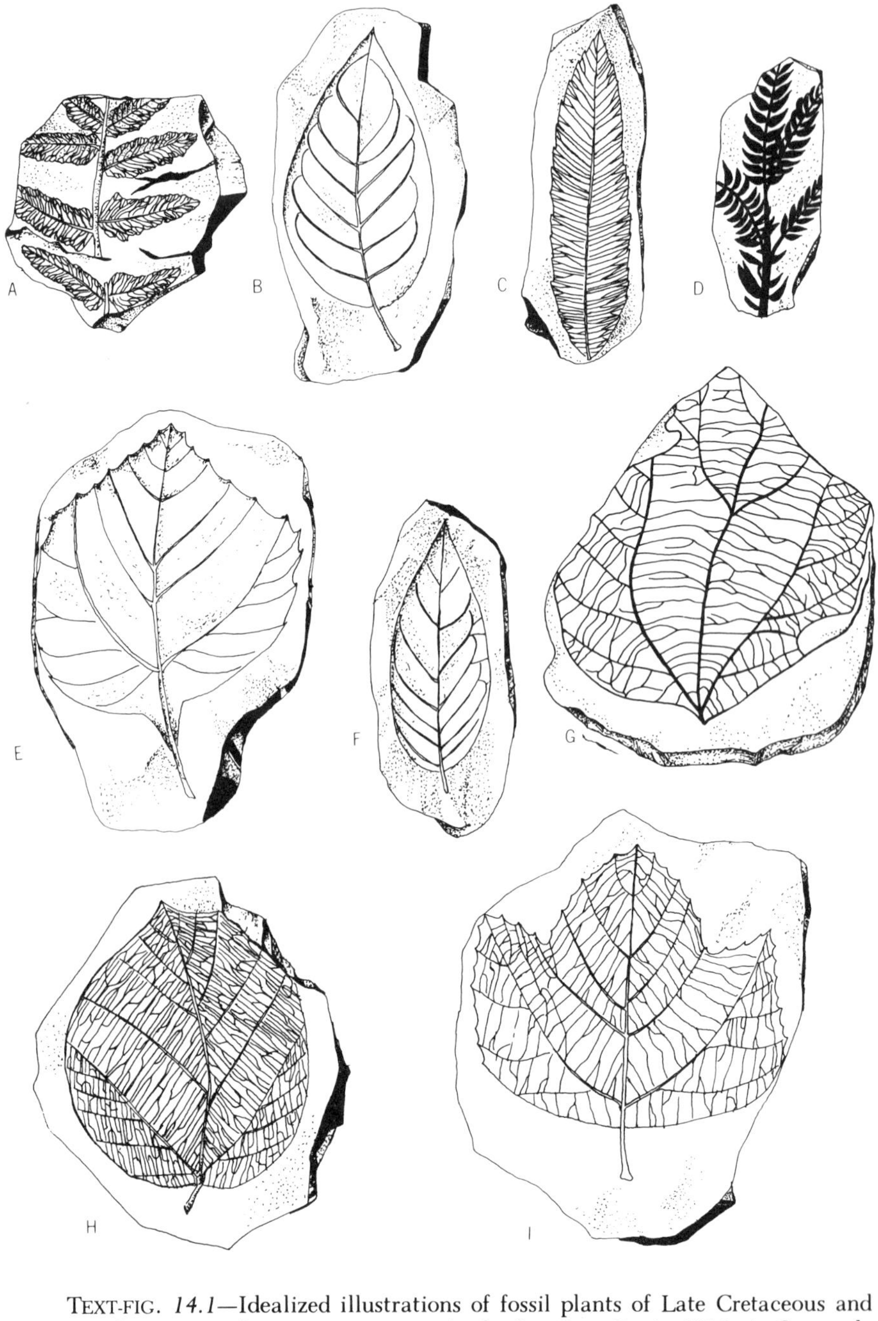

TEXT-FIG. *14.1*—Idealized illustrations of fossil plants of Late Cretaceous and Paleocene age having occurrences in the San Juan Basin, N.M. *A*, *Osmunda* sp. similar to *Osmunda hollicka* from the Fruitland Formation; *B*, *Carya antiquorum* (Fruitland Formation); *C*, *Allantodiopsis erosa* (Animas Formation); *D*, *Sequoia cuneata* (Menefee Formation); *E*, *Platanus raynoldsi* (Fruitland Formation); *F*, *Magnolia* sp. (Fruitland Formation); *G*, *Ficus planicostata* (Fruitland Formation); *H*, *Ficus* sp., similar to *Ficus planicostata* (Fruitland Formation); *I*, *Cissus marginata* (Menefee Formation). All × 0.5.

Table 14.1—Flora of the Fruitland-Kirtland Formations

Source of data: A. Knowlton (1916), B. Lee (1917) near Dulce, N.M., C. O'Sullivan et al. (1972), D. Kues et al. (1977), E. This report.

Filicophyta
- Osmundaceae
  - *Osmunda hollicki* E
  - *Cladophlebis* sp. D
- Schizaeaceae
  - *Anemia hesperia* A, B
  - *Anemia* sp. A
- Cyathaceae
  - *Cyathea pinnata* D
- Polypodiaceae
  - *Asplenium neomexicana* A
  - *Asplenium? coloradense* E
  - *Onoclea neomexicana* A
- Salvinaceae
  - **Salvinia* sp. D

Coniferophyta
- Araucariaceae
  - *Araucaria* sp. D, E
  - *Araucaria longifolia* A, B
- Cupressaceae
  - *Brachyphyllum macrocarpum* B
- Taxodiaceae
  - *Sequoia reichenbachii* A, B, C
  - *Sequoia obovata* A
  - *Sequoia cuneata (= Metasequoia cuneata)* C
  - *Cunninghamites pulchellus* B
  - *Cupressinoxylon* sp. E
- Coniferales incertae sedis
  - *Podozamites* sp. D

Anthophyta (Angiosperms)
- Monocotyledonae
  - Sparganiaceae
    - *Sparganium* sp. D
  - Cyperaceae
    - **Cyperacites* sp. D
  - Palmae
    - *Sabalites imperialis* A, B, D
    - *Phoenocites?* sp. D
    - *Palmoxylon* sp. A D, E
    - *Palmoxylon* sp. B E
    - *Palmoxylon* sp. C E
    - *Rhizopalmoxylon* sp. D
  - Araceae
    - *Pistia corrugata* A
  - Pontederiaceae
    - *Heteranthera cretacea* A
  - Cannaceae
    - **Canna? magnifolia* D

Dicotyledonae
- Salicaceae
  - *Salix baueri* A
  - *Salix* sp. A
- Juglandaceae
  - **Carya antiquorum* D, E
    - *(= Juglans nigella)*
- Fagaceae
  - *Quercus baueri* A
  - *Quercophyllum* sp. D
- Moraceae
  - *Ficus squarrosa?* A
  - *Ficus rhamnoides* A
  - *Ficus planicostata* A, B, C, E
  - *Ficus praetrinervis* B
  - *Ficus starkvillensis?* B
  - *Ficus curta* A
  - *Ficus wardii* B
  - *Ficus baueri* A, E
  - *Ficus* sp. A, D
- Nymphaeceae
  - *Nelumbo* sp. A
- Cercidiphyllaceae
  - *Cercidiphyllum* sp. E
- Menispermaceae
  - *Menispermites* sp. A
    - (= *Ficus* sp.)
- Magnoliaceae
  - *Magnolia cordifolia* E
  - **Magnolia berryi* D
- Lauraceae
  - *Laurus baueri* A
  - *Laurus coloradensis* A
  - *Cinnamomum sezannense* A
    - *(= Ficus eucalyptifolia)*
  - **Laurophyllum* sp. D
- Saxifragaceae
  - *Ribes neomexicana* A
- Platanaceae
  - **Platanus nobilis* D
  - **Platanus raynoldsi* D
- Leguminosae
  - *Leguminosites? neomexicana* A
  - *Leguminosites* sp. E
- Rhamnaceae
  - **Rhamnus goldianus* D, E
  - *Zizyphus?* sp. B
- Vitaceae
  - **Vitis lobata* D

**Cissus marginata* D
Dillenaceae
Dillenites cleburni E
Myrtaceae
**Myrtophyllum torreyi* A, D, E
Myrtophyllum neomexicanum A
Cornaceae
Cornus coloradense E
Caprifoliaceae
**Viburnum antiquum* D
Bignoniaceae
Dombeyopsis obtusa C
Incertae sedis
Phyllites neomexicanus A
Phyllites petiolatus A
Pterospermites undulatus A
Pterospermites sp. A
Carpites baueri A
Carpites cf. *lancensis* E
*unidentified flowers and fruit D

*Specimens of these species were collected by Kues et al. (1977) from near the base of the Ojo Alamo Sandstone. This locality is now considered to be in the Kirtland Shale (Kues, oral communication, 1979).

Among the conifer leaf compressions, *Araucaria* sp. (Pl. 14.1, fig. 1) is a frequent form in these floras and occurs commonly in coal mines in the San Juan Basin. Although *Sequoia cuneata* and *Brachyphyllum macrocarpum* have not, as yet, been collected in large numbers, they are considered important trees in the peat-forming swamp community of this age (Parker, 1976). Specimens of *Sequoia reichenbachii* and *S. obovata* were noted in the flora by Knowlton (1916) to be poorly preserved but "clearly *Sequoia*."

Leaves assignable to *Podozamites* were originally thought to be cycadean but are now considered to be related to the conifers, possibly the living genus *Agathis*, which is a member of the araucarian family. Species of *Agathis* are currently restricted to the Southern Hemisphere.

Among the monocotyledoneous plants only the families Sparganiaceae, Cyperaceae, Palmae, Araceae, and Pontederiaceae are represented in the Fruitland–Kirtland floras. The palms, comprising four different genera, are not as varied as might be expected. *Sabalites* (a fan palm; Pl. 14.2, fig. 2), like its possible modern correlative *Sabal*, may be a lowland coastal form. The feather palms are represented by *Phoenocites*, a genus which probably includes the ancestor of the date palms. According to Berry (1924), *Phoenocites* is a sand-inhabiting form. Although this genus and its modern allies require abundant water, he postulated that these plants needed a long growing season under clear, hot, sunny skies.

Silicified remains of palm axes are exceedingly common in some areas of the Fruitland Formation, particularly near Cuba, N.M. (Pl. 14.1, figs. 2, 3). Although these remains are often fragmentary, those near Cuba occur as columnar stem bases in growth position. The roots from these bases extend outward into the paleosoils in which they originally grew. Specimens of the four species of *Palmoxylon* from this formation are well preserved. Significantly, fossilized palm bases are rather rare in the Cretaceous of North America. Only one base, which is from the Upper Cretaceous of New Jersey, has been previously reported (Stevens, 1912).

The living members of the Araceae are cosmopolitan, although most are found in the tropics of South America, Africa, and southeast Asia (Berry, 1924). *Pistia corrugata*, a member of this family, was reported from the Fruitland Formation by Knowlton (1916). The living species of *Pistia (P. stratiotes)*, commonly called water lettuce, is a floating plant. This is probably the same environment in which the fossil species grew.

The Pontederiaceae is represented in this flora by the genus *Heteranthera* (pickleweed). Modern species of this genus grow in mud or shallow water with creeping, ascending or floating stems. Knowlton (1916) noted that his species *H. cretacea* from the Fruitland Formation agreed closely with the living form *H. limosa*. The living species ranges throughout eastern United States and tropical America.

Other monocotyledons in the Fruitland–Kirtland floras are a species of the burreed (*Sparganium*) and *Cyperacites*.

The Dicotyledonae are the predominate plants in the Fruitland–Kirtland floras. The willows (*Salix*) and figs (*Ficus*; Pl. 14.3, figs. 4, 5) are well represented. The Magnoliaceae has two species in these floras. One of these, *Magnolia cordifolia* (Pl. 14.3, fig. 1) was combined into *Ficus uncata* by Brown (1962). He mentioned that this species resembled some species of *Combretum* and *Magnolia* and that his own allocation to the genus *Ficus* was dubious. Therefore, we elect to leave this species under *Magnolia* until something more certain is demonstrated. *Juglans nigella* was recombined by Brown (1962) into *Carya antiquorum* (Text-fig. 14.1B). He stated that it is difficult to distinguish between leaflets of *Carya* and *Juglans* as well as between their fruits; the latter he assigned to the ubiquitous genus *Juglandicarya* sp. The genera *Rhamnus* (Pl. 14.2, fig. 6) and *Zizyphus*(?) were present in the Basin during Fruitland–Kirtland time. Living species of *Zizyphus* grow predominately in the Indo-Malaysian region although a few species occur in all tropical areas. *Rhamnus cleburni* was combined into *Dillenites cleburni* (Text-fig. 14.2C) by Dorf (1942). As noted by Dorf (1938), it is difficult to see any resemblance between this fossil species and any modern species of *Rhamnus*, whereas it is similar to several modern genera of the Dilleniaceae. The continued reference to *Rhamnus* would be misleading; therefore, his assignment seems advisable. The genus *Dombeyopsis* is the only member of the Bignoniaceae in the Fruitland–Kirtland floras. It is supposedly the ancestral form for the genus *Dombeya* in the modern African flora (Berry, 1924). The family Lauraceae has three species in these floras. There is a relatively narrow leaf species of the camphor tree *(Cinnamomum)* and two species of laurels *(Laurus)*. The Fagaceae is represented by a single species of oak *(Quercus)*. The extinct fagaceous genus *Dryophyllum* that is fairly common in other Upper Cretaceous floras of the Rocky Mountains has not been observed in the Fruitland–

1

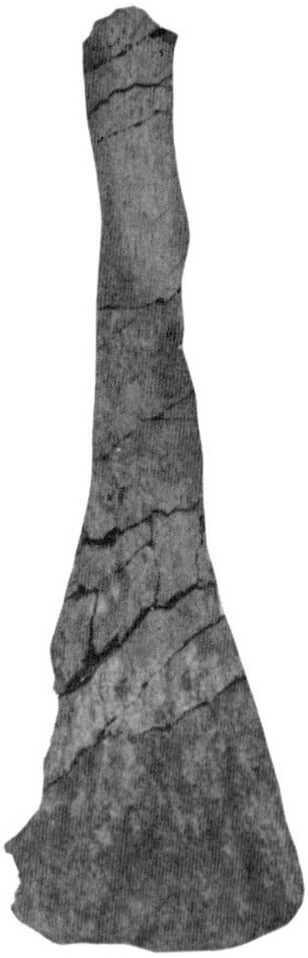

2

3

EXPLANATION OF PLATE 14.1

FIG. *1*—*Araucaria* sp. This specimen was collected from the highwall (Fruitland Formation) at San Juan Mine, Farmington, N.M. (By -0501; × 0.80).

2–3—*Palmoxylon* sp. *2*, palm axis collected from the Fruitland Formation near Cuba, N.M. (B177-1; × 0.07); *3*, palm axis collected from the Fruitland Formation near Cuba, N.M. (B177-2; × 0.08).

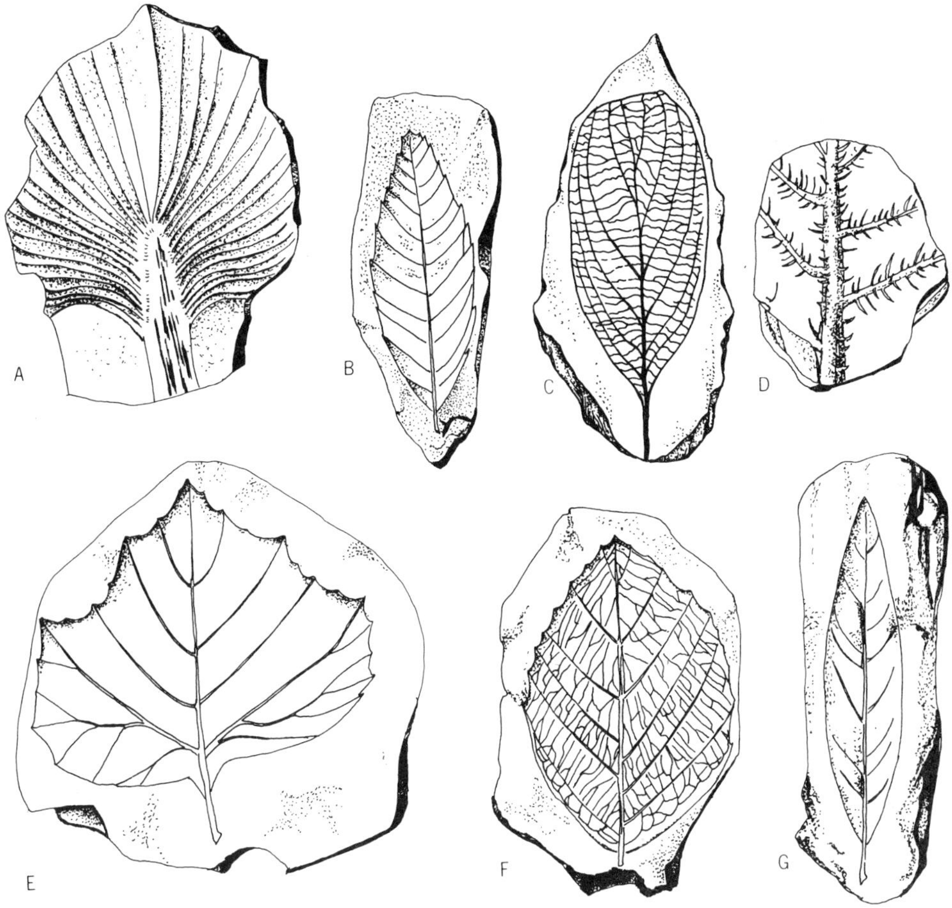

TEXT-FIG. *14.2*—Idealized illustrations of fossil plants of Late Cretaceous age having occurrences in the San Juan Basin, N.M. *A*, *Sabalites* sp., similar to *Sabalites imperialis* from the Fruitland Formation; *B*, *Carya* sp. similar to *Carya antiquorum* from the Fruitland Formation; *C*, *Dillenites cleburni* (Fruitland Formation); *D*, *Araucaria* sp. (Menefee Formation); *E*, *Platanus nobilis* (Kirtland Shale); *F*, *Viburnum antiquum* (Fruitland Formation); *G*, *Laurophyllum* sp. (Fruitland Formation). All × 0.5.

Kirtland floras. The Leguminosae (legume family) includes two species of *Leguminosites* (Pl. 14.3, fig. 3), whereas Saxifragaceae has only one species from these floras, which Knowlton (1916) assigned to *Ribes*. Other members of the flora are *Myrtophyllum*, *Cercidiphyllum* (Katsura Tree; Pl. 14.2, figs. 4, 5), *Menispermites*, a member of the moonseed family (Menispermaceae), and *Cornus* (Dogwood).

On a portion of an unidentified fossil dicotyledon leaf from the Fruitland Formation in the Bisti area, a leaf mine is preserved (Pl. 14.2, fig. 3). These are relatively rare in the fossil record. They are caused by an insect that, in its larval stage, lives between the upper and lower epidermal layers of dicotyledonous leaves. It eats the mesophyll of the leaf and leaves winding, S-shaped tunnels, known as leaf mines.

The two species of *Platanus* in the flora also occur in other Upper Cretaceous and Paleocene floras and may be of paleoecological significance. Parker (1976) considered *Platanus raynoldsi* (Text-fig. 14.1E) as an element of the bottomland community, but when associated with *Cissus marginata* and *Myrtophyllum torreyi*, as in this flora, it may represent one of the subordinate trees in the peat-forming community.

Anderson (1960) described a suite of palynomorphs from the Kirtland Shale near Cuba, New Mexico, consisting principally of dicotyledonous pollen. Spores were uncommon in both variety and number, and the coniferous grains were absent. The dominant dicotyledon in the florule was *Proteacidites thalmanni*. Other common dicotyledons were *Cupanieidites* and *Tilia*. A second suite was later described by Anderson (in Baltz et al., 1966) from a locality near Ojo Alamo in the western part of the basin. This suite was not at all similar to the first one, as it was dominated by polypodiaceous spores and a peculiar monosulcate grain with echinate-clavate sculpture. Conifer pollen was also common, whereas dicotyledon grains were uncommon. The florule includes *Liliacidites leei*, *L. hyalacinatus?*, and *Proteacidites thalmanni*.

The Fruitland flora grew in and around long, relatively narrow coastal swamps which formed a belt behind the barrier beaches of the sea (Fassett & Hinds, 1971). Individual swamps ranged in size from small bogs to inundated areas of more than 150 km^2. Water lilies, water lettuce, and pickleweed grew in the open water and marshy areas (Text-fig. 14.3). Several species of *Sequoia* lived in the swamps, rooted in the peaty soils along with several other woody plants, including the palms. These were the plants which seem to have contributed most to the accumulation of peat on these coastal plains. The ferns lived as part of the understory in these swamp forests. On the better drained bottomlands between the swamps grew the oaks and the hickories. Willows may have lived along the meandering rivers on the coastal plain, between the swamps and bottomlands. A few larger rivers crossed from the south and west, carrying sand to the beaches and fine clays and silt to the open sea. On the landward side of the swamps were broad, low flood plains (Fassett & Hinds, 1971). The climate was probably warm temperate to subtropical.

According to Dickinson et al. (1968) an important floristic change occurs near the top of the Kirtland Shale. A dramatic impoverishment of the flora as indicated by the palynomorphs took place, with the loss of 14

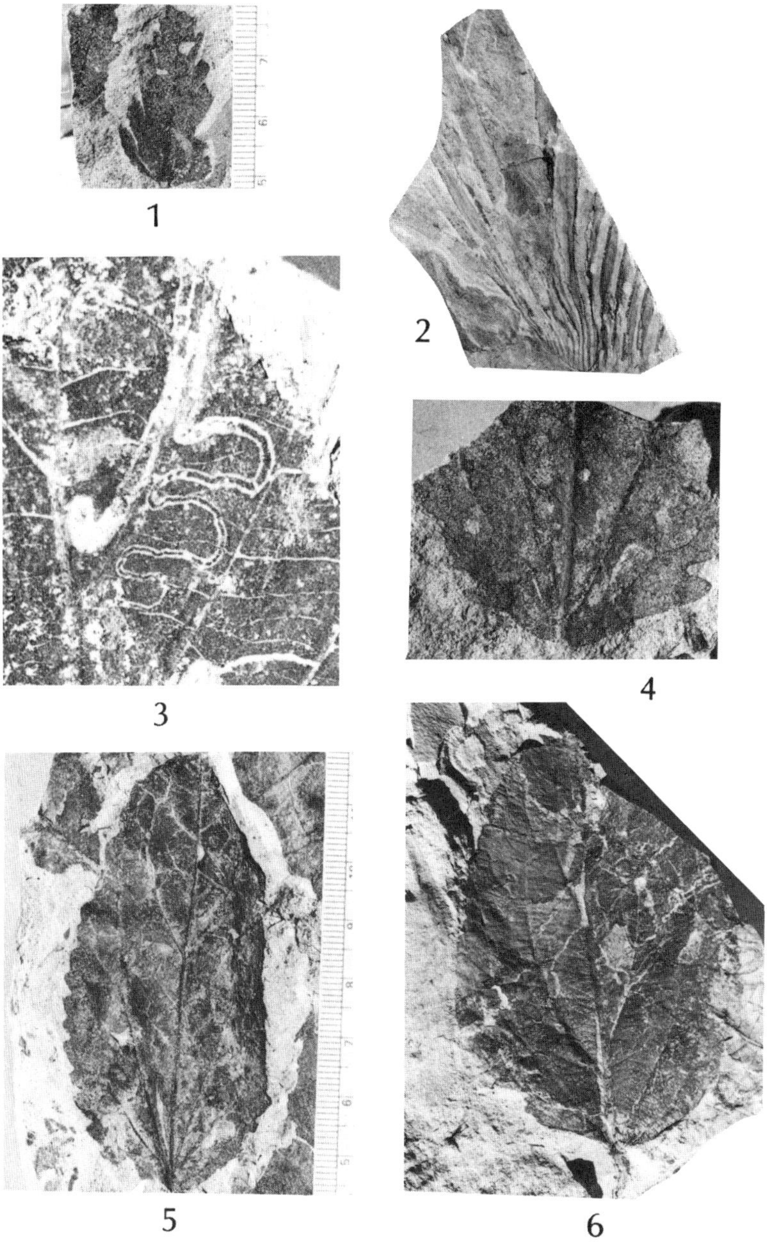

EXPLANATION OF PLATE 14.2

FIGS. *1–6*—Various specimens collected from the Fruitland Formation in the Bisti Badlands, N.M. All are × 0.8 except for fig. 3, which is × 1.2. *1*, *Asplenium? coloradense*; *2*, *Sabalites* (?) sp.; *3*, Leaf mine; *4*, *Cercidiphyllum* sp.; *5*, *Cercidiphyllum?* sp.; *6*, *Rhamnus goldianus*.

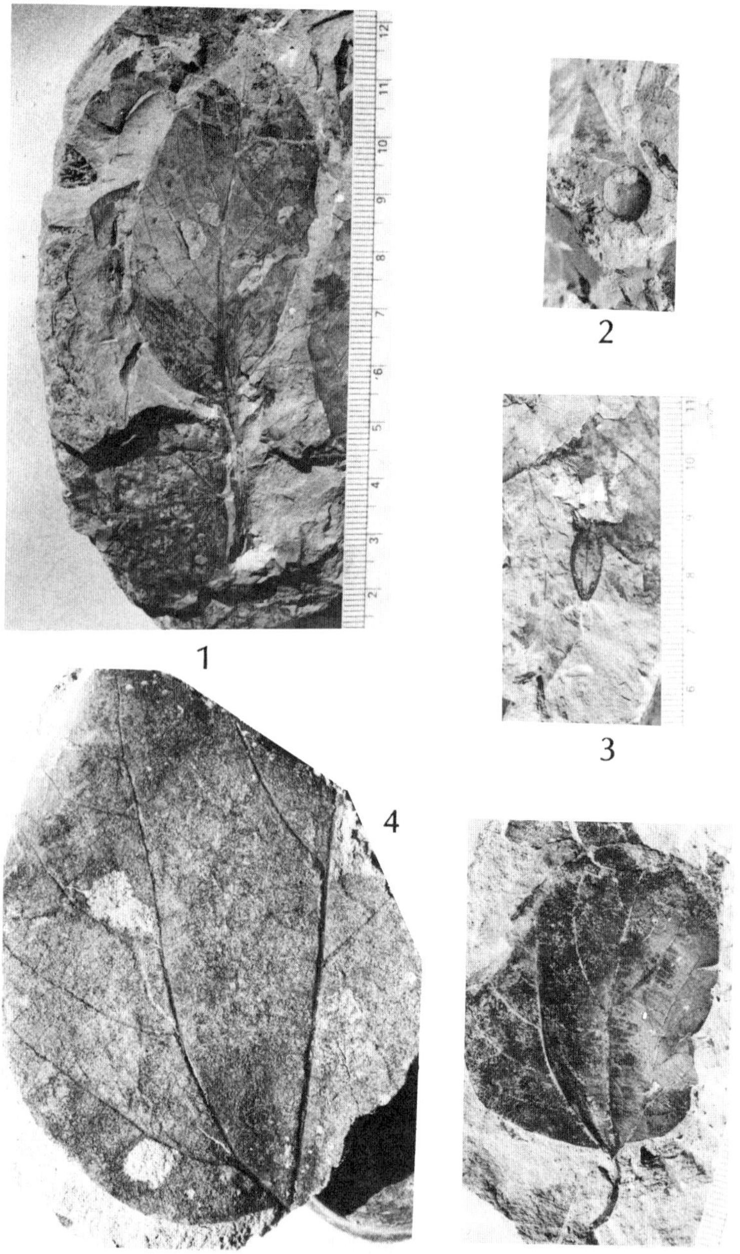

EXPLANATION OF PLATE 14.3

FIGS. *1–5*—Specimens from the Fruitland Formation, collected from the Bisti Badlands, N.M. All are × 0.8. *1*, *Magnolia cordifolia*; *2*, *Carpites* sp.; *3*, *Leguminosites* sp.; *4*, *Ficus praelatifolia*; *5*, *Ficus planicostata*.

TEXT-FIG. *14.3*—Reconstruction of the San Juan Basin during the time the Fruitland Formation was being deposited. The duckbill dinosaur, *Parasaurolophus*, lived amid these forests. The ceratopsian dinosaur *Pentaceratops* and the soft-shelled turtle *Trionyx* are at the right of the illustration.

species of angiosperms and fern taxa from the Fruitland Formation and the Kirtland Shale. In their place, abundant saccate conifer pollen of the Pinaceae and the Podocarpaceae suddenly appear along with certain new angiosperm groups. The change from what was undoubtedly a basin flora in the Fruitland and lower part of the Kirtland to an impoverished flora of possible upland environment at the top of the Kirtland suggests that an uplift occurred in the area during late Kirtland time.

PALEOCENE

Ojo Alamo Sandstone.—The flora of the Ojo Alamo Sandstone includes 8 species based on megafossils and 17 based on palynomorphs. Some of the megafossils are very fragmentary and only tentatively identified. The megafossils come from three localities south of Farmington (Bauer, 1916; Reeside, 1924). The flora includes two ferns, one monocotyledon, and five dicotyledons (Table 14.2). *Anemia*- and *Pteris*-like ferns are represented by fragmentary specimens, whereas the monocotyledoneae are represented by a fragmentary "palm leaf." The scarcity of monocots in this flora may not be real. Berry (1919) suggested that it in fact may be due to the lack of differentiation of the monocot foliage into lamina and petiole, which precludes the regular shedding of their leaves as in dicots. Therefore, the monocot leaves hung on and were flayed by the wind until they became almost unrecognizable.

The occurrence of two species of soapberry *(Sapindus)* in the flora could be important, as the genus may represent a Paleocene index fossil in the Rocky Mountains (Anderson, 1960). This genus has not been reported from Cretaceous floras of the western United States. Anderson (1960) studied a palynological florule from the Ojo Alamo Sandstone near Cuba, New Mexico and another from a locality near Ojo Alamo, in which podocarpaceous pollen was extremely abundant. This abundance suggests that the florules may have been decidedly more upland or temperate. However, modern *Podocarpus* is restricted to mountainous areas of about 2000 m elevation. Thus, the *Podocarpus* pollen in the florules may have been washed in from higher elevations. Anderson (1960) assumed that local uplift, secular change, or locally xeric conditions contributed to the temperate aspect of this florule.

In a second florule Anderson studied from the Cuba area, *Quercus* and palmlike pollen were dominant, whereas a mixed assemblage of upland and lowland types formed the subdominant flora. Some of these probably represent the more upland, xeric, or riparian environments surrounding the depositional site of this florule.

The flora of the Ojo Alamo Sandstone appears to have grown in lowland habitats under warm, humid climatic conditions.

Nacimiento and Animas Formations.—The floras of the Nacimiento and Animas Formations are considered together because the formations are more or less correlative and intertongue with each other (Baltz, 1967, fig. 27). Although the base of the Animas is lower than that of the Nacimiento, the plant fossils which were previously described from the Animas Formation (Knowlton, 1924) appear to have been collected from higher parts of the formation in strata equivalent to the Nacimiento.

Plant fossils were first collected from the Nacimiento Formation in the Barrel Spring Arroyo area and from a locality about 4.5 km south of Farmington. These fossils were identified by Knowlton (in Reeside, 1924, p. 38, 42). Later, additional fossil plant materials were reported from the Nacimiento by Kues et al. (1977). The Animas plant fossils were collected from several localities in southwestern Colorado.

There appear to be at least 58 species in the Nacimiento–Animas flora

Table 14.2—Flora of the Ojo Alamo Sandstone

Source of data: Bauer (1916) and/or Reeside (1924). These fossils were attributed to the Ojo Alamo Sandstone by O'Sullivan et al. (1972).

Filicophyta
- Polypodiaceae
  - *Anemia*-like
  - *Pteris*-like

Anthophyta
- Monocotyledoneae
  - Palmae
    - "palm-leaf?"
- Dicotyledoneae
  - Salicaceae
    - "willow-like leaf"
  - Moraceae
    - *Ficus* sp.
  - Platanaceae
    - *Platanus* or *Aralia*
  - Sapindaceae
    - *Sapindus* cf. *S. angustifolia*
    - *Sapindus* cf. *S. affinis*

(Table 14.3). The flora is dominated by angiosperms, especially the dicotyledons. The monocotyledons are represented by a single species, the ferns by two species, and *Equisetum* by a single species. No conifers have been reported from this flora.

The walnut family (Juglandaceae) is well represented in the flora, as it includes three species of walnut *(Juglans)*, one of which, *J. schimperi*, occurs widely in Paleocene and Eocene floras. The mulberry family (Moraceae) is purportedly one of the best represented families in the flora, containing possibly nine species most of which are attributed to figs *(Ficus)*. *Ficus* is also the dominant genus among the living Moraceae and includes about 60% of the existing species of this family. The Moraceae are distinctly tropical and warm temperate types. They are most abundant in the oriental tropics, although the dominant genus *Ficus* is widespread (Berry, 1916).

The laurel family (Lauraceae), the plane tree family (Platanaceae), and the honeysuckle family (Caprifoliaceae) are also present in the flora. Three genera of the buckthorn or coffeeberry family (Rhamnaceae) *Rhamnus*, *Paliurus*, and *Zizyphus* have been recognized in the flora. All three genera are common in both living and fossil floras. At present *Zizyphus* consists mostly of large shrubs and some small trees in the tropics of America, Africa, and Australia. The genus *Paliurus* (Jerusalem thorn) includes at least two living species which range from southern Asia to China and Japan.

Table 14.3—Floras of the Nacimiento and Animas Formations

Source of data: A. Animas Formation, N. Nacimiento Formation, Rp. Reeside (1924; Puercan plants), Rt. Reeside (1924; Torrejonian plants), Kn. Knowlton (1924), K. Kues et al. (1977).

- Equisetophyta
  - Equisetaceae
    - *Equisetum* sp. N, K
- Filicophyta
  - Schizaeaceae
    - *Anemia elongata* A, Kn
  - Polypodiaceae
    - *Allantodiopsis erosa* A, Kn
- Anthophyta
  - Monocotyledoneae
    - Palmae
      - *Sabalites grayana* A, Kn
  - Dicotyledoneae
    - Salicaceae
      - *Salix reesidei* A, Kn
      - *Populus* cf. *P. cuneata* N, Rp
    - Juglandaceae
      - *Juglans berryana* A, Kn
      - *Juglans schimperi* A, Kn
      - *Juglans? innominata* A, Kn
    - Fagaceae
      - *Fagus crossi* A, Kn
      - *Castanea intermedia* A, Kn
      - *Quercus* sp. N, Rp
    - Moraceae
      - *Ficus occidentalis* N, Rp, A, Kn
      - *Ficus* sp. N, K, A, Kn
      - *Ficus planicostata* A, Kn
      - *Ficus subtruncata* A, Kn
      - *Ficus denveriana* A, Kn
      - *Ficus monodon* A, Kn
      - *Ficus affinis* A, Kn
      - *Artocarpus lessigiana* N, Rt, A, Kn
      - *Artocarpus* sp. indet. N. Rp
    - Nymphaeaceae
      - *Nelumbo lakesiana* A, Kn
    - Magnoliaceae
      - *Magnolia hilgardiana* A, Kn
      - *Magnolia berryi* A, Kn
    - Lauraceae
      - *Laurus primigenia* A, Kn
      - *Laurus socialis* A, Kn
      - *Laurophyllum coloradensis* N, K
      - *Lindera obtusata* A, Kn
      - *Cinnamomum salicoides* A, Kn
      - *Persea brossiana* A, Kn
      - *Nectandra lacifolia* A, Kn
      - *Malapoenna sessiliflora* A, Kn
    - Hammelidaceae
      - *Liquidamber cucharas?* N, Rt
    - Platanaceae
      - *Platanus* cf. *P. haydenii* N, Rp
      - *Platanus aceroides* N, Rt, A, Kn
      - *Platanus nobilis* N, K
      - *Platanus raynoldsii* N, K, A, Kn
    - Rosaceae
      - *Crataegus? gigantea* A, Kn
    - Sapindaceae
      - *Sapindus obtusifolius* A, Kn
    - Rhamnaceae
      - *Rhamnus goldianus* N, Rt, A, Kn
      - *Paliurus zizyphoides* N, Rp, Rt
      - *Zizyphus fibrillosus* A, Kn
      - *Zizyphus lanceolatus* A, Kn
    - Vitaceae
      - *Vitis lobata* A, Kn
      - *Cissus marginata* A, Kn
    - Dilleniaceae
      - *Dillenites cleburni* A, Kn
    - Nyssaceae
      - *Nyssa? racemosa* A, Kn
    - Cornaceae
      - *Cornus?* sp. N, K
    - Bignoniaceae
      - *Dombeyopsis obtusa* N, Rt
    - Caprifoliaceae
      - *Viburnum lakesii?* N, Rp
      - *Viburnum* sp. N, Rp
      - *Viburnum rotundifolium* A, Kn
      - *Viburnum speciosum* A, Kn
      - *Viburnum tilioides* A, Kn
    - Incertae sedis
      - *Phyllites pagosensis* A, Kn
      - *Phyllites innominata* A, Kn

The palynomorphs in the Nacimiento Formation at a locality near Ojo Alamo and at two localities near Cuba, New Mexico were investigated by Anderson (1960; in Baltz et al., 1966). He found that dicotyledonous pollen was dominant in both number and variety at both localities and concluded that the physical environment of one florule was a mixed upland and lowland assemblage, whereas the other was from a low swampy area.

The composition of the Nacimiento–Animas flora suggests that it probably grew in swampy humid areas under a warm to hot climate. These conclusions are suggested by such examples as figs, palms, and laurels, as they are commonly associated in the tropics. Palms now exist under warm temperate conditions. Nearly all the Lauraceae are tropical, but several species extend into the southern United States. The genus *Cinnamomum* is especially characteristic of warm temperate rain forests. *Zizyphus* is mainly tropical, but withstands the winters of Louisiana.

EOCENE

San Jose Formation.—A small flora of leaves and leafy shoots was recently collected from the Regina member of the San Jose Formation near Santos Peak in the northeastern part of the San Juan Basin (Ash et al., 1980). This flora is particularly significant because it is the first to be reported from this formation. It includes a total of 15 species representing ferns, conifers, and flowering plants (Table 14.4).

Impressions of leaf fragments of the fossil fern *Acrostichum hesperium* are abundant in the flora. This species has very large leaves. The living species *Acrostichum aureum*, with which the fossil was compared by Newberry (1898), has leaves that are often as much as 1.5 m in length, with large pinnules that may be sterile in their lower part and fertile in the upper. It inhabits tropical areas throughout the world, particularly coastal areas. Possibly *A. hesperium* had leaves which were nearly as large. *A. hesperium* or a similar species occurs in the Eocene Claiborne and Green River floras.

The only conifer in the flora is *Taxodium olriki*. It is represented by more or less pointed needles that are inserted on the twig with their petioles parallel to the axis of the twig. This species probably grew close to swamps.

The angiosperms in this flora include willow *(Salix)*, several *Pterocarya*-like fruits and catkins, and the water plant *Paleonelumbo*. Among the Lauraceae in this flora, *Persea coriacea* is a species that also occurs in the Green River flora. This species is similar to the foliage of the living species *P. borbonia* of the southeastern United States and *P. cinerascens*,

Table 14.4—Flora of the San Jose Formation (from Ash et al., 1980)

Filicophyta	Lauraceae
Marattiaceae	*Cinnamomum* cf. *hesperium*
Danaea? sp.	*Lindera obtusata*
Polypodiaceae	cf. *Persea coriacea*
Acrostichum hesperium	Sapindaceae
	Sapindus dentoni
Coniferophyta	Myrtaceae
Taxodiadeae	*Eugenia americana*
Taxodium olriki	Araliaceae
	Araliophyllum sp.
Anthophyta	Leguminosae
Dicotyledoneae	*Leguminosites* sp.
Salicaceae	Incertae sedis
Salix sp.	*Carpites* sp.
Juglandaceae	
Pterocarya-like catkin	
Pterocarya-like fruits	
Nymphaceae	
Paleonelumbo macroloba?	

found in New Mexico (MacGinitie, 1969). Living members of *Lindera* (Benzoin) comprise approximately 80 species. Of this number, all but one or two are confined to warm temperate and tropical Asia from Japan and central China southward to Java. The occurrence of the water plant *Paleonelumbo* in the flora also suggests that this flora grew around swamps or a lake.

The species of soapberry, *Sapindus dentoni*, in the San Jose flora was previously described from the Green River flora. It has also been reported from the Eocene, middle and upper Claiborne and Jackson floras or the Mississippi embayment.

Eugenia americana of this flora resembles living *E. fluviatilis* or the rose apple *E. jambos* native to the tropics or subtropics of southeast Asia (MacGinitie, 1969). MacGinitie noted that *E. arnaceaeformis* from the Oligocene Florissant beds of Colorado may be an evolutionary derivative of *E. americana* which in turn may have been derived from *Myrtophyllum walderi* of the Dakota flora. *Araliophyllum* as defined by MacGinitie (1969) is comparatively rare in the San Jose flora and in the contemporary Green River flora.

It may be concluded that the San Juan Basin during Eocene time was in general a humid, forested region as none of the living corollaries of the fossil plants are typically associated with dry environments (Text-fig. 14.4). Most appear to characterize regions of ample rainfall and some probably grew in or near swampy areas.

TEXT-FIG. *14.4*—The large pantodont *Coryphodon* in a reconstruction of the environment of the Eocene San Jose Formation.

CONCLUSIONS

The floras of the San Juan Basin are abundant and varied. However, more extensive collecting and study is necessary before they will be completely understood. Although these floras have been known for some time, it is interesting to note that they have been collected and studied largely for stratigraphic purposes and have appeared simply as floral lists. Even Knowlton's descriptions in the only detailed publications on the floras of this area are very brief. The San Juan Basin floras are of importance because they can help us understand changes in the paleoenvironment of this region between Late Cretaceous and Eocene times. Also, they enable us to observe floral changes in this region over the same period of time, about 60 million years.

ACKNOWLEDGMENTS

The authors acknowledge with thanks the assistance of Mr. Leslie Smith of Brigham Young University in preparing the drawings for this paper. We also wish to express our appreciation to Dr. Coleman Robison of the Bureau of Land Management, Albuquerque, N.M., for reviewing the manuscript, and to the Bureau of Land Management for the use of some of the information from their reports.

REFERENCES

Anderson, J. A. R. 1961. The ecology and forest types of peat swamp forests of Sarawak and Brunei in relation to their silviculture. Unpublished Ph.D. dissertation, Univ. of Edinburg.

Anderson, R. Y. 1960. Cretaceous-Tertiary palynology eastern side of San Juan Basin, New Mexico. N.M. Bur. Mines Mineral. Res. Mem. 6, 59 p.

Ash, S. R. and C. B. Read. 1976. North American species of *Tempskya* and their stratigraphic significance. U.S. Geol. Surv. Prof. Paper 874, 42 p.

Ash, S. R., S. G. Lucas, and W. D. Tidwell. 1980. Paleontological Survey San Juan Planning Unit and Rio Puerco Resource Area, N.M. Unpublished report for U.S. Bur. Land Management, Albuquerque Office, 200 p.

Baltz, E. H., Jr. 1967. Stratigraphy and regional tectonic implications of part of Upper Cretaceous and Tertiary rocks, east-central San Juan Basin, New Mexico. U.S. Geol. Surv. Prof. Paper 552, 101 p.

Baltz, E. H., Jr., S. R. Ash, and R. Y. Anderson. 1966. History of nomenclature and stratigraphy of rocks adjacent to the Cretaceous-Tertiary boundary, western San Juan Basin, New Mexico. U.S. Geol. Surv. Prof. Paper 524D, 23 p.

Bauer, C. M. 1916. Contributions to the geology and paleontology of San Juan County, New Mexico; pt. 1. Stratigraphy of a part of the Chaco River Valley. U.S. Geol. Surv. Prof. Paper 98:271–278.

Berry, E. W. 1916. The lower Eocene floras of southeastern North America. U.S. Geol. Surv. Prof. Paper 91, 481 p.

————. 1919. Upper Cretaceous floras of the eastern gulf region in Tennessee, Mississippi, Alabama, and Georgia. U.S. Geol. Surv. Prof. Paper 112, 177 p.

———. 1924. The Middle and Upper Eucene floras of southeastern North America. U.S. Geol. Surv. Prof. Paper 92, 206 p.

Brown, R. W. 1962. Paleocene flora of the Rocky Mountains and Great Plains. U.S. Geol. Surv. Prof. Paper 375, 119 p.

Delevoryas, T. 1959. Investigations of North Americans cycadeoids: *Monanthesia*. Am. J. Bot. 46:657–666.

Dickinson, R. G., E. G. Leopold, and R. Marvin. 1968. Late Cretaceous uplift and volcanism on the north flank of the San Juan Mountains, Colorado. Colo. Sch. Mines Quart. 6:125–148.

Dilcher, D. L. 1971. A revision of the Eocene flora of southeastern North America. The Palaeobot. 20(1):7–18.

———. 1974. Approaches to the identification of angiosperm leaf remains. Bot. Rev. 40:1–157.

Dorf, E. 1938. Upper Cretaceous floras of the Rocky Mountain region; I, Stratigraphy and paleontology of the Fox Hills and Lower Medicine Bow Formations of southern Wyoming and northwestern Colorado. Carnegie Inst. Wash., D. C., Publ. 580:1–78.

———. 1942. Upper Cretaceous floras of the Rocky Mountain region; II, Flora of the Lance Formation at its type locality, Niobrara County, Wyoming. Carnegie Inst. Wash. D. C., Publ. 580:83–159.

Doyle, J. A., and L. J. Hickey. 1976. Pollen and leaves from the Mid-Cretaceous Potomac Group and their bearing on early angiosperm evolution, p. 139–205. *In* C. B. Beck (ed.), Origin and early evolution of the Angiosperms. Columbia Univ. Press, New York.

Eames, A. J. 1936. Morphology of vascular plants, lower groups. McGraw-Hill, New York, 433 p.

Fassett, J. E., and J. S. Hinds. 1971. Geology and fuel resources of the Fruitland Formation and Kirtland Shale of the San Juan Basin, N.M. and Colorado. U.S. Geol. Surv. Prof. Paper 676:1–76.

Gregory, H. E. 1917. Geology of the Navajo Country—a reconnaissance of parts of Arizona. U.S. Geol. Surv. Prof. Paper 93, 161 p.

Hickey, L. J., and J. A. Wolfe. 1975. The bases of angiosperm phylogeny: Vegetative morphology. Ann. Mo. Bot. Gard. 62(3):538–589.

Knowlton, F. H. 1916. Contributions to the geology and paleontology of San Juan County, N.M.; pt. 4. Flora of the Fruitland and Kirtland Formations. U.S. Geol. Surv. Prof. Paper 98:327–333.

———. 1924. Flora of the Animas Formation. U.S. Geol. Surv. Prof. Paper 134:71–112.

Kues, B. S., J. W. Froehlich, J. A. Schiebout, and S. G. Lucas. 1977. Paleontological survey, resource assessment, and mitigation plan for the Bisti–Star Lake area, northwestern N.M. Unpublished report for U.S. Bur. Land Management, Albuquerque Office, 1523 p.

Lee, W. T. 1912. Stratigraphy of the coal fields of northern central New Mexico. Geol. Soc. Am. Bull. 23:471–686.

———. 1917. Geology of the Raton Mesa and other regions in Colorado and New Mexico. U.S. Geol. Surv. Prof. Paper 101:9–221.

MacGinitie, H. D. 1969. The Eocene Green River flora of northwestern Colorado and northwestern Utah. Univ. Calif. Publ. Geol. Sci. 83, 202 p.

Newberry, J. S. 1861. Geological Report. Report on the Colorado River of the west by Lieut. J. C. Ives, part III (36th Congr., 1st sess.), 154 p.

———. 1898. The later extinct floras of North America: U.S. Geol. Surv. Mon. 35, 295 p.

Northrop, S. A. 1962. New Mexico's fossil record. New Mexico Quart. (supplement) 32:5–74.

O'Sullivan, R. B., C. A. Repenning, E. C. Beaumont, and H. G. Page. 1972. Stratigraphy of the Cretaceous rocks and the Tertiary Ojo Alamo Sandstone Navajo and Hopi Indian Reservations Arizona, New Mexico, and Utah. U.S. Geol. Surv. Prof. Paper 521-E, 65 p.

Parker, L. R. 1976. The paleoecology of the fluvial coal-forming swamps and associated floodplain environments in the Blackhawk Formation (Upper Cretaceous) of central Utah. Brigham Young Univ. Geol. Stud. 22(3):99–116.

Parker, L. R., and J. K. Balsley. 1979. Paleoecology of delta and coastal plain plant communities in the Upper Cretaceous Blackhawk Formation of Utah. Am. J. Bot. Misc. Publ. 157:35.

Penfound, W. T. 1952. Southern swamps and marshes: Bot. Rev. 18:413–446.

Reeside, J. B., Jr. 1924. Upper Cretaceous and Tertiary formations of the western part of the San Juan Basin of Colorado and New Mexico. U.S. Geol. Surv. Prof. Paper 134:1–70.

Roth, J. L., and D. L. Dilcher. 1979. Investigations of angiosperms from the Eocene of North America: stipulate leaves of the Rubiaceae including a probable polyploid population. Am. J. Bot. 66:1194–1207.

Shaler, M. K. 1907. A reconnaissance survey of the western part of the Durango–Gallup coal field of Colorado and New Mexico. U.S. Geol. Surv. Bull. 316:375–426.

Simpson, J. H. 1852. Journal of a military reconnaissance from Santa Fe, N.M. to the Navajo Country. Rept. Sec. War with Reconn. of routes from San Antonio to El Paso, 31st Cong., 1st Sess., Senate Ex. Dec. 64:146–148.

Stevens, N. E. 1912. A palm from the Upper Cretaceous of New Jersey. Am. J. Sci. 34:421–436.

Tidwell, W. D. 1975. Common fossil plants of western North America. Brigham Young Univ. Press, Provo, 197 p.

Tidwell, W. D., G. F. Thayn, and J. L. Roth. 1976. Cretaceous and early Tertiary floras of the Intermountain area. Brigham Young Univ. Geol. Stud. 22(3):77–98.

Wieland, G. R. 1928. Associate in Paleontology, reports on cycadeoids. Carnegie Inst. Wash. D. C. Yearb. No. 27:390–391.

————. 1930. Researches in paleobotany: the Mesa Verde cycadeoids. Carnegie Inst. Wash. D. C. Yearb. No. 29:243–247.

————. 1934. Fossil cycads, with special reference to *Raumeria reichenbachiana* Geopport sp. of the Zwinger of Dresden. Palaeontogr. 79B:85–130.

————. 1936. Cycadeoid investigations and field work. Carnegie Inst. Wash. D. C. Yearb. No. 34:333–340.

Part 5
Paleoecology

15 Dinosaur communities of the San Juan Basin: A case for lateral variations in the composition of Late Cretaceous dinosaur communities

Spencer G. Lucas
Department of Geology and Geophysics and
Peabody Museum of Natural History, Yale University

"...most contemporary authorities recognize the principle of dinosaurian habitat zonation in a broad sense."

—Langston (1974, p. 93)

INTRODUCTION

During the last decade there has been heated debate over the physiology of dinosaurs. Functional morphological, histological, and paleoecological evidence has been presented to support the view that dinosaurs were not the ponderous lethargic ectotherms reconstructed by earlier students, but rather they were sleek agile endotherms whose metabolism resembled that of modern birds and mammals more than that of modern reptiles. The paleoecological evidence, first introduced by Bakker (1972), entails the quantitative reconstruction of dinosaur trophic dynamics. This approach, further elaborated by Bakker, Farlow, and Beland and Russell, involves the acceptance of numerous assumptions about the structure and composition of dinosaur communities and assumes that the fossil record faithfully preserves these communities. Many feel that the acceptance of some of these assumptions is not warranted, and that the reconstruction of dinosaur trophic dynamics thus oversteps the limits imposed by the inadequacy of the fossil record (see reviews by Benton, 1979, and Ostrom, 1980). But, in spite of this, such a novel approach to dinosaur paleoecology has had a heuristic effect in generating further interest in the subject to either confirm or refute and replace its results.

This novel approach has coincided with and, in part, reawakened interest in collecting dinosaurs, particularly with the object of acquiring fossils from a precisely known sedimentary context so as to have taphonomic, biostratinomic, and distributional data for use in paleoecological studies. A quantum jump in our knowledge of dinosaurs, dinosaur occurrences, and their sedimentary context has resulted. This has been particularly true of the Late Cretaceous (Campanian–Maastrichtian) of North America and Asia, largely due to the work of Dodson, Gradzinski, Langston, Morris, Rozhdestvensky, D. A. Russell, and many others.

The Late Cretaceous dinosaur faunas of the San Juan Basin, known principally from fieldwork early in this century by Barnum Brown, C. W. Gilmore, and C. H. Sternberg, have been somewhat neglected in these new developments. The claim to fame of the San Juan Basin Upper Cretaceous traditionally lay in possessing the peculiar hadrosaur *Parasaurolophus*, the endemic ceratopsian *Pentaceratops* and the last of the North American sauropods, *Alamosaurus sanjuanensis*. Because of the rarity of complete or nearly complete skeletons, the Upper Cretaceous beds of the San Juan Basin have not previously been considered a particularly promising fossil field.

Recent restudy of these beds, however, is beginning to show otherwise and has revealed a much more diverse fauna than hitherto expected. Unlike most other Upper Cretaceous dinosaur-bearing beds, those of the San Juan Basin comprise a long uninterrupted sequence of terrestrial deposits containing dinosaur remains that were preserved in two significantly different environments of deposition. The San Juan Basin record thus allows study of how the composition of fossil assemblages containing dinosaurs varies with respect to depositional environments in the same geographical area. The study of the differences between two Late Cretaceous dinosaur communities reconstructed from San Juan Basin dinosaur assemblages from two different depositional environments is the object of this paper. These dinosaur assemblages provide evidence that there were lateral variations in the composition of dinosaur communities, that different habitats ranging from coastal deltas and interdeltaic plains to inland basinal floodplains would have been populated by different dinosaurs.

COMMUNITIES, PHYSICAL ENVIRONMENTS AND ECOSYSTEMS

The reconstruction of communities, physical environments and ecosystems undertaken here employs the following chain of reasoning:

1) The distribution of fossils in the sediments forms the basis for the recognition of fossil assemblages, "any group of fossils from a suitably restricted stratigraphic interval and geographic locality" (Fagerstrom, 1964, p. 1198). Two broad fossil assemblages are utilized herein: one

from the Upper Cretaceous Fruitland Formation in the western Central San Juan Basin, and the other from the Naashoibito Member of the overlying Kirtland Shale, also in the western Central Basin.

2) These assemblages come from relatively thin sedimentary sequences which may represent relatively short intervals of time (ASSUMPTION 1).[1] The assemblages thus are assumed to represent two past communities (ASSUMPTIONS 2 and 3). A past community is ideally construed as "any assemblage of populations living in a prescribed area or physical habitat; it is an organized unit to the extent that it has characteristics additional to its individual and population components and functions as a unit through coupled metabolic transformations" (Odum, 1971, p. 140). This does not mean that it is assumed that such a past community is completely represented by the fossil assemblage, but rather that the assemblage *qualitatively* represents a significant part of at least the vertebrate members of the past community (ASSUMPTION 3). Each community identified is named after a characteristic vertebrate herbivore.

3) The identified communities are depicted by food webs which are hypothesized patterns of the trophic interactions and habitat preferences of their members (ASSUMPTIONS 4, 5 . . . N). Some elements of the fossil assemblage do not appear in the food web, reflecting their exclusion from the community based on my judgment that they represent allochthonous elements of the assemblage (ASSUMPTION 4). An obvious problem with the food webs is that they betray our ignorance about the feeding specificity of extinct organisms and thus tend to both obscure and complicate the "actual" trophic relationships. Thus, for example, the feeding specificity of a piscivorous turtle on only certain types of fishes is unknown. Such a turtle is necessarily portrayed in the web as potentially feeding on most (or all) fishes in its habitat, thus obscuring the possiblity of specific feeding preferences and complicating the web by adding energy vectors. A similar argument applies to organisms whose trophic relationships change through ontogeny since the web necessarily casts all members of the community into one trophic scheme, generally that of their adult stages.

Despite these drawbacks, I believe that food webs are more useful than some recent attempts to portray past communities in matrices compartmentalized according to habitat and feeding preference (e.g., Sloan, 1970, fig. 1; Sahni, 1972, fig. 16).[2] These are static, although admit-

1. All assumptions listed in this section are discussed in Appendix 15.1.

2. One logical outgrowth of this compartmentalization has been the identification of megaterrestrial, microterrestrial, semiaquatic and aquatic "communities" (e.g., Sloan, 1970) each of which is a subdivision of a community (as defined here) based on body size and/or substrate preference. Throughout this paper I informally refer to such subdivisions as "associations."

tedly somewhat less speculative portrayals of community organization that belie the complexity and dynamism inherent to terrestrial communities. Food webs seem to depict better (as well as can be done on a two dimensional surface) this complexity and dynamism.

4) Coincident with the reconstruction of the communities is the reconstruction of their corresponding past physical environments. The determination of the broad physical environment (for example, "inland floodplain") is based initially on the sediments (ASSUMPTION N + 1), not utilizing fossil data so as to avoid circularity when later comparing variations in the communities with respect to variations in the physical environments. Feedback between fossil data and sedimentary data is pertinent in further elaborating certain aspects of the past environment to arrive at a dim, though palpable, picture of the past ecosystem (= the nexus formed by community + physical environment) (ASSUMPTION N + 2).

5) Analysis of variations in community composition with respect to different physical environments, however, necessitates separating the communities from the sedimentary environments. All communities (assemblages) compared here are Campanian-Maastrichtian in age and are assumed to have been more or less coexistent (ASSUMPTION N + 3). Thus the variable of time, which can confound such a comparison, is eliminated from the analysis.

6) A pattern is identified (that different environments contained differently constituted dinosaur communities) in the San Juan Basin record and is compared with what is known of some other Late Cretaceous dinosaur assemblages. This suggests, in a broad way, that the pattern of variation seen in the San Juan Basin was also present in other parts of North America and in part of Asia during the Late Cretaceous. Some implications of the broad recognition of this pattern are discussed, but no explanation of how the pattern came about is offered.

The above chain of reasoning thus proceeds from the discernment of low level patterns (fossil assemblages, sedimentary facies) through the acceptance of assumptions to build scenarios (past communities, past physical environments) and culminates in the identification of a higher level pattern of the distribution of the scenarios with respect to each other. If this higher level pattern is accepted, a hypothesis of the mechanism(s) by which it was generated leads to another series of assumptions that are beyond the scope of this paper.

GEOLOGIC SETTING

Introduction.—The Upper Cretaceous terrestrial deposits of the San Juan Basin, the Fruitland Formation and Kirtland Shale, represent sediments deposited chiefly by rivers during the interval between the last retreat of the epicontinental sea from northwestern New Mexico during the Campanian and the deposition of the Ojo Alamo Sandstone at the onset of the Tertiary as a response to the uplift of the San Juan Mountains (Fassett & Hinds, 1971; Powell, 1973). Although there is an extensive outcrop belt of the Fruitland and Kirtland Formations in the San Juan Basin (Fassett & Hinds, 1971, pl. 1), most knowledge of their contained fossils comes from a relatively limited area in the western Central Basin (Text-fig. 15.1). In this area both formations are best exposed and appear to be most fossiliferous, and it is thus in this area that virtually all productive collecting has gone on.

Structure.—The San Juan Basin is a broad subcircular downwarping of the Earth's crust. It is usually construed as being some 40 to 50 thousand km^2 in extent, bounded on the west by the Defiance Uplift and Carrizo Mountains near the New Mexico–Arizona border, on the east by the Nacimiento Uplift and "Monocline" and extending north from the Zuni and Lucero Uplifts near Grants and Gallup to the San Juan Uplift near Durango in southern Colorado (Kelley, 1950). The basin, as thus defined, includes a section some 4,500 m thick of sedimentary rocks ranging in age from Cambrian to Holocene. All outcrops of the Fruitland and Kirtland Formations are within Kelley's (1950) "Central Basin," a much smaller area approximately demarcated by the outcrop belt of the Pictured Cliffs Sandstone (Text-fig. 15.1).

The structure of the Central Basin is relatively simple with bedding dipping gently (3–5°) towards the basin axis. The axis trends roughly southeast to northwest (Baltz, 1967, fig. 5) and passes northeast of the geographic center of the basin, giving it a distinct asymmetry. Some faults are present locally in the southern and western Central Basin, but major faults and folds are confined to its eastern margin. There the Laramide deformation, which, among other things, resulted in the Nacimiento Uplift, has produced a complex structural picture (Baltz, 1967).

Although the tectonic evolution of the San Juan Basin may have begun as early as the late Paleozoic (Kelley, 1950), it was not until the latest Cretaceous and early Tertiary that the present basin was formed (Kelley, 1950; Baltz, 1967; Tweto, 1975). Sometime during the deposition of the Kirtland Shale a changeover occurred from sediment sources primarily south and west of the basin to those north and east in the form of the rising Laramide geanticline (Baltz, 1965, 1967; Fassett & Hinds, 1971; Tweto, 1975).

Rock-stratigraphy and biostratigraphy.—The rock-stratigraphic sequence from the Lewis Shale to Nacimiento Formation (Text-fig. 15.2) has mostly been worked out by U.S. Geological Survey stratigraphers in this century (notably Bauer, 1916; Reeside, 1924; Dane, 1936, 1946; Baltz et al., 1966; Baltz, 1967; Fassett & Hinds, 1971; O'Sullivan et al., 1972) with some important contributions by other workers (especially

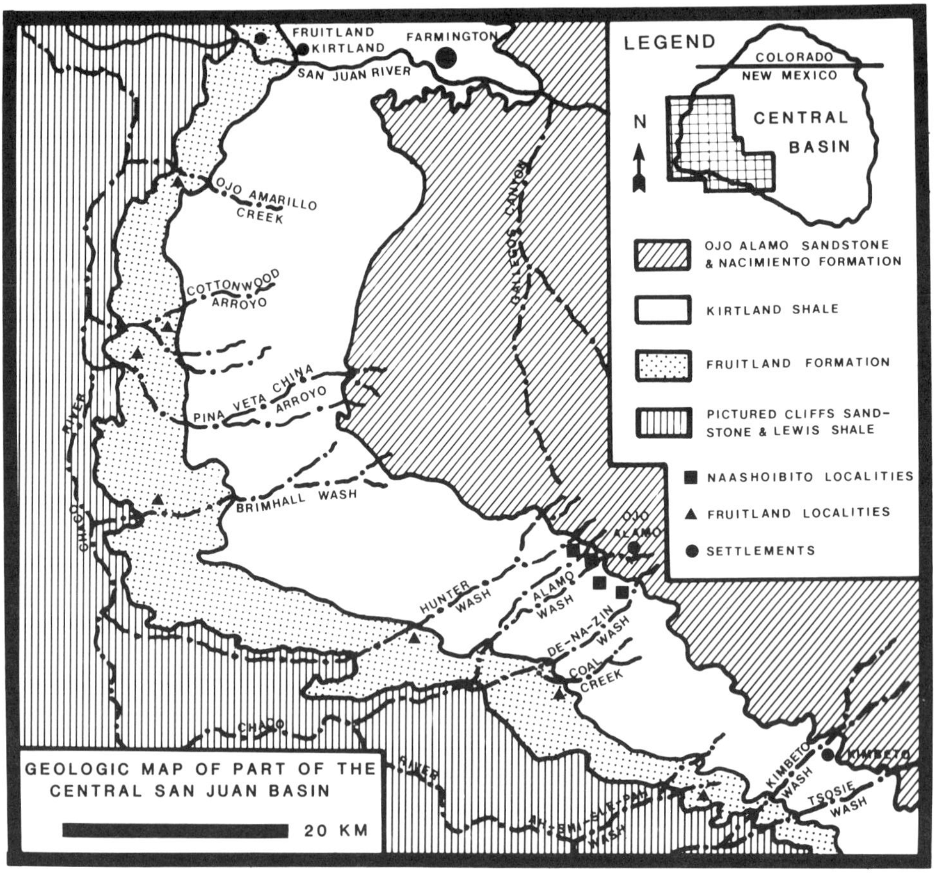

TEXT-FIG. *15.1*—Geologic map of part of the western Central San Juan Basin showing major tributaries of the Chaco River mentioned in the text and general location of most known fossil localities in the Fruitland Formation and Naashoibito Member of the Kirtland Shale. Geology after Reeside (1924, pl. 1) and Fassett & Hinds (1971, pl. 1); fossil localities after Bauer (1916) and Kues et al. (1977).

Brown, 1910; Sinclair & Granger, 1914; Simpson, 1959), all building on the pioneering work in the last century of Cope, Holmes, Newberry and Shaler. Not surprisingly, most remaining stratigraphic problems center on the rocks immediately adjoining the Cretaceous–Tertiary boundary, especially the Animas, McDermott and Ojo Alamo Formations. Problems with the Animas and McDermott largely stem from a lack of detailed paleontological and stratigraphic study in the northern part of the basin, an area where structural complexities and limited outcrop obscure the stratigraphic picture.

The problem of the Ojo Alamo began with Brown's (1910) original and somewhat nebulous use of the term (and Sinclair & Granger's [1914] subsequent upward extension of Brown's) "Ojo Alamo beds." The result has been a polemical literature divided between those who restrict definition of the Ojo Alamo Sandstone mainly to a widespread conglomeratic sandstone (Baltz et al., 1966) and those who prefer to consider the Ojo Alamo as two conglomeratic sandstones with a shale-soft sandstone sequence sandwiched in between (Clemens, 1973; Fassett, 1973; Powell, 1973; Butler et al., 1977). These latter workers thus profess to retaining some allegiance to Brown's original proposal and the later work of Sinclair & Granger (1914) and Bauer (1916), although they have not yet reached a consensus on what nomenclature is most appropriate.

I here follow and urge the acceptance of the restricted definition of the Ojo Alamo Sandstone of Baltz et al. (1966) because my field experience has shown it to be the most readily mappable and persistent unit and thus the most useful definition of this rock-stratigraphic unit. The objection raised by some paleontologists (e.g. Clemens, 1973) that in so doing the "Ojo Alamo Fauna" (Gilmore, 1919) is placed in a rock unit other than the Ojo Alamo should not be a deterrent to this procedure. Although Gilmore's use of a rock-stratigraphic name for a faunal unit was common practice in his day, it was an unfortunate choice that is best forgotten. Gilmore's "Ojo Alamo Fauna" came from what is now termed the Naashoibito Member of the Kirtland Shale; the fauna of the Naashoibito Member is best called the Alamo Wash local fauna (Lehman, this volume).

The biostratigraphy of the Lewis Shale to Nacimiento Formation sequence, like the rock-stratigraphy, is beset with some problems, although for the most part the age of these units is well established. Ammonites from the Lewis Shale and Pictured Cliffs Sandstone securely place these rocks in the Campanian along the western edge of the Central Basin, the rocks becoming younger in age toward the northeast (Cobban, 1973; Cobban et al., 1974). The lower part of the Fruitland Formation intertongues with the Pictured Cliffs and is thus of Campanian age in the western Central Basin. The dinosaur fauna (Gilmore, 1919) supports this; corroboration is also lent by pollen (Tschudy, 1973) and the somewhat less precise evidence of megafossil plants (Knowlton, 1916) and nonmarine invertebrates (Stanton, 1916). The mammals of the Hunter Wash local fauna (Clemens, 1973) do not contradict such an age, but they do indicate that the Fruitland–Kirtland sequence may range up into the early Maastrichtian.

The presence of a major unconformity at the base of the Ojo Alamo Sandstone has been documented by several workers (Baltz et al., 1966; Baltz, 1967; Fassett & Hinds, 1971; O'Sullivan et al., 1972; Powell, 1972, 1973; Lucas & Rigby, 1979). The evaluation of the temporal significance of this unconformity has been one of the most trying problems of San Juan Basin biostratigraphy. Lehman (this volume) suggests

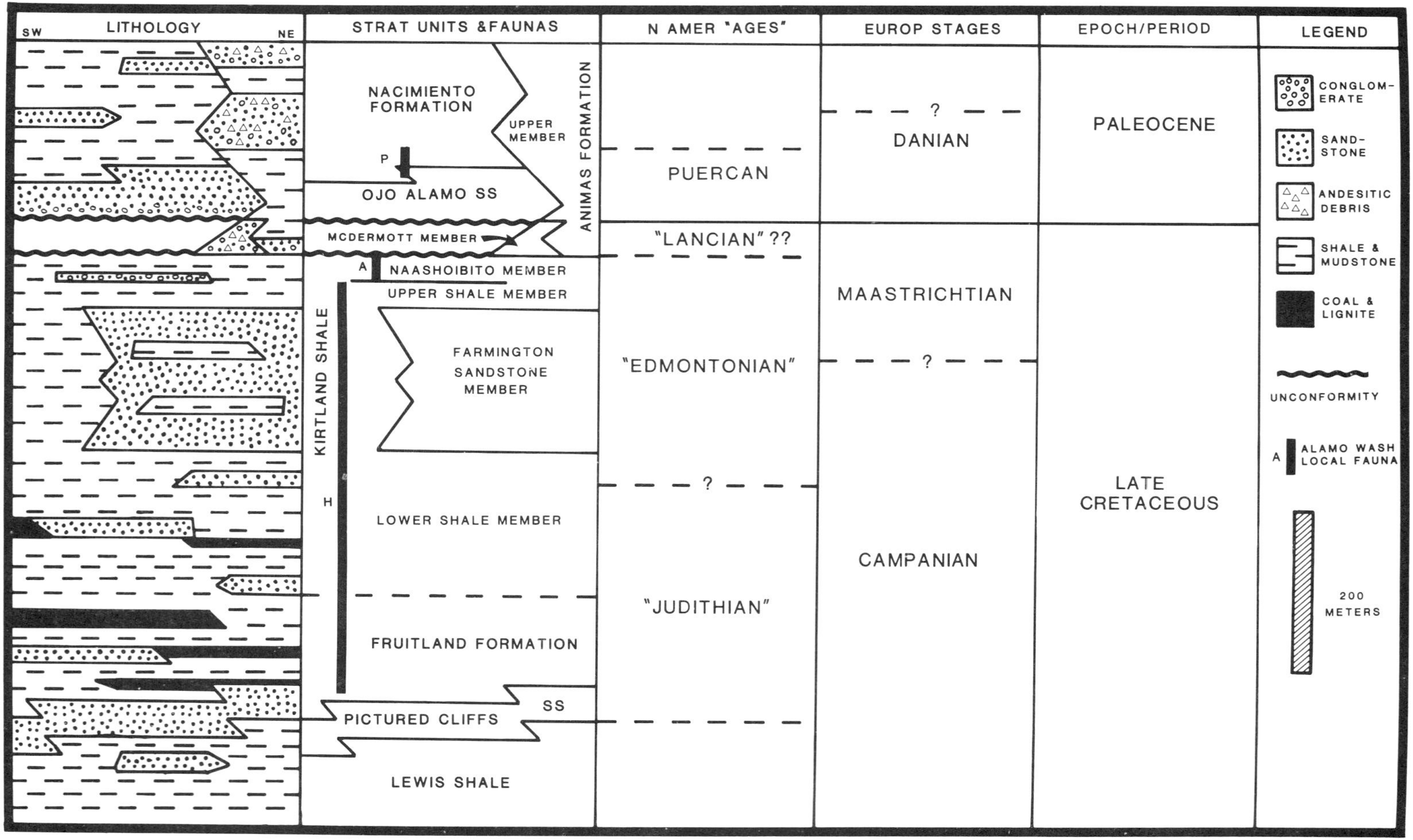
LITHOLOGY
SW
NE
STRAT UNITS &FAUNAS
N AMER "AGES"
EUROP STAGES
EPOCH/PERIOD
LEGEND
NACIMIENTO FORMATION
UPPER MEMBER
ANIMAS FORMATION
P
OJO ALAMO SS
MCDERMOTT MEMBER
A
NAASHOIBITO MEMBER
UPPER SHALE MEMBER
KIRTLAND SHALE
FARMINGTON SANDSTONE MEMBER
H
LOWER SHALE MEMBER
FRUITLAND FORMATION
PICTURED CLIFFS
SS
LEWIS SHALE
PUERCAN
"LANCIAN"??
"EDMONTONIAN"
?
"JUDITHIAN"
?
DANIAN
MAASTRICHTIAN
?
CAMPANIAN
PALEOCENE
LATE CRETACEOUS
CONGLOM-ERATE
SAND-STONE
ANDESITIC DEBRIS
SHALE & MUDSTONE
COAL & LIGNITE
UNCONFORMITY
A
ALAMO WASH LOCAL FAUNA
200 METERS

that the Alamo Wash local fauna, and hence Naashoibito Member of the Kirtland Shale, is a Lance equivalent, but earlier workers (Gilmore, 1919, 1946; Baltz, et al., 1966) have considered it pre-Lance, particularly because it contains the hadrosaurs *Kritosaurus*[3] and *Parasaurolophus*. Because the evidence remains equivocal (Lucas & Rigby, 1979), no solution to this problem is proposed here, and I thus adopt a conservative stance in considering the Alamo Wash local fauna as pre-Lancian (Text-fig. 15.2). This suggests that a significant portion of Maastrichtian time is not represented by sediments preserved in the southern and western Central Basin, and this is the most striking break in the Lewis–Nacimiento sequence.

The age of the Ojo Alamo Sandstone is Paleocene, as indicated by pollen (Baltz, et al., 1966; Fassett & Hinds, 1971), its intertonguing relationship with the Paleocene Nacimiento Formation (Baltz et al., 1966; O'Sullivan & Beikman, 1963; Lucas & Rigby, 1979), and the presence of Paleocene mammals in one of its tongues (Rigby & Lucas, 1977; Lucas & Rigby, 1979; Rigby, this volume). Thus the boundary between rocks of Cretaceous and Tertiary age in the San Juan Basin is at an uncomformity,[4] a situation that would please nineteenth century stratigraphers

3. Baird and Horner (1977), followed by Brett-Surman (1978) and Horner (1979) considered *Kritosaurus* a synonym of *Hadrosaurus*. In this paper, however, I retain the name *Kritosaurus* pending substantiation of this synonymy as well as to maintain continuity with earlier literature on the San Juan Basin in which the name *Kritosaurus* has figured prominently since Brown (1910) coined it for a skull he found near Ojo Alamo.

4. Powell (1973, p. 118–119) reported that "mammal and reptile bone scraps were found as low as three feet above the base of the Kimbeto Member [= Ojo Alamo Sandstone of this paper] and as much as 30 feet above the base of the Kimbeto Member" at a site "in the NW¼ sec. 32 T25N, R12W off a small tributary of Hunter's Wash." Although he pointed out that these fossils were "waterworn and undiagnostic" (p. 119), Powell named this assemblage the "Monisco local fauna" (probably a *lapsus*, intending *Moncisco* for nearby Moncisco Wash). He further asserted that "the Kimbeto Member of the Ojo Alamo

TEXT-FIG. *15.2*—Rock-stratigraphic and biostratigraphic framework for the late Campanian-early Paleocene sedimentary sequence in the San Juan Basin from which come the fossils that form the basis of the communities reconstructed in this paper. Vertical scale is keyed to the average thickness of lithologic units. Lithology is schematic, and lithologic units essentially follow Baltz (1967, fig. 27). Faunas (see Matthew, 1937; Clemens, 1973; Powell, 1973; Lehman, this volume) are: A = Alamo Wash local fauna, H = Hunter Wash local fauna, P = Puerco local fauna(s). North American "ages" and their tentative correlation with European stages are from Wood et al. (1941) and L. S. Russell (1975).

but that makes twentieth century stratigraphers uneasy. The lower part of the Nacimiento Formation is Paleocene in age and bears the "type" fauna of the early Paleocene (Puercan) land mammal "age" (Wood et al., 1941). The Animas Formation is mostly Paleocene in age, although its basal McDermott Member ranges into the Cretaceous (Reeside, 1924; Baltz et al., 1966); more rock-stratigraphic and biostratigraphic study of the Animas Formation is needed.

I refrain from presenting an absolute time scale for the San Juan Basin Late Cretaceous because such a scale would have to be inferred from evidence that is mostly extrabasinal or that I consider unreliable. The late Campanian is approximately 70 to 74 million years ago, and the end of the Cretaceous is about 65 million years ago (Van Hinte, 1978). No radiometric dates within this interval are given because no datable rocks are known to be present in the Lewis–Nacimiento sequence except in the northern part of the Basin (Dickinson et al., 1968), and their precise stratigraphic and temporal relationships to rocks further south are presently somewhat unclear. A recent date of 66.1 ± 4.7 million years (Lindsay et al., 1979) from a reworked ash in the Kirtland Shale is the only date from the western Central Basin, and it has such a large margin of error and is based on so small a sample that its reliability is questionable.

The only other attempt at a nonbiostratigraphic correlation of the Lewis–Nacimiento sequence is the recent magnetostratigraphic work of Butler et al., (1977) and Lindsay et al., (1978, 1981) on the interval from the Fruitland through Nacimiento Formations in the western Central Basin. Their polarity column could be used to infer absolute dates for this interval. However, these workers did not fully take into account the fundamentally discontinous nature of continental sedimentation and

---

Sandstone must be considered Cretaceous at least 20 feet above its base, and Paleocene where dinosaur remains can no longer be found" (Powell, 1973, p. 119).

I recently examined the locality described by Powell in the SE¼ NW¼ NW¼ sec. 32 T25N, R12W and confirmed his observations by finding several *in situ* dinosaur bone fragments about 2 m above the base of the Ojo Alamo (Text-fig. 15.8f). One of these fragments, placed in the University of New Mexico Department of Geology collections (UNM TOA-1), is typical in being a well-abraded angular chunk of bone, large enough (5 cm long and 6 cm wide) to be considered dinosaurian. However, the Ojo Alamo at this locality contains much gravel and petrified wood, including a prone fossil log more than 10 m in length (Text-fig. 15.8e). This evidence of high energy of deposition supports the hypothesis that these bone fragments are reworked, and I thus consider their presence insufficient evidence to justify moving the Cretaceous–Tertiary boundary up to within the Ojo Alamo Sandstone. Even though such bone fragments may occur as much as 30 ft above the base of the Ojo Alamo, this cannot be considered definitive evidence against their being reworked, since in some areas the Ojo Alamo has channeled down through 60 or more ft of the underlying Kirtland Shale (Baltz et al., 1966, plate 1).

In this paper I do not use Powell's "Monisco local fauna" because I feel that naming as a local fauna an undiagnostic assemblage of reworked bone fragments is hardly justifiable.

failed to recognize the potential gap in their polarity column represented by the unconformity at the base of the Ojo Alamo Sandstone; they thus assumed a continuous, unbroken section and correlated it with the type Late Cretaceous–early Tertiary polarity sequence from Gubbio, Italy (e.g., LaBrecque et al., 1977). This correlation indicates that what Butler et al. (1977) call anomaly 31 in the San Juan Basin corresponds to most of the Fruitland Formation, and thus the Fruitland–Kirtland sequence is of late Maastrichtian age. Not only does this paleomagnetic correlation contradict the biostratigraphic evidence, but if it is correct, it will necessitate a major retiming of the paleogeographic history of the Western Interior during the Late Cretaceous (McGookey et al., 1972) as well as a retiming of some Laramide tectonic events in the southern Rocky Moutains (Tweto, 1975).

The Upper Cretaceous sedimentary sequence in the San Juan Basin is typical of most continental deposits in being characterized by numerous breaks in sedimentation, ranging from ubiquitous bedding plane paraconformities to at least one major regionally angular unconformity. It is *not* a continuous record of Late Cretaceous–early Tertiary sedimentation, and this should always be borne in mind when drawing conclusions from it.

ADEQUACY OF THE RECORD

Previous studies.—The history of study of the Fruitland and Kirtland Formations and their fossil biota has already been discussed by Reeside (1924), Baltz et al. (1966), Fassett (1973), Clemens (1973), Lehman (this volume), Mateer (this volume), Simpson (this volume), and others. Major works on the stratigraphy and paleontology of the Fruitland and Kirtland are: Cope (1885), Brown (1910, 1914), Hay (1910), Sinclair & Granger (1914), Bauer (1916), Knowlton (1916), Stanton (1916), Gilmore (1916, 1919, 1921, 1922, 1935), Reeside (1922, 1924), Osborn (1923), Wiman (1930, 1931, 1932, 1933), Sternberg (1932), Dane (1936, 1946), Anderson (1960), Ostrom (1961, 1963), Baltz et al. (1966), Baltz (1967), Dickinson et al. (1968), Powell (1969, 1972, 1973), Fassett & Hinds (1971), Tschudy (1973), Fassett (1973, 1974), Clemens (1973), Mateer (1976, this volume), Kues et al. (1977, 1980), Armstrong-Ziegler (1978), Rowe et al. (this volume), Lehman (this volume), and Tidwell et al. (this volume). Rather than recount the history of this study here, I discuss several points that are of importance in considering the adequacy of the study of the Fruitland and Kirtland Formations and their fossils, and thus the appropriateness of using these rocks and fossils for paleoecological reconstructions. While seemingly much work has been done and the record here is remarkably rich in places, much remains unknown.

Sedimentology.—No paleoecological study can ignore the sedimentary context in which fossils occur. But, as Powell (1972, p. 42) recently put it, "previous studies of the geology of formations near or adjacent to the Cretaceous–Tertiary boundary in the San Juan Basin have dealt largely with non-quantitative stratigraphic problems; sedimentary structures have usually been described in very little detail. Quantitative measurements of such parameters as crossbed directions, variation in grain size or pebble imbrication are almost non-existent." Thus, although there has been a wealth of stratigraphic study of the Upper Cretaceous rocks of the San Juan Basin, there have been few sedimentological studies aimed at reconstructing depositional environments. Consequently, the reconstructions of sedimentary environments presented here are based largely on my observations made during the course of fossil collecting, section measuring, and mapping. These observations are coupled with my synthesis of previously published observations, most of which are general descriptions of lithologies, measured sections, geologic maps, and incidental discussions of depositional environments. The environments reconstructed in this paper are, therefore, fairly general, underscoring the need for more detailed sedimentological studies in the San Juan Basin. Such studies are slowly appearing (Powell, 1972, 1973; Erpenbeck & Flores, 1979; Flores, 1979) and hopefully will gain momentum in the future.

Paleobotany.—Without doubt the most poorly understood fossils in the San Juan Basin are those of plants. With the exception of Tidwell et al. (this volume), only Knowlton (1916) has published significantly on the megafossil plants of the Fruitland and Kirtland Formations. Plants are briefly discussed in Sinclair & Granger (1914), Reeside (1924), and some of the older literature, mostly listing identifications and comments by Knowlton. The palynology of the Fruitland, Kirtland, and Ojo Alamo has received somewhat more attention, principally by Anderson (1960; Baltz et al., 1966), Leopold (Dickinson et al., 1968), and Tschudy (1973; in Fassett & Hinds, 1971), but much remains to be done here as well.

This neglect of the Late Cretaceous fossil plants of the San Juan Basin is surprising in view of the abundance and generally high quality of preservation of floral material here. Excellent leaf localities, at which there are many whole leaves with preserved cuticle, are present in the Fruitland and Kirtland. Large petrified logs, often well enough preserved to discern cell structure, are locally abundant in the Fruitland and Ojo Alamo.

The lack of extensive paleobotanical study renders detailed paleoclimatic interpretations impossible, and only very general statements about paleoclimates are presented in the ensuing discussion. Lack of paleobotanical study has also prohibited detailed reconstruction of the vegetational component of the communities considered here.

Invertebrate paleontology.—The marine invertebrate faunas of the San Juan Basin, including those of the Lewis Shale and Pictured Cliffs Sandstone, have been studied by many workers, and important paleoecological observations on the fauna of the Lewis Shale were recently presented by Mannhard (1976). In contrast, the Late Cretaceous nonmarine invertebrates (mostly unionids and freshwater gastropods) have received little study, only Stanton (1916) and Reeside (1924), even though they are locally abundant, especially in the Fruitland Formation. To my knowledge, only indirect evidence of insects has ever been reported from the Lewis–Nacimiento sequence: leaf mines from the Fruitland Formation (Tidwell et al., this volume) and insect borings preserved in petrified wood from the Nacimiento Formation (Brues, 1936). This evidence, coupled with the fact that most insect orders (including ants and termites) are known at least as far back as the middle Campanian (Carpenter, 1977), has led me to assume the presence of a diverse and abundant insect fauna in the communities reconstructed, an insect fauna presumably not unlike that found in comparable environments at present. Many of these insects were important primary consumers who provided a food source for insectivorous and omnivorous vertebrates, thus occupying a key position in the food web. As the unionids and gastropods are not as important in this regard, the relative lack of study of these forms is less of a hindrance than might first be supposed.

Vertebrate paleontology.—Since the major concern of this paper is with vertebrates, the discussion of the adequacy of present knowledge of their fossil record forms a leitmotif throughout the ensuing discussion and need not be elaborated upon here. Suffice it to say that although the Late Cretaceous vertebrate fossil record of the San Juan Basin has been known and studied for about a century, much remains to be done, especially with regard to the taphonomy and biostratinomy of the vertebrate occurrences and the collection and study of microvertebrates (mammals, lizards, amphibians etc.). The most complete synthesis of what has been done and what is known of this record is Kues et al. (1977), and I owe a heavy debt to this diverse compendium for aiding in the assimilation of the data upon which this paper is based.

FRUITLAND ECOSYSTEM

Introduction.—Reconstruction of the ecosystem present during the time of deposition of the Fruitland Formation is principally based on the rocks and fossils from a relatively small area in the western part of the Central Basin near the eastern border of the Navajo Indian Reservation, especially along Hunter Wash, Coal Creek and Ah-shi-sle-pah Wash (Text-fig. 15.1). This area has produced the overwhelming majority of

TEXT-FIG. *15.3*—View of badlands of the Fruitland Formation along the south branch of Hunter Wash looking north-northeast across Sec. 34, T24N, R13W and beyond. The dark sandstone capping the butte in the center foreground corresponds to Dane's (1936, p. 113) "stratigraphically highest brown sandstone" and has been used by some workers to arbitrarily mark the Fruitland-Kirtland contact.

Fruitland fossils and contains the most extensive and spectacular outcrops of the formation (Text-fig. 15.3). Much collecting, utilizing a variety of methods (surface prospecting, quarrying, screen washing), has yielded a very diverse assemblage of fossil plants, invertebrates, and vertebrates which form the basis for the community reconstructed below.

Physical environment.—As the Cretaceous epicontinental sea retreated from northwestern New Mexico during Campanian time ("R3" or regressive phase of the Clagget cycle: Weimer, 1960; McGookey et al., 1972), rivers flowed northeastward towards the regressing shoreline. The sediments deposited along the coastal reaches of these rivers and the swamps and ponds of their poorly drained floodplains constitute the Fruitland Formation. These sediments intertongue extensively with the littoral sands and shales of the Pictured Cliffs Sandstone which, in turn, grade into the deeper marine deposits of the Lewis Shale.

Fruitland sediments were derived principally from volcanically active highlands southwest of the San Juan Basin (Fassett & Hinds, 1971) and are a relatively thin (up to 170 m) sequence of interbedded sandstone, shale, mudstone, silt and lignitic to bituminous coal (Text-fig. 15.3). Many of the sandstones are medium- to coarse-grained, moderately well sorted subarkoses[5] (Burgener, 1953), most of which are cross-bedded and have a shoestring geometry typical of point bar deposits of meandering streams. Some sheet sands are present, probably representing splay deposits. The shales and mudstones are generally gray, buff, or olive in color; most represent the overbank deposits of the meandering streams. The silts and some shales form local lenses that were probably pond deposited, though no extensive lakes are evident. The frequent and sometimes thick (up to about 10 m) coals were deposited in extensive swamps, confirming that deposition of the Fruitland Formation took place in a predominately reducing environment.

In the area where it has been most intensively studied, the Fruitland is extremely fossiliferous. The shales and mudstones contain abundant fossil leaves of various angiosperms, conifers and ferns. Fossil logs (Text-fig. 15.4a) and *in situ* stumps are also common and include at least two fossil "forests," one almost exclusively of taxodiaceous conifers (Text-fig. 15.4b) and the other of palms (Tidwell et al., this volume). Oyster beds and coquinas of nonmarine invertebrates (unionids and gastropods) are sometimes found in the shales and mudstones (Text-fig. 15.4c). Vertebrate remains are generally fragmentary, predominately isolated dinosaur bones (Text-fig. 15.4d) or partial turtle shells, and are found in all lithologies, though a concentration in the sandstones is noticeable. A few relatively complete dinosaur skeletons and skulls have been recovered (Osborn, 1923; Ostrom, 1961, 1963; Rowe et al., this volume). Some rich accumulations of small vertebrate remains have also been discovered, usually in gravelly pockets at the base of channel sandstones (Text-fig. 15.4e & f).

In general, the Fruitland Formation was deposited in a spectrum of environments ranging from strict delta plain (Erpenbeck & Flores, 1979) to interdeltaic and resembles deposits formed under similar conditions in the Western Interior during the Late Cretaceous (Weimer & Land, 1975). The more palpable picture that emerges is of a swampy and heavily vegetated coastline. The coastal belt of swamps during the time of Fruitland deposition was probably about 10 to 25 km wide, as suggested by the present geometry of the coal deposits (Fassett & Hinds, 1971). Estuaries and river courses near the shoreline were brackish and inhabited by *Ostrea* and various selachians, a condition typical of such coastlines today. A subtropical climate with possible wet to dry seasonality is

---

5. All petrologic terminology used in this paper follows Folk (1974); descriptive terms for sedimentary structures and environments (e.g., "delta plain," "meandering stream," etc.) follow standard usage as recently compiled in the glossary of Friedman and Sanders (1978).

A

B

C

D

E

F

suggested by, among other things, the presence of palms and ferns (Knowlton, 1916; Mahabale, 1954) and presence of various lower vertebrates (especially certain salamanders and turtles). The luxuriant flora probably supported a diverse and abundant fauna of insects and vertebrate herbivores who, in turn, supported a number of primary (and inevitably secondary) carnivores. Sedimentary influx in a poorly drained setting and high resident biomass contributed to the preservation of a large number of fossils. The disarticulated vertebrate remains in part reflect transport, but scavenging and decomposition were probably also significant contributing factors. Though the elements of the Fruitland assemblage that are especially rare and fragmentary, like dromaeosaurs and pachycephalosaurs, are evidently allochthonous, most elements are abundant and occasionally represented by complete or nearly complete remains, and thus probably represent organisms that actually lived along the 10 to 25 km wide swampy coastline during late Campanian time.

Parasaurolophus community.—The vertebrate fauna preserved in the coastal deposits of the Fruitland Formation is a taxonomically diverse array of fishes, amphibians, lizards, a snake, turtles, crocodiles, dinosaurs and mammals (Table 15.1). More collecting and study will un-

TEXT-FIG. *15.4*—Typical fossil occurrences in the Fruitland Formation. *A*, prone fossil log at University of New Mexico (UNM) locality 77-10 on the south branch of Hunter Wash. Rock hammer is 28 cm long; *B*, *in situ* petrified stump, probably of a taxodiaceous conifer (D. Tidwell, oral communication, 1979), one of a "forest" of such stumps in the headlands of Coal Creek, UNM locality 77-799. Rock hammer is 28 cm long; *C*, part of a coquina bed of unionids and gastropods at UNM locality 77-812 in the headlands of Coal Creek. Diameter of camera lens cap is 5.5 cm.; *D*, part of a hadrosaur tail eroding out of an arroyo bottom in the headlands of Coal Creek. Rock hammer is 28 cm long; *E*, Dane's (1936, p. 113) "stratigraphically highest brown sandstone" marking the top of the Fruitland Formation just above UNM locality 77-467 (see *F*). Rock hammer is 28 cm long; *F*, UNM locality 77-467 on the south branch of Hunter Wash. This locality is at the base of the white sandstone below the caprock in fig. *E*. Small "pebbles" around the camera lens cap are actually bone fragments, fish scales and teeth of crocodiles, rays, and dinosaurs. Screen washing at this locality has also produced teeth of multituberculates and therian mammals, as well as abundant lower vertebrate remains. Diameter of camera lens cap is 5.5 cm.

Table 15.1—Fossil vertebrates from the Fruitland Formation and Kirtland Shale (exclusive of the Naashoibito Member). Acronyms following names indicate sources in which the taxa were first reported: AZ = Armstrong-Ziegler (1978); C3 = Clemens (1973); G5 = Gilmore (1935); G6 = Gilmore (1916); G9 = Gilmore (1919); K7 = Kues et al. (1977); O1 = Ostrom (1961); Ob3 = Osborn (1923); P9 = Powell (1969); W0 = Wiman (1930); W1 = Wiman (1931); W2 = Wiman (1932); W3 = Wiman (1933). Baenids have been revised following Gaffney (1972); tyrannosaurids follow D. Russell (1970a); mammals follow Clemens et al. (1979). The stratigraphic ranges of some turtles previously listed as occurring in the Fruitland Formation (e.g., Wiman, 1933; Gaffney, 1972) have been revised following Mateer (this volume). Taxonomic names in quotation marks are probably invalid. Essentially all mammals reported by Clemens (1973) were collected from the Fruitland Formation (W. Clemens, oral communication, 1979), but some may also have come from the Kirtland; uncertainty about the exact boundary between the Fruitland and Kirtland in some areas has prevented resolution of this problem at present.

	FRUITLAND	KIRTLAND
Chondrichthyes		
Selachii		
Hybodontidae		
Lonchidion selachos (AZ)	X	
Isuridae		
Isurus (Oxyrhina) sp. (AZ)	X	
Orectolobidae		
Squatirhina americana (AZ)	X	
Squatirhina sp. (AZ)	X	
Squatinidae		
gen. & sp. indet. (G5)		X
Batoidea		
Dasyatidae		
Myledaphus bipartitus (AZ)	X	
Myledaphus sp. (G6)		X
Pristidae		
Ischyrhiza avonicola (AZ)	X	
Osteichthyes		
Acipenseriformes		
Acipenseridae		
Acipenser eruciferus (AZ)	X	
Polyodontidae		
Paleopsephurus wilsoni (AZ)	X	
Amiiformes		
Amiidae		
Amia fragosa (AZ)	X	
Lepisosteiformes		
Lepisosteidae		
Lepisosteus occidentalis (AZ)	X	
Lepisosteus sp. (G6)	X	X
Elopiformes		
Phyllodontidae		
Paralbula casei (AZ)	X	
Perciformes		
Sciaenidae		
Platacodon nanus (AZ)	X	

Amphibia		
Anura		
Discoglossidae		
Scotiophryne pustulosa (AZ)	X	
Pelobatidae		
?*Eopelobates* sp. (AZ)	X	
Urodela		
Prosirenidae		
Prodesmodon ?copei (AZ)	X	
Batrachosauroididae		
Opisthotriton kayi (AZ)	X	
?Urodela		
"*Cuttysarkus mcnallyi*" (AZ)	X	
Reptilia		
Testudines		
Baenidae		
"*Baena*" *nodosa* (G6, G5)	X	X
"*Baena*" *ornata* (G5)		X
Baena sp. (G5)		X
Boremys pulchra (G5, P9)	X	X
Neurankylus eximius (G6)		X
Thescelus insiliens (W3)		X
Compsemys sp. (AZ)	X	
Dermatemydidae		
Adocus bossi (G9, G5)	X	X
Adocus kirtlandius (G9, P9)	X	X
Adocus sp. (G9)	X	
Basilemys nobilis (W3)		X
?*Basilemys* sp. (AZ)	X	
Trionychidae		
Aspideretes ovatus (G5)		X
Aspideretes vorax (G5)		X
Aspideretes sp. (G9)	X	
Plastomenus robustus (G9)		X
Plastomenus sp. (G6)		X
Platypeltis sternbergi (W3)		X
Trionyx sp. (AZ)	X	
Sauria		
Teiidae		
Leptochamops denticulatus (AZ)	X	
Chamops segnis (AZ)	X	
Anguidae		
cf. *Gerrhonotus* sp. (AZ)	X	
gen. & sp. indet. (AZ)	X	
Serpentes		
Aniliidae		
Coniophis cosgriffi (AZ)	X	
Crocodilia		
Goniopholidae		
Goniopholis kirtlandicus (W2)		X
Crocodylidae		
Brachychampsa sp. (G6, AZ)	X	X
Crocodylus sp. (G9, K7)	X	X

Leidyosuchus sp. (AZ)	X	
?*Thoracosaurus* sp. (AZ)	X	
Saurischia		
Coeluridae		
Paronychodon lacustris (AZ)	X	
?Coeluridae		
gen. & sp. indet. (AZ)	X	
Ornithomimidae		
gen. & sp. indet. (K7)	X	
?Dromaeosauridae		
gen. & sp. indet. (AZ)	X	
Tyrannosauridae		
Albertosaurus sp. (G5)		X
"Deinodon horridus" (AZ)	X	
gen. & sp. indet. (G6, G9)	X	X
Ornithischia		
Ceratopsidae		
Chasmosaurus sp. (G5)		X
Monoclonius? sp. (G6)	X	
Pentaceratops fenestratus (W0)		X
Pentaceratops sternbergii (Ob3, G5)	X	X
Hadrosauridae		
"Kritosaurus" navajovius (G6, K7)	X	X
Parasaurolophus cyrtocristatus (O1)	X	
Parasaurolophus tubicen (W1)		X
gen. & sp. indet. (G6, K7)	X	X
Pachycephalosauridae		
gen. & sp. indet. (K7, AZ)	X	
Ankylosauridae		
gen. & sp. indet. (K7)	X	
Nodosauridae		
gen. & sp. indet. (G5)		X
Mammalia		
Multituberculata		
Neoplagiaulacidae		
Mesodma sp. (C3)	X	
Mesodma? sp. (AZ)	X	
new gen.? & new sp(p).? (AZ)	X	
?Neoplagiaulacidae		
Cimexomys, cf. *C. judithae* (C3)	X	
Ptilodontidae		
cf. *Kimbetohia campi* (C3)	X	
Cimolodontidae		
Cimolodon sp. (C3)	X	
Eucosmodontidae		
new gen. & sp. (C3)	X	
Family *incertae sedis*		
Essonodon? sp. (AZ)	X	
Marsupialia		
Didelphidae		
Alphadon cf. *marshi* (C3)	X	X
Alphadon?, new sp. (C3)	X	
cf. *Peradectes* sp. (AZ)	X	

Pediomyidae	
Pediomys cf. *cooki* (C3)	X
Eutheria	
"Leptictidae"	
Gypsonictops new sp. (C3)	X
Palaeoryctidae	
cf. *Cimolestes* sp. (C3)	X
Eutheria	
order indet. (C3)	X

doubtedly add to this diversity, but most of the major groups that were present already appear to be known (though birds are conspicuously absent), and the fauna of the Fruitland broadly resembles the best sampled contemporaneous faunas from the Western Interior (e.g., L. S. Russell, 1964). Fruitland megafossil plants are summarized by Knowlton (1916) and nonmarine invertebrates by Stanton (1916) and Reeside (1924). These groups are in dire need of restudy and are probably more diverse than indicated. I here term the community of organisms that occupied the swampy coastline during the time of Fruitland deposition the *Parasaurolophus* community, choosing a characteristic hadrosaur as the basis of the eponym. Based on autecological hypotheses of the feeding and habitat preferences of organisms preserved in the Fruitland fossil assemblage (Appendix 15.2), a hypothetical food web has been constructed (Text-fig. 15.5). Remains of ankylosaurs, pachycephalosaurs and dromaeosaurs are so rare and fragmentary that I judge them to be allochthonous elements of the assemblage and thus exclude them from the *Parasaurolophus* community and food web. Within this food web the following are worthy of note:

1) A large and complicated association of mostly euryhaline fishes forms a striking part of the community. Many of these fishes have close modern analogs that are either piscivores or generalized carnivores (Appendix 15.2). Teleosts are relatively rare in the Fruitland assemblage and of low diversity, which may result from the fact that many teleost groups had not yet diversified and/or entered North American coastal environments by the Late Cretaceous (Estes, 1970). In the rarity of teleosts and in its general composition, the aquatic association of the *Parasaurolophus* community resembles other aquatic associations represented by assemblages in Late Cretaceous coastal deposits known from the Western Interior (e.g., Estes, 1964; Estes et al., 1969; Estes & Berberian, 1970).

2) Many of the fishes, particularly freshwater forms, were a major food source for turtles and crocodiles of a semiaquatic association that also included amphibians, though they probably fed mainly on terrestrial and aquatic arthropods. This association of the *Parasaurolophus* community also closely resembles other semiaquatic associations represented by assemblages of the Western Interior Late Cretaceous (e.g.,

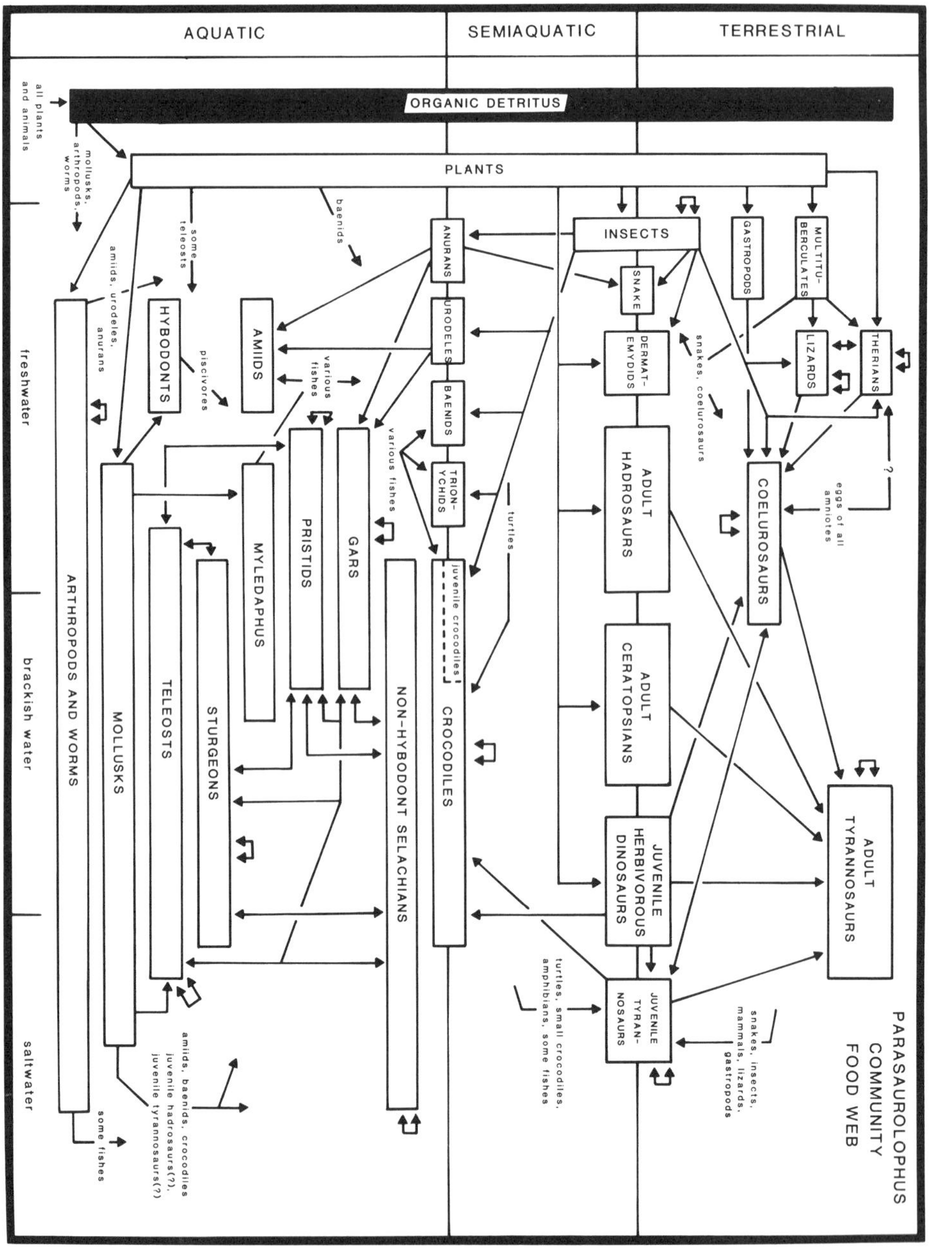

AQUATIC
SEMIAQUATIC
TERRESTRIAL
ORGANIC DETRITUS
all plants and animals
PLANTS
mollusks, arthropods, worms
freshwater
brackish water
saltwater
some teleosts
baenids
amids, urodeles, anurans
piscivores
various fishes
ARTHROPODS AND WORMS
MOLLUSKS
HYBODONTS
TELEOSTS
STURGEONS
AMIDS
MYLEDAPHUS
PRISTIDS
GARS
NON-HYBODONT SELACHIANS
ANURANS
URODELES
BAENIDS
TRION-YCHIDS
juvenile crocodiles
CROCODILES
turtles
INSECTS
SNAKE
DERMAT-EMYDIDS
GASTROPODS
MULTITU-BERCULATES
LIZARDS
THERIANS
snakes, coelurosaurs
COELUROSAURS
eggs of all amniotes
?
ADULT HADROSAURS
ADULT CERATOPSIANS
JUVENILE HERBIVOROUS DINOSAURS
ADULT TYRANNOSAURS
JUVENILE TYRAN-NOSAURS
turtles, small crocodiles, amphibians, some fishes
snakes, insects, mammals, lizards, gastropods
amids, baenids, crocodiles juvenile hadrosaurs(?), juvenile tyrannosaurs(?)
some fishes
PARASAUROLOPHUS COMMUNITY FOOD WEB

Estes, 1964, 1970; Estes et al., 1969; Estes & Berberian, 1970). Indeed, a semiaquatic association of turtles (principally baenids and trionychids), crocodiles, salamanders, anurans, and champsosaurs (absent in the Fruitland) is a recurring "complex" (*sensu* Olson, 1966) in most Late Cretaceous and early Tertiary (Paleocene-Eocene) assemblages of Western North America (e.g., Estes, 1964, 1970, 1976).

3) In the terrestrial zone there is an association of small vertebrates, mostly lizards and mammals, that may have been partly or mostly arboreal. Its trophic resource base probably was mostly terrestrial plants and insects. Numerically predominate as fossils are the marsupials and ptilodontoid multituberculates, quite similar to other associations of "typical latest Cretaceous" mammals in the Western Interior (Van Valen & Sloan, 1977; Clemens & Archibald, 1980).

4) What Sloan (1970) termed a "megaterrestrial community" is a somewhat self-contained trophic nexus of adult dinosaurs. The resource base of this association was in semiaquatic and terrestrial plants eaten by hadrosaurs and ceratopsians, who in turn were the primary food source for the top carnivores of the *Parasaurolophus* community, the tyrannosaurs.

An important consideration in the reconstruction of dinosaur food webs is the recognition that in the course of its ontogeny a dinosaur passed through a large range of sizes, from weights of a few grams or kilograms on hatching to fully adult sizes of several thousand kilograms (Case, 1978). Thus, at different growth stages a dinosaur, especially a carnivorous dinosaur, would have occupied significantly different positions in the food web (Farlow, 1976a; Tracy, 1976).[6] Recognition of

6. This, of course, assumes that there was little or no parental care of juvenile dinosaurs, or at least that parental care of juvenile dinosaurs was not as extensive as in most birds and mammals. Horner & Makela (1979), however, have recently reported what they believe to be a "nest" of juvenile hadrosaurs, suggesting that some sort of parental care occurred among hadrosaurs. It seems likely that parental care similar to that seen in modern crocodilians (Neill, 1971) may have occurred in these, as well as other dinosaurs. However, it is just as likely that at a relatively early age and small size a young dinosaur would have "entered" the food web and had to procure its own food.

TEXT-FIG. *15.5*—Hypothesized food web of the *Parasaurolophus* community. Animals are subdivided according to inferred substrate preference. Relative size of boxes and names of organisms is of no significance. Arrows indicate inferred direction of flow of trophic energy (see Appendix 15.2). "Insects" and "arthropods and worms" are not known from body fossils in the Fruitland Formation but are assumed to have been present (see text and Appendix 15.2). Compare with Text-fig. 15.9.

these changes in feeding have been incorporated into the food web (Text-fig. 15.5) by including juvenile tyrannosaurs and juvenile herbivorous dinosaurs as entities separate from their adult counterparts. This lessens the apparent uncoupling of the mammal-lizard terrestrial association, adult dinosaur association, semiaquatic association and aquatic association by highlighting animals that bridged the gaps. Consequently, juvenile tyrannosaurs at some stage in their growth may have eaten mammals, lizards, turtles, frogs, insects, in short, any small animal (Farlow, 1976a), and may also have provided a food source for adult tyrannosaurs (Tracy, 1976). Some juvenile herbivorous dinosaurs conceivably competed with herbivorous mammals for the same plant items; such "diffuse competition" may have contributed to the dinosaur's extinction (Van Valen & Sloan, 1977). Probably some form of juvenile dinosaur-mammal competition occurred throughout the Mesozoic.

The Fruitland dinosaur assemblage is numerically dominated by hadrosaurs and ceratopsians (Table 15.2) represented by a wide variety of remains, ranging from complete skeletons to isolated bones and fragments. In contrast, carnosaurs are exceedingly rare except for fairly common occurrences of isolated teeth. Throughout this paper I eschew quantitative analysis of this sort of data, though others have used such data in the quantitative reconstruction of dinosaur trophic dynamics (e.g., Bakker, 1972; Farlow, 1976b; Beland & Russell, 1979). Predator-prey ratios derived from dinosaur fossil assemblages are not necessarily an accurate reflection of previous community compositions (see Appendix 15.1, ASSUMPTION 3). However, I do believe that the magnitude of such figures, interpreted in the light of the relative completeness of the remains, is of some significance. Thus, carnosaurs were quite rare in the Fruitland swamp belt as indicated by their few and fragmentary remains, mostly of isolated teeth either shed during replacement, lost in feeding, and/or transported into these deposits. In contrast, hadrosaurs and ceratopsians were quite common in the Fruitland swamp belt, as suggested by their abundance and the relative completeness of some of their skeletal remains.

5) Olson (1966, 1971) has argued that in the evolution of terrestrial communities there was a progression from communities where the trophic base lay in aquatic plants ("Type I Community") to those where the base lay in terrestrial invertebrates ("Type II Community") to those where the trophic base lay in terrestrial vertebrates feeding on terrestrial vegetation ("Type III Community"). It appears that all three types of trophic bases were present in the *Parasaurolophus* community. Thus, the semiaquatic association discussed above was based, through its ultimate dependence on aquatic herbivores (e.g., various molluscs, arthropods and fishes), on aquatic plants. In the terrestrial zone, the mammal-lizard association was based on terrestrial plants and insects and the adult

dinosaur association was based largely on terrestrial plants and vertebrates. This is not to say that there are three types of community here; the significant amount of trans-association feeding, particularly by juvenile dinosaurs, links all of the associations together as one community. Rather, it suggests that by the Late Cretaceous a community in a deltaic setting was a multilayered nexus of all three types of community identified by Olson.

THE KIRTLAND BLUR

Throughout much of the Central Basin, beds of the Fruitland Formation grade upward into the Kirtland Shale, a thick (up to 500 m) and

Table 15.2—Census of dinosaur specimens from the Fruitland Formation and the Naashoibito Member of the Kirtland Shale. The numbers are based on specimens in the collections of the University of New Mexico (Albuquerque), Museum of Northern Arizona (Flagstaff), U.S. National Museum (Washington), American Museum of Natural History (New York), University of Kansas (Lawrence), University of California Museum of Paleontology (Berkeley), University of Arizona Laboratory of Paleontology (Tucson), Field Museum of Natural History (Chicago) and University of Oklahoma Stovall Museum (Norman). They also include some specimens still in the field (unpublished field notes in the University of New Mexico Department of Geology). The vast majority of specimens are isolated bones, but a few complete skulls and nearly complete skeletons are included. An impression of the quality of the preserved material can be gained by consulting Brown (1910), Gilmore (1916, 1919, 1922, 1935), Osborn (1923), Wiman (1930, 1931), Ostrom (1961, 1963), Mateer (1976), Kues et al. (1977), Rowe et al. (this volume) and Lehman (this volume). Except for pachycephalosaurs, isolated teeth are not included in this census. No claim is made here that these figures are an absolutely complete census of all dinosaur specimens ever collected from or present in the Fruitland Formation and Naashoibito Member. Confident conversion of these figures to minimum number of individuals is difficult to impossible because of the isolated nature of most remains and imprecise locality data accompanying most early collections. These figures thus are *not* to be used for the quantitative reconstruction of dinosaur trophic dynamics. Nevertheless, the magnitude of the figures may give a general impression of the relative abundances of some of the dinosaurs in the *Parasaurolophus* and *Alamosaurus* communities.

	FRUITLAND FORMATION		NAASHOIBITO MEMBER	
Dinosaurs	number of specimens	percentage	number of specimens	percentage
Hadrosaurs	80	54%	10	27%
Ceratopsians	30	20%	10	27%
Ankylosaurs	2	1%	5	14%
Pachycephalosaurs	2	1%	--	--
Ornithischia (indet.)	23	15%	2	5%
Carnosaurs	1	1%	4	11%
Coelurosaurs	10	7%	--	--
Dromaeosaurs	1	1%	1	3%
Sauropods	--	--	5	14%
TOTALS	149	± 100%	37	± 101%

variable sequence of fluvial sediments (Text-fig. 15.2). The contact between the two formations is often difficult to define with certainty in the field, and their distinctiveness was originally based on the greater percentage of coal in the Fruitland with arbitrary contacts chosen "at the top of the stratigraphically highest brown sandstone" (Dane, 1936, p. 113) or at the highest extensive coal bed (Fassett & Hinds, 1971) (Text-fig. 15.4e). Throughout much of the Central Basin, particularly along its southern and eastern margin, these criteria do not suffice to separate the Fruitland from the Kirtland, and these units are mapped together as the "undivided Fruitland and Kirtland Formation" (Baltz, 1967) (Text-fig. 15.6f).

As was the case with the Fruitland, knowledge of the past physical environment and biota of the Kirtland is based largely on a relatively small area in the western part of the Central Basin (Text-fig. 15.1), although fossils have been found throughout the entire outcrop belt of the formation. Unfortunately this knowledge is limited and too diffuse to allow the reconstruction of a community. Thus, the picture of the Kirtland ecosystem (ecosystems?) painted here is very broad brush, predominately one of transition:

1) The Fruitland–Kirtland transition is an excellent example of Walther's Law of the Succession of Facies. As the Cretaceous epicontinental sea regressed, a sequence of drab shales, mudstones and subarkosic sandstones, the lower shale member of the Kirtland, was deposited along the courses of rivers that flowed to the sea, proximal to the deposits that produced the Fruitland Formation. These more inland floodplains were better drained (less swampy) than those at the coast, accounting for the lesser amount of coal formed. After the sea had left northwestern New Mexico, and these lower shale Member deposits had prograded completely over the Fruitland Formation, there apparently was a major change in source area and sedimentation with the deposition of the Farmington Sandstone Member. The predominately medium- to coarse-grained arkoses of the Farmington Sandstone, with their interbedded shales, represent a large meandering (and in parts, braided) river system that, because of its thickening and coarsening northward (Reeside, 1924), was evidently shed off a rising highland to the north of the basin. Thus began the Laramide tectonics that dominated subsequent deposition in the San Juan Basin through the Eocene (Tweto, 1975). The high energy deposits of parts of the Farmington Sandstone give way laterally to and are overlain by meandering stream deposits of extensive overbank shales and shoestrings of meandering channel sandstones. The environment during the deposition of the upper shale member was, thus, that of a complex river system wholly free from marine influence. Silver's (1950, p. 112) suggestion that there was a "Late Cretaceous

basin of deposition lying somewhat to the west of the present area of Tertiary outcrop," a Kirtland paleobasin, although recently denied by Fassett and Hinds (1971), may be of some merit in explaining the thickness and paleocurrent trends of much of the Kirtland.

2) A floral change near the top of the Kirtland in southern Colorado, where there is an abrupt increase in coniferous pollen (Dickinson et al., 1968), mirrors the changeover to a more inland and upland environment. The megafossil plants of the Kirtland show a drastic drop in diversity (from 32 species in the Fruitland to 9 in the Kirtland, using Knowlton's taxonomy), but I hesitate to dwell on this change since I have little faith in the adequacy of the sampling or the taxonomy it is based on. The flora of the Kirtland Shale is in great need of further collection and restudy.

3) The vertebrate fauna of the Kirtland (Table 15.1) exclusive of the Naashoibito Member is inadequately sampled (and preserved?!). Particularly striking is the virtual absence of a lower vertebrate microfauna, reflecting a lack of both screenwashing and study of material acquired in association with the mammals. Dinosaur remains are reasonably common, though often highly localized and fragmented (Text-fig. 15.6a & b). There are only slight indications of transition here, transition from the coastal *Parasaurolophus* community to the inland *Alamosaurus* community. Thus, for example, ankylosaurs, known from a few isolated dermal scutes in the Fruitland, are somewhat better represented in the Kirtland, but admittedly only by isolated bones and dermal scutes. Such a difference may, however, only reflect the accidents of preservation and/or sampling.

In general, as Table 15.1 indicates, the fauna of the Kirtland Shale is a taxonomically depauperate subset of the fauna of the Fruitland. It is likely that further collecting and study would increase their similarity, although above the lower shale member the absence of various saltwater fish (selachians, etc.) and oysters would probably never be eradicated, simply because these beds are wholly freshwater in derivation. The indistinctness of the Kirtland fauna is also reflected by Clemens' (1973) designation of the Hunter Wash local fauna to include vertebrates from both the Fruitland and the lower part of the Kirtland. There is as yet no evidence for a distinct fauna, or community, in the lower ¾ of the Kirtland Shale. I thus refer to the interval represented by the lower shale, Farmington Sandstone and upper shale members of the Kirtland, and their poorly sampled biota, as the "Kirtland Blur," an indistinct and hazy record of the transition from the coastal swamps and hadrosaur–ceratopsian dominated *Parasaurolophus* community of the Fruitland to the *Alamosaurus* community of the inland floodplains of the Naashoibito Member of the Kirtland.

A

B

C

D

E

F

NAASHOIBITO ECOSYSTEM

Introduction.—In the uppermost Kirtland Shale, an unusual and limited rock facies and fauna is found in the Naashoibito Member and Alamo Wash local fauna (Baltz et al., 1966; Lehman, this volume). The Naashoibito has been mapped in only a limited but quite fossiliferous area in the western part of the Central Basin (Text-fig. 15.1). Thus knowledge of the paleoenvironment and biota at the time of its deposition is based mostly on this region.

Physical environment.—Deposits of the Naashoibito Member are a thin (up to 30 m) sequence of interbedded sandstone, shale and siltstone, underlain in many places by a thin (up to 5 m) conglomeratic sandstone (Baltz et al., 1966) (Text-fig. 15.7). The Naashoibito is separated from the underlying upper shale member of the Kirtland by this conglomerate, the "lower conglomerate" of Bauer (1916), which was the base of his "Ojo Alamo Sandstone." In many places this conglomerate

TEXT-FIG. *15.6*—Strata of and typical fossil occurrences in the Kirtland Shale. *A*, channel sandstones amidst overbank shales in the upper part of the Kirtland Shale (upper shale member?) at Kansas University (KU) New Mexico locality 16 (see *B*) on the north bank of Tsosie Wash; *B*, badly weathered and fragmented skull of a hadrosaur at KU New Mexico locality 16 along the north bank of Tsosie Wash. Length of rock hammer is 28 cm.; *C*, "Cut-bank failure" of a channel of the Ojo Alamo Sandstone (*sensu* Baltz et al., 1966) into underlying shales of the Naashoibito Member of the Kirtland Shale. The arrow indicates the contact between sandstone of the Ojo Alamo lateral to shales of the Naashoibito. The Naashoibito shales contain abundant fossil leaves (University of New Mexico (UNM) locality 77-227; see *D*). Rock hammer is 28 cm long; *D*, pieces of monocot leaves, probably palm, from the Naashoibito shales at UNM locality 77-227 (see *C*) along the west bank of a tributary of Kimbeto Wash. Dime for scale; *E*, partial femur of a sauropod, presumably *Alamosaurus sanjuanensis*, in the Naashoibito Member of the Kirtland Shale at UNM locality 77-492 along the north bank of Willow Wash. Rock hammer is 28 cm long; *F*, undivided Fruitland and Kirtland Formations, underlain by the Pictured Cliffs Sandstone (bench near base of cliff in center of picture) and overlain by the Ojo Alamo Sandstone (cap rock of mesa) along the south edge of Mesa Portales in the SW¼, Sec. 9, T19N, R2W. The sandstone body indicated by the arrow was considered a lower sandstone tongue of the Ojo Alamo Sandstone by Fassett & Hinds (1971, fig. 12) but is probably more reasonably considered as a large sandstone body in the upper part of the Kirtland Shale (Baltz, 1967). There is no clear separation of the Kirtland Shale and Fruitland Formation in this section, which is about 120 m thick.

TEXT-FIG. *15.7*—View of the Naashoibito Member of the Kirtland Shale and overlying Ojo Alamo Sandstone in the SE¼, Sec. 12, T24N, R12W just southeast of the headlands of Alamo Wash. Gravel in the foreground is weathered down conglomerate of Bauer's (1916) "lower conglomerate," the base of the Naashoibito Member of the Kirtland Shale as defined by Baltz et al. (1966). Arrow points to erosional remnant of the Ojo Alamo Sandstone (*sensu* Baltz et al., 1966) overlying variegated purple, blue, pink, and gray shales interbedded with gray white sandstones that are typical of the Naashoibito in this area.

has weathered to form a thin bench of pebbles that are predominately quartz and quartzite although other rock types, especially chalcedony and various volcanics, are common (Baltz et al., 1966). In other places the conglomerate grades into a fine- to coarse-grained sandstone which is irregularly bedded and has but few conglomeratic lenses (Baltz et al., 1966). The sequence overlying the conglomerate has a high proportion of fine- to coarse-grained sandstones that represent channel deposits of meandering and, possibly in places, braided streams. The shales are predominately silty, often sandy, and represent the overbank deposits of these streams. They display diverse colors, ranging from drab grays, buffs, and olives to striking variegated sequences of purples, blues, and reds. Occasional small ponds are indicated by limited lenses of siltstone and claystone. Fossil occurrences in the Naashoibito are less common but similar to those described for the Fruitland, especially in the occurrence of partial turtle shells and isolated dinosaur bones (Text-fig. 15.6e). Few nonmarine invertebrates and plants, however, have been recovered (except for one outstanding locality, Text-fig. 15.6c & d).

The stratigraphic relationships of the Naashoibito and upper shale Members of the Kirtland are very complex, and are characterized by the interfingering of variable shale and sandstone facies. A facies of variegated purple, blue, and red shale with a high percentage of interbedded sandstone characterizes the Naashoibito Member in and around its type locality (Baltz et al., 1966), and occurs sporadically in various parts of the basin (Text-fig. 8a), though often no "lower conglomerate" is discernible. In contrast, a facies of drab gray, buff, and olive shale with a lower percentage of interbedded sandstone comprises the uppermost part of the Kirtland throughout much of the basin, giving the impression that the upper shale member, not the Naashoibito Member, lies directly under the Ojo Alamo Sandstone in these areas. In addition, as Tom Lehman showed me, there is at least one place where the variegated shale-high percentage sandstone facies occurs below the "lower conglomerate"; this is along the west flank of "South Mesa" (Clemens, 1973).

As a consequence of these variations, the depositional picture of the upper part of the Kirtland is complex and deserves more detailed study. It is possible, though, to assert that all these deposits are fluvially derived, and Powell (1972) has claimed (without substantiation) that the mean cross-bedding trend of the "lower conglomerate" is 210°, indicating a source area to the north-northeast of this unit, and possibly for the remainder of the Naashoibito as well. The "lower conglomerate" and sandstones of the Naashoibito probably reflect the tectonic spasms of continued uplift north of the San Juan Basin, presaging the major uplift that later produced the Ojo Alamo Sandstone (Powell, 1973). Variations in the color of the shales may be due to variations in paleotopography and drainage or fluctuations in rainfall or both. The drab sequences may have been deposited in poorly drained topographic lows during periods of greater rainfall; the variegated sequences may have been deposited on relatively better drained ("oxidizing") highs during drier periods. The picture of the environment during the time of deposition of the Naashoibito Member in the San Juan Basin, then, is one of broad, and possible topographically diverse floodplains adjacent to a volcanically active and rising mountain chain inland from the sea.

Alamosaurus community.—The Alamo Wash local fauna (Lehman, this volume) is in much need of further collecting and study. Nevertheless, the dinosaurs are relatively well known (Gilmore, 1916, 1919, 1921, 1922; Mateer, 1976; Lehman, this volume) and have enabled the

A

B

C

D

E

F

reconstruction of a community, called the *Alamosaurus* community[7] after its largest herbivore, the sauropod *Alamosaurus sanjuanensis.* Its hypothesized food web (Text-fig. 15.9) is much less complex than that of the *Parasaurolophus* community due to the lower diversity (especially of small forms) of the Naashoibito fossil assemblage (Lehman, this volume). Though some Fruitland taxa will never be found in the Naashoibito Member, particularly those tied to marine or brackish water like certain sharks and oysters, there is every reason to expect a comparable diversity of mammals, lizards, and amphibians to appear with further collecting. Until this occurs, few comparisons of the *Parasaurolophus* and *Alamosaurus* communities are possible at this level.

A semiaquatic association dominated by turtles and crocodilians is

7. The term "*Alamosaurus* community" here strictly refers to the community represented by the fossil assemblage from the Naashoibito Member of the Kirtland Shale in the San Juan Basin. This contrasts with Sloan's (1970, p. 433) use of the same term to refer to a "southerly *Alamosaurus–Pentaceratops–Parasaurolophus* community," thus lumping together the assemblages of the Naashoibito Member of the Kirtland Shale (New Mexico), North Horn Formation (Utah), and Javelina Formation (Texas).

TEXT-FIG. *15.8*—Strata of and typical fossil occurrences in the Ojo Alamo Sandstone (*sensu* Baltz et al., 1966). *A*, uppermost beds of the Kirtland Shale (presumably of the Naashoibito Member, though no lower conglomerate has been discerned in this area) overlain by the Ojo Alamo Sandstone in the NE¼, Sec. 3, T22N, R9W along the north bank of Tsosie Wash. Note differential weathering of the Ojo Alamo; *B*, the *alamo* (= poplar) tree at Ojo Alamo, a spring located in the NE¼, Sec. 7, T24N, R11W along the north bank of Alamo Wash (the spring is *not* shown on the U.S. Geological Survey Alamo Mesa East topographic map). The tree is the same tree that is to the left of the building in Bauer's (1916, pl. 68B) photograph and is flanked by pillars and benches of the Ojo Alamo Sandstone; *C*, cross-bedded strata of the Ojo Alamo Sandstone near "Barrel Spring" in the NW¼, Sec. 16, T24N, R11W along the south bank of De-na-zin Wash (the spring is *not* shown on the U.S. Geological Survey Alamo Mesa East topographic map). Rock hammer is 28 cm long; *D*, pebbles, principally of jasper, quartz, and chalcedony, weathered from conglomeratic strata of the Ojo Alamo Sandstone near "Barrel Spring." Rock hammer is 28 cm long; *E*, prone fossil log over 10 m long, weathered out of the Ojo Alamo Sandstone along a tributary of Hunter Wash (see footnote 4). The root base of the log is to the man's left; *F*, fragments of dinosaur bone (indicated by arrows), weathering out of the Ojo Alamo Sandstone near the fossil log shown in fig. *E*. See Powell (1973) and footnote 4 (this paper) for discussion of the significance of these remains. Diameter of camera lens cap is 5.5 cm.

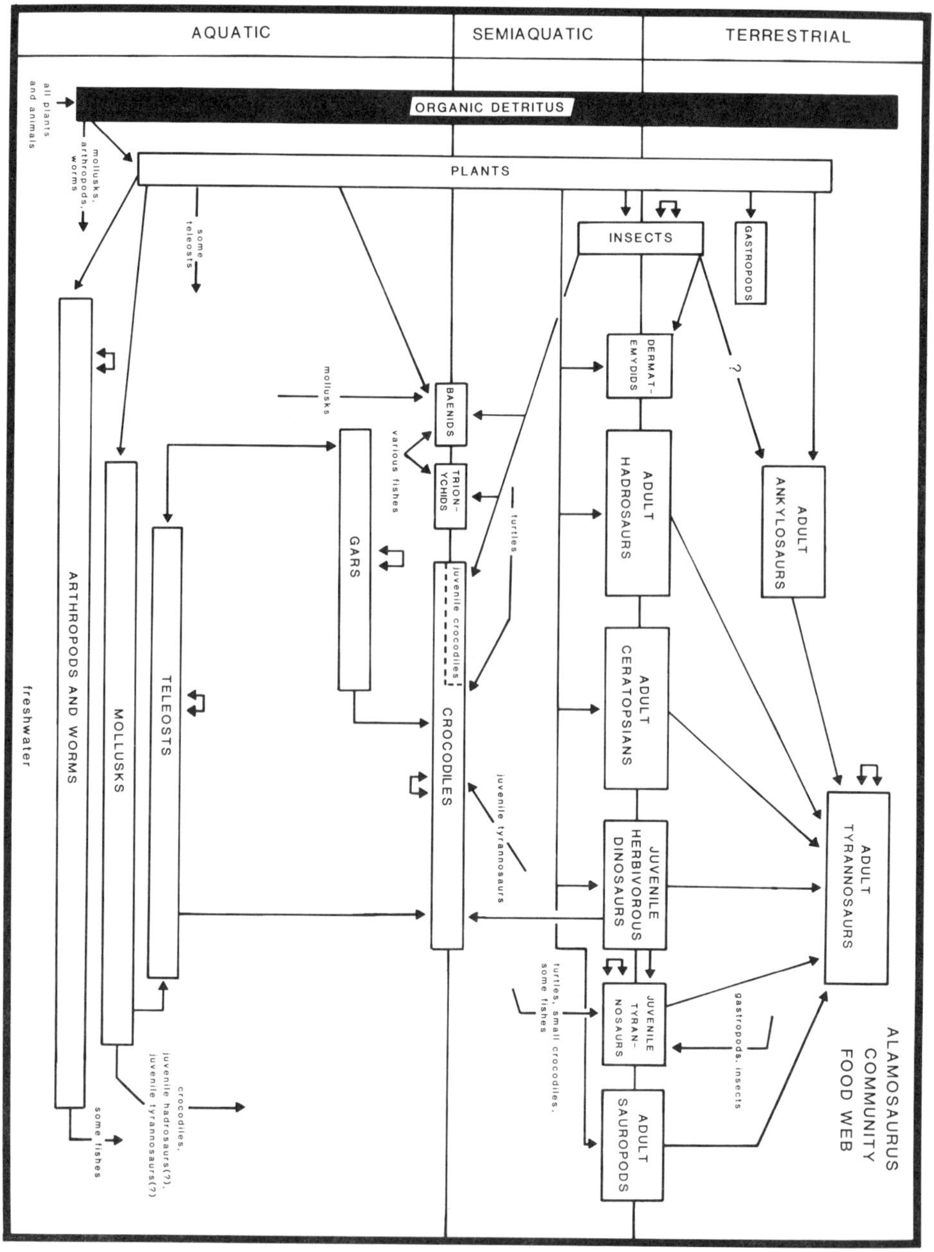

AQUATIC
SEMIAQUATIC
TERRESTRIAL
all plants and animals
ORGANIC DETRITUS
mollusks, arthropods, worms
PLANTS
some teleosts
INSECTS
GASTROPODS
DERMAT-EMYDIDS
?
mollusks
BAENIDS
various fishes
TRION-YCHIDS
turtles
GARS
juvenile crocodiles
ADULT HADROSAURS
ADULT ANKYLOSAURS
ARTHROPODS AND WORMS
MOLLUSKS
TELEOSTS
CROCODILES
ADULT CERATOPSIANS
juvenile tyrannosaurs
JUVENILE HERBIVOROUS DINOSAURS
ADULT TYRANNOSAURS
freshwater
JUVENILE TYRAN-NOSAURS
turtles, small crocodiles, some fishes
gastropods, insects
crocodiles, juvenile hadrosaurs(?), juvenile tyrannosaurs(?)
some fishes
ADULT SAUROPODS
ALAMOSAURUS COMMUNITY FOOD WEB

present in the *Alamosaurus* community. It broadly resembles the semi-aquatic association of the *Parasaurolophus* community; the differences are due to the changeover from marine to freshwater conditions (loss of marine fish and crocodiles), probable sampling biases (absence of amphibians), and taxonomy (the jump in turtle diversity, due largely to taxonomic splitting by Hay, 1908, 1910). Similar arguments apply to the aquatic association, which is very limited in the *Alamosaurus* community principally because of sampling effects. Both the aquatic and semi-aquatic associations already greatly resemble those of early Tertiary time (Matthew, 1937; Estes, 1970, 1976).

The only differences between the *Parasaurolophus* community and *Alamosaurus* community that can be discussed with any confidence are the differences in the composition of their dinosaur associations:

1) The Naashoibito fossil assemblage is numerically dominated by hadrosaurs and ceratopsians (Table 15.2), whose remains range from partial skulls to isolated bone fragments. This suggests a predominance of these dinosaurs in the *Alamosaurus* community, as was the case in the *Parasaurolophus* community.

2) There is a greater percentage of carnosaurs in the Naashoibito assemblage than in the Fruitland assemblage due to the occurrence of a few limb elements in addition to the usual isolated teeth. The numbers are small, and no rarefaction is available to assess whether the sample is simply too small to reflect actual abundances; a few preservational accidents or idiosyncracies of collecting could possibly have accounted for all the carnosaur bones in the Naashoibito. Thus, there is suggestive, but not very reliable, evidence that there was a greater proportion of carnosaurs present in the *Alamosaurus* community than there was in the *Parasaurolophus* community.

3) The pattern of carnosaur occurrence is mirrored by ankylosaurs; several isolated bones of nodosaurid ankylosaurs are known from the Naashoibito (Lehman, this volume). These are judged by me to be autochthonous, justifying the inclusion of ankylosaurs in the *Alamosaurus* community food web (Text-fig. 15.9). However, since the number

TEXT-FIG. *15.9*—Hypothesized food web of the *Alamosaurus* community. Animals are subdivided according to inferred substrate preference. Relative size of boxes and names of organisms is of no significance. Arrows indicate inferred direction of flow of trophic energy (see Appendix 15.2). "Insects," "arthropods and worms," "mollusks," and "gastropods" are not known from body fossils in the Naashoibito Member of the Kirtland Shale but are assumed to have been present (see text and Appendix 15.2). Compare with Text-fig. 15.5.

of ankylosaur specimens is small, the evidence of their presence and relative abundance is suggestive at best.

4) The typical Fruitland ceratopsian *Pentaceratops* is represented in the Naashoibito by several specimens, including part of a juvenile skull (Lehman, this volume). *Torosaurus* also may be present in the Naashoibito (Lehman, this volume). There is thus a slight difference in the types of ceratopsians that were present in the *Alamosaurus* and *Parasaurolophus* communities.

5) No pachycephalosaurs or coelurosaurs have been reported from the Naashoibito, and only the slightest remains of dromaeosaurs have been recovered (Lehman, this volume). Given the dearth of screenwashing that has been carried out in the Naashoibito, these circumstances probably result from sampling biases. The small size and frailty of the remains of most of these smaller dinosaurs also would have worked against their preservation in the high energy deposits typical of much of the Naashoibito. That no abraded skull caps of pachycephalosaurs have been recovered is peculiar, and may suggest that these animals were not present in the drainage basin in which the Naashoibito was deposited. A single scapula may record the presence of a hypsilophodontid in the Naashoibito (Lehman, this volume), but identification of this specimen is tentative, and thus its paleoecological significance is difficult to judge.

The differences between the dinosaur associations of the *Alamosaurus* and *Parasaurolophus* communities are correlated with differences in physical environment, a correlation that may not be spurious. Indeed, it suggests that during the Late Cretaceous there were differences in the composition of coastal and inland communities in the San Juan Basin reflecting different habitat preferences of different dinosaurs. Hadrosaurs and ceratopsians appear to have been relatively eurytopic, whereas carnosaurs, ankylosaurs and sauropods apparently preferred inland habitats. The data on coelurosaurs, dromaeosaurs and pachycephalosaurs are less conclusive, but their virtual absence from the *Parasaurolophus* and *Alamosaurus* communities may suggest that they preferred upland habitats away from centers of deposition on coastal and inland floodplains.

LATERAL VARIATIONS IN OTHER LATE CRETACEOUS DINOSAUR COMMUNITIES

Introduction.—It is logical to inquire whether the pattern of lateral variation in dinosaur communities seen in the San Juan Basin is evident elsewhere during the Late Cretaceous, thus establishing the pattern as a more general one. To answer this question it is necessary to identify Late Cretaceous assemblages that represent communities that lived in both coastal and inland environments. I limit this discussion to assemblages in

North America and eastern Asia. Although many Late Cretaceous assemblages are known from other regions, such as India (Prasad, 1968) Africa (Lapparant, 1957) and South America (Bonaparte, 1978), they are generally of low diversity, and their sedimentary context is poorly known in most cases, prohibiting confident assessment of the composition of the communities they represent as well as the past physical environments that these communities inhabited.

Coastal assemblages.—There is a wealth of Campanian-Maastrichtian age assemblages in North America that represent dinosaur communities that lived in coastal deltaic-interdeltaic settings similar to that represented by the deposits of the Fruitland Formation. These include the faunas of the Oldman, Judith River, Scollard, Hell Creek, St. Mary River and Lance Formations, among others (e.g., L. S. Russell, 1964; D. A. Russell, 1967, 1975; Sloan, 1970; Dodson, 1971; Sahni, 1972; Langston, 1975a, b). All of these assemblages are similar to the Fruitland assemblage in the numerical dominance of hadrosaurs and ceratopsians and rarity of carnosaurs, coelurosaurs, pachycephalosaurs, and dromaeosaurs. Some differ from the Fruitland in the occasional occurrence of relatively complete skeletons and skulls of ankylosaurs (Brown, 1908; Coombs, 1978). Other elements of these assemblages—fishes, amphibians, lizards, turtles, mammals—are also broadly similar to those found in the Fruitland Formation, as noted earlier. The evidence thus seems to indicate that dinosaur communities very similar to the *Parasaurolophus* community were present throughout coastal areas in North America during the Late Cretaceous.

Inland assemblages.—There are relatively few well known Late Cretaceous dinosaur assemblages from inland deposits. In North America these are from the Javelina Formation in Big Bend, Texas (Lawson, 1972, 1976) and the lower part of the North Horn Formation in east-central Utah (Gilmore, 1946; Spieker, 1946). These assemblages are relatively small but record the presence of autochthonous hadrosaurs, ceratopsians, carnosaurs, and the sauropod *Alamosaurus.* They lack ankylosaurs, coelurosaurs, pachycephalosaurs, and dromaeosaurs. These absences may be due to both sampling and preservational biases, although these factors may not account for the absence of ankylosaurs. What is known of the non-dinosaurian vertebrates of the North Horn and Javelina reveals elements similar to those of the Naashoibito, although there are some peculiar forms not present in the Naashoibito, such as the giant pterosaur in the Javelina (Lawson, 1975) and the lizards and tiny spiked crocodilian *Pinacosuchus* in the North Horn (Gilmore, 1942a, b). Thus some support, particularly with respect to sauropod distribution, is lent to the idea that communities similar to the *Alamosaurus* community were present in other inland areas of the Western Interior during Late Cretaceous.

In Asia, the Mongolian Nemegt Formation is composed of braided stream deposits that were laid down well within the interior of an eastern Asian proto-continent during the Late Cretaceous (Gradzinski, 1970; Gradzinski et al., 1977; Vakhrameev et al., 1978). The Nemegt assemblage thus represents organisms that lived even further inland than those of the Naashoibito, North Horn and Javelina in North America. It is numerically dominated by carnosaurs, coelurosaurs and ankylosaurs whose remains range from numerous complete or nearly complete skeletons to isolated bone fragments (Gradzinski, 1970). There are autochthonous remains of dromaeosaurs (Barsbold, 1974), pachycephalosaurs (Maryanska & Osmolska, 1974), and sauropods (Borsuk-Bialynicka, 1977), but hadrosaurs are uncommon, except for the "Dragon's Cave," a local pond(?) lense that contained the remains of at least seven individuals of *Saurolophus* (Efremov, 1955). Ceratopsians are absent, reflecting their probable absence in Asia during the latest Cretaceous (Colbert, 1948).

D. A. Russell (1970b) and Beland & Russell (1979) have explained this unusual abundance of carnosaur remains by postulating that the Nemegt Formation constituted a predator trap. Presence of a predator trap has also been suggested to explain the numerous *Allosaurus* remains in the Cleveland–Lloyd Quarry in Utah (Stokes, 1961), but it seems unlikely that a predator trap could account for the abundance of carnosaur remains in the fossiliferous part of the Nemegt Formation, a thickness of many meters of sediment cropping out over many square kilometers (Gradzinski et al., 1977). Rather, it seems that a large number of carnosaurs lived, or at least died, in the area now called the Nemegt Basin of Mongolia, and their remains were preserved in fluvial deposits now called the Nemegt Formation. These carnosaurs were the top carnivores of an inland dinosaur community, feeding on sauropods, ankylosaurs, and occasional hadrosaurs.

If the Nemegt assemblage accurately reflects the community outlined above, and communities similar to the one it represents were present in other far inland areas during the Late Cretaceous, it is probable that in moving from coastal to inland environments there was a change from communities dominated by hadrosaurs and ceratopsians to those dominated by carnosaurs, coelurosaurs, and ankylosaurs (Text-fig. 15.10). Sauropods seem to have been limited to these more inland communities. From this perspective, the *Alamosaurus* community of the San Juan Basin can be seen as transitional between the coastal *Parasaurolophus* community and the far inland community represented by the Nemegt assemblage (Text-fig. 15.10).

Rarity of ankylosaurs.—The suggestion that ankylosaurs preferred inland habitats, based on their distribution in the San Juan Basin and abundance and completeness of preservation in the Nemegt, is impaired

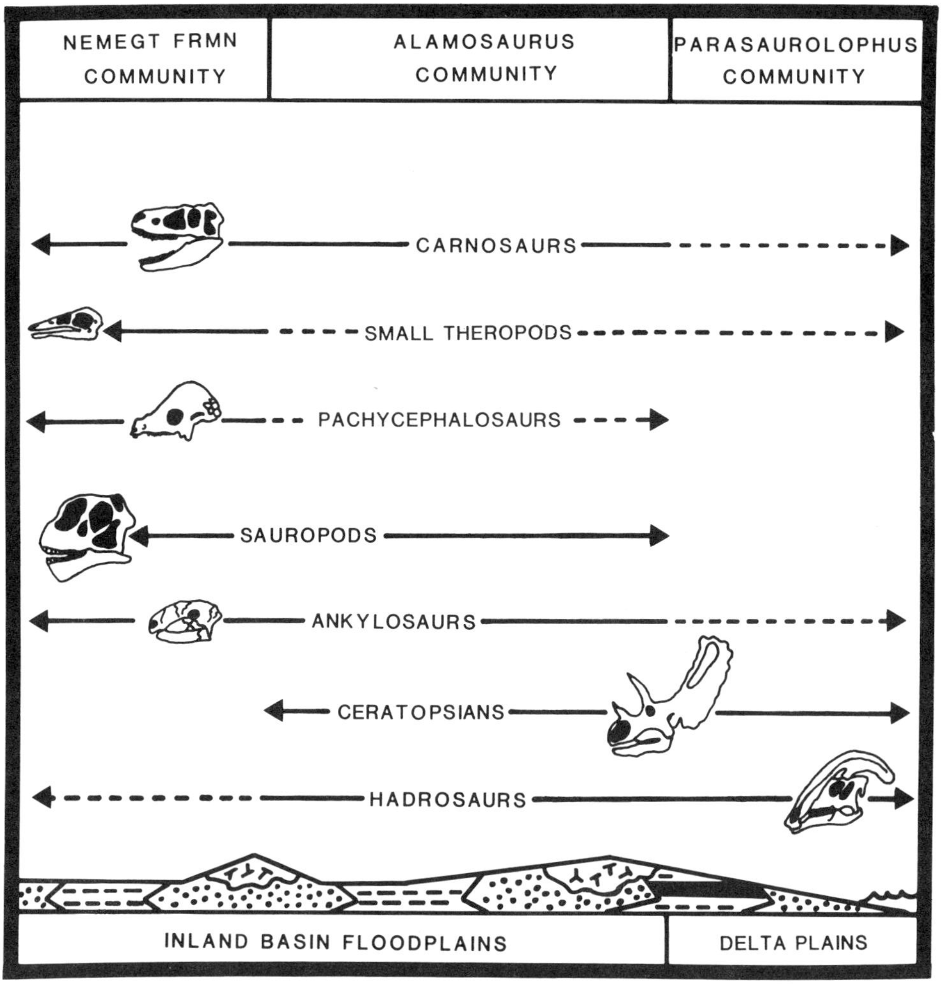

TEXT-FIG. *15.10*—Dinosaur habitat preferences during the Late Cretaceous (Campanian-Maastrichtian), as inferred in this paper from the distribution of fossils and sedimentary environments in the San Juan Basin (New Mexico), Nemegt Basin (Mongolia), and several North American Late Cretaceous fossil assemblages (see text). Solid line indicates normal habitat of dinosaur group; dashed lines indicate habitats in which a given group of dinosaurs was probably rare. Skulls not drawn to scale. Compare with Text-fig. 15.11.

by their absence in the North Horn and Javelina and the presence of some excellent (though not generally abundant) specimens in coastal deposits like the Lance. The rarity of ankylosaurs in most deposits may indicate that these animals were generally low in numerical abundance and/or that they were upland forms that didn't frequent centers of deposition such as coastal deltas, interdeltaic plains or inland flood plains. Unfortunately, the ankylosaur sample from inland habitats in North America is so limited that it is difficult to discern confidently one consistent pattern of their distribution in North America and Asia, if such a pattern even existed at all.

Ceratopsians and ecological vicars.—The presence of ecological vicars, one form in a coastal community being "replaced" by an ecologically analogous form in an inland community, is a logical outgrowth of the recognition of lateral variations in dinosaur communities. Such replacements may have operated at various taxonomic levels. Thus, for example, although the North Horn and Javelina faunas both are acknowledged to be Lancian in age (Spieker, 1946; Lawson, 1976) the common Lancian ceratopsian *Triceratops* is not present. Instead, the large fenestrate *Torosaurus* is present (Gilmore, 1946; Colbert & Bump, 1947; Lawson, 1976), suggesting that it may be an inland ecological vicar of *Triceratops*. (The extremely rare occurrence of *Torosaurus* in the coastal Lance and Hell Creek deposits [Colbert & Bump, 1947] may represent the rare individual that wandered into and died in a coastal environment.) The probable presence of *Torosaurus* in the Naashoibito (Lehman, this volume) supports the hypothesis that it preferred inland habitats. However, its occurrence with *Pentaceratops* militates against arguing for a simple replacement of the "coastal ceratopsian" *Pentaceratops* by *Torosaurus*. The presence of a juvenile specimen of *Pentaceratops* in the Naashoibito (Lehman, this volume) is intriguing but slender evidence that ceratopsian breeding grounds may have been inland. The distribution of ceratopsians thus deserves further investigation, but it does suggest the presence of some ecological replacement. Similar replacement among hadrosaurs may also have occurred (Efremov, 1954; Russell & Chamney, 1967; Langston, 1975b; Horner, 1979), but the data are likewise insufficient to support firm conclusions.

Alternative interpretations.—The distributions of various dinosaurs, here ascribed to lateral variations in dinosaur community composition, have also been explained by various evolutionary and biogeographic factors. It would be fatuous to argue that lateral variation in community composition was the only factor reflected by Late Cretaceous dinosaur distributions; surely various historical factors contributed to producing these distributions. Nevertheless, the recognition of lateral variations in community composition adds an ecological factor to traditionally used historical factors in explaining dinosaur distribution.

The distribution of North American Late Cretaceous sauropods is one example where an explanation based on historical factors may better be replaced (or, at least, supplemented) by an explanation based on lateral variations in community composition. Sloan (1970, p. 432) suggested that the appearance of the sauropod *Alamosaurus* in the Naashoibito and what he presumed were correlative deposits of the North American Late Cretaceous[8] reflects a migration "either from Europe along the north shore of Tethys or by crossing the Tethys from South America." Since potential sauropod migrators are known from the South American (e.g., Bonaparte, 1978) and European (e.g., Ostrom, 1970) Cretaceous, Sloan's suggestion is plausible, though it requires a local extinction of sauropods in North America after the time of Cloverly and Glen Rose deposition (i.e., post early Albian, Ostrom, 1970; Langston, 1974) and their subsequent reintroduction in Maastrichtian time. I alternatively argue that sauropods were in North America throughout the Cretaceous, but their absence from Albian to Maastrichtian time is due to the absence of preservation and/or study of the inland fluvial sediments that represent the environments where the sauropods lived. Rather than having to propose ad hoc a local extinction of sauropods in North America and their subsequent reentry into the continent at a later date, it seems more parsimonious to ascribe their absence to a lack of sampling (or preservation) of sediments representing preferred sauropod habitats. Possibly, however, both biogeographic and ecologic factors were at work, with the reintroduced sauropods only gaining a foothold in the inland environments. In such a way, the possibility of different habitat preferences of different dinosaurs can either supplant or supplement traditional explanations of dinosaur distribution based on biogeographic and evolutionary factors.

SOME CONSEQUENCES

Introduction.—Some of the consequences of recognizing that there probably were lateral variations in the composition of Late Cretaceous dinosaur communities have already been pointed out in the above discussion of possible explanations of Late Cretaceous sauropod distribution in North America. Biostratigraphic correlations based on dinosaurs as well as biogeographic and evolutionary scenarios explaining dinosaur distribution and evolution should be reappraised in the light of this

8. In addition to the Kirtland, North Horn, and Javelina Formations, Late Cretaceous North American sauropods are only definitely known from the Evanston Formation in Wyoming (D. Lawson, written communication, 1978). The Evanston lies unconformably on major thrust faults of the Sevier orogenic belt and is a predominately coarse clastic unit that was deposited inland from the sea (e.g., Tracy & Oriel, 1959; Armstrong, 1968). A supposed occurrence of a sauropod in the Lance Formation cited by Sloan (1970, p. 431) is based on unreliable provenance (J. McIntosh, oral communication, 1978).

recognition. It is possible that much of what is known of Late Cretaceous dinosaur occurrences (for example, the abundance of carnosaurs in the Nemegt) may reflect an ecological pattern (carnosaurs preferred inland habitats) rather than just an historical or idiosyncratic pattern (the Nemegt is a predator trap). The concept of lateral variations in dinosaur communities also has significance for the quantitative reconstruction of dinosaur trophic dynamics and the explanation of the ultimate extinction of the dinosaurs.

Dinosaur trophic dynamics.—The quantitative reconstruction of dinosaur trophic dynamics, taken by some as strong evidence for (or against) dinosaur endothermy (e.g., Bakker, 1972; Farlow, 1976b; Beland & Russell, 1979), assumes, among other things, that the relative abundance of carnosaurs in a dinosaur community was based entirely on the available standing crop of herbivorous prey. Thus, primacy is given to biotic interactions in determining community composition. That lateral variations in dinosaur community composition appear to be correlated with variations in physical environments implies that abiotic and/or more complex biotic (vegetational?) factors were also important in determining dinosaur community composition. Thus, for example, swampy coastlines may have been unfavorable environments for large bipedal carnosaurs to successfully hunt potentially quadrupedal hadrosaurs and fully quadrupedal ceratopsians; a biped is likely to be at a locomotory disadvantage to a quadruped in a predominately subaqueous and densely vegetated setting. Consequently, carnosaurs were rare in coastal habitats where their chances of successful hunting were low, and abundant in inland (i.e., drier and more open) habitats where their speed and agility were better suited to successfully capturing prey. If this were the case, then a much more complex picture of predatory-prey interaction and relative abundance emerges than has been assumed by those who have reconstructed dinosaur trophic dynamics.[9]

Dinosaur extinction.—The recognition of differences in dinosaur community composition linked to differences in physical environments also complicates understanding and explanation of the extinction of dinosaurs. Dinosaur extinction should be envisioned as the extinction of different dinosaur communities. An explanation must be sought to explain not only the disappearance of coastal communities like the *Parasaurolophus* community, but also the disappearance of inland communities like the *Alamosaurus* community or the community represented by the Nemegt assemblage must be explained.

Most recent explanations of dinosaur extinction (Van Valen & Sloan, 1977; Bakker, 1977; D. A. Russell, 1979) concentrate on the extinction of coastal dinosaur communities. The extinction of these communities may

9. Farlow (1980) has reached a similar conclusion from a different viewpoint.

largely be explained by the simple disappearance of their habitat, due to a major marine regression, an oft proposed agent of mass extinction (Ager, 1976). Thus, the extinction of the *Parasaurolophus* community in the San Juan Basin simply resulted from the regression of the sea, which removed its habitat from northwestern New Mexico during the late Campanian. This, of course, does not ultimately explain the extinction of coastal dinosaur communities throughout the world, since there have always remained coastal habitats for such communities. However, it does point out that much of the Late Cretaceous record of dinosaurs thought to be significant for determining the cause of their extinction merely records the extinction of communities whose disappearance was insured by the disappearance of favorable habitats resulting from marine regression.

The more challenging, and heretofore little discussed, aspect of dinosaur extinction concerns what factors eliminated the inland dinosaur communities of the Naashoibito, North Horn, Javelina and Nemegt. Bakker (1977), largely working with data from coastal dinosaur communities, suggested that the Late Cretaceous regression coincided with a decrease in the world-wide level of orogeny (Haug Effect), resulting in decreased geographic-topographic diversity which lessened barriers for speciation in such a way that extinction rates overtook origination rates and insured the dinosaurs' extinction. That this cannot be a general explanation of dinosaur extinction is demonstrated by the fact that some inland habitats, such as those in the San Juan Basin, were experiencing an increase in geographic-topographic diversity during the Late Cretaceous. The Naashoibito Member of the Kirtland Shale, for example, was deposited on topographically diverse floodplains adjacent to an uplifting mountain chain. Speciation rates among dinosaurs, using Bakker's assumptions, should have been on the rise in such an environment.

The disappearance of the *Alamosaurus* community in the San Juan Basin occurs at a major facies change, recording the deposition of the Ojo Alamo Sandstone (Text-fig. 15.8 a, b, c & d) by an extensive braided river system fed by coarse-grained detritus shed off the San Juan Uplift (Powell, 1972, 1973). From a local standpoint, it thus appears that the physical environment of the *Alamosaurus* community disappeared; this can "explain" its extinction. But what of a section like the North Horn Formation in Utah where no obvious change in sedimentary regime coincides with the disappearance of the dinosaurs? Early Paleocene mammals immediately overlie dinosaurs in this section, and the sedimentary sequence is remarkably similar in both its Cretaceous and Tertiary parts (Gazin, 1941; Spieker, 1946, 1960). The question, then, of what factors exterminated inland dinosaur communities remains perhaps the greatest riddle of them all. I offer no hypothesis here but merely reiterate that there is evidence that different dinosaur communities lived

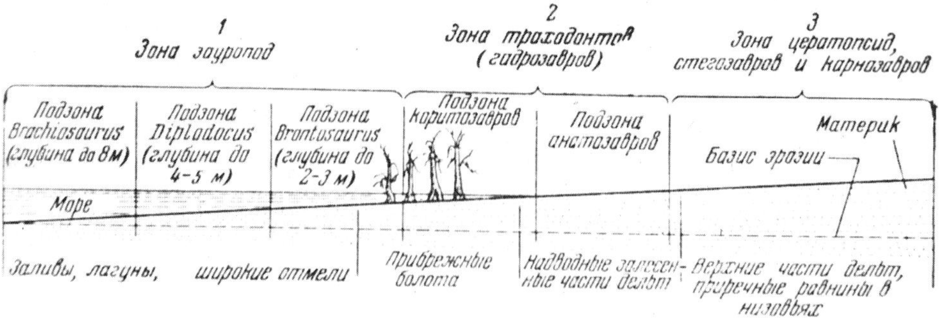

TEXT-FIG. *15.11*—Dinosaur habitat zonation during the Mesozoic proposed by Efremov (1954). Three zones are depicted: 1, sauropod zone in the coastal bays and nearshore shallows of the sea. This zone is subdivided into a *Brachiosaurus* subzone (water up to 8 m deep), a *Diplodocus* subzone (water up to 4–5 m deep) and a *Brontosaurus* subzone (water up to 2–3 m deep); 2, "Trachodont" (hadrosaur) zone in the coastal swamps and distal delta. This zone is subdivided into a more coastal corythosaur subzone and a more terrestrial anatosaur subzone, reflecting Efremov's conviction that there were different habitat preferences among lambeosaurine ("crested") and hadrosaurine ("flat-headed") hadrosaurs; 3, ceratopsian, stegosaur, and carnosaur zone on inland riverine plains of proximal deltas. From Efremov (1954, fig. 1). Compare with Text-fig. 15.10.

in coastal and inland settings during the terminal Cretaceous, and any satisfactory explanation(s) of dinosaur extinction will have to account for the disappearance of all dinosaurs of all types of communities in all environments.

CONCLUSION: EXTENSION OF THE PATTERN

The concept of lateral variations in dinosaur community composition is not new. That different types of dinosaurs preferred different habitats has either been explicit or implicit in virtually all of the literature on dinosaur autecology. However, Efremov (1954) was the first, as far as I know, to explicitly develop the concept of dinosaur habitat zonation (Text-fig. 15.11). Though Efremov's zonation was based largely on his own interpretations of physical environments and dinosaur occurrences in the Nemegt Basin (some of which were incorrect, see Gradzinski, 1970), he claimed generality for his zonation, applying it to the entire Mesozoic. Some of his results (e.g., carnosaurs preferred inland habitats) coincide with mine, but in others (e.g., sauropods preferred coastal habitats according to Efremov) we differ. Langston (1974) has recently discussed dinosaur habitat zonation with regard to early Cretaceous dinosaur assemblages, and his conclusions are a mixed lot of agreements and disagreements with those of Efremov (1954) and myself.

It seems likely that some sort of lateral variations in dinosaur community composition were present throughout the Mesozoic. But, it is not likely that the pattern of variation was the same everywhere throughout the Mesozoic. A pattern that seems consistent for well known Campanian–Maastrichtian occurrences has been identified here and should not necessarily be expected to be simply extensible back through the Mesozoic. Such extension brings in the factor of time, early eliminated in this analysis, and thus creates a more difficult problem by introducing the factors of evolutionary changes in the dinosaurs as well as changes in the nature of physical environments due to plate tectonics and other long term causal agents. These difficulties, however, do not invalidate the importance of the concept of lateral variations in dinosaur communities. As a general pattern such variations are not to be denied; it is the specific pattern of variation at a given time, its causes, and its consequences that are very difficult to identify. But only through the identification of such patterns and their further investigation will a better understanding of dinosaur ecology and all of its ramifications be attained.

ACKNOWLEDGMENTS

This paper is based largely on information gathered by me while employed by the University of New Mexico to work on a survey of San Juan Basin fossils for the U.S. Bureau of Land Management (Kues et al., 1977). Numerous people, too many to mention by name here, have helped me in my studies of the stratigraphy and paleontology of the Fruitland, Kirtland and Ojo Alamo Formations. For assistance in the field I especially thank my good friends Jeff Froehlich, Barry Kues, Tom Lehman, Mike O'Neill, Pete Reser, Keith Rigby, Jr. and Costas Tsentas. For permission to examine collections of Fruitland–Kirtland fossils I thank W. Clemens and J. H. Hutchison (University of California Museum of Paleontology), R. Emry (U.S. National Museum), E. Gaffney (American Museum of Natural History), B. Kues (University of New Mexico), E. H. Lindsay (University of Arizona) and L. Martin (University of Kansas). For information on Fruitland–Kirtland fossil collections I have been unable to examine I thank J. O. Farlow, N. Mateer and J. K. Rigby, Jr.

In the field and in correspondence, Elmer Baltz has provided me with much useful insight into the problems of San Juan Basin stratigraphy. Sid Ash, Leo Hickey, Lee Parker, and Don Tidwell have likewise taught me much about the paleobotany of the basin. Tom Lehman has allowed me free access to the results of his work on the dinosaurs and stratigraphy of the Naashoibito Member of the Kirtland Shale, and I am also grateful to him for many interesting discussions.

Elmer Baltz, Jim Farlow, Barry Kues, Tom Lehman, Niall Mateer, John Ostrom, and Keith Rigby, Jr. reviewed all or part of the manuscript, and their comments have improved its content and the clarity of its presentation; all errors of fact and extravagant interpretations it contains are, of course, my responsibility, not theirs. Virginia Moore typed the final version of the manuscript, and Robert Schoch, Keith Rigby, Jr., and several members of the Albuquerque Office of the Bureau of Land Management helped with the preparation of the illustrations. This paper is dedicated to the memory of Ivan Antonovich Efremov.

REFERENCES

Ager, D. V. 1976. The nature of the fossil record. Proc. Geol. Assoc. 87:131–159.

Anderson, R. Y. 1960. Cretaceous–Tertiary palynology, eastern side of the San Juan Basin, New Mexico. New Mex. Bur. Mines and Mineral Res. Mem. 6:59 p.

Archibald, J. D. 1977. Fossil Mammalia and Testudines of the Hell Creek Formation, and the geology of the Tullock and Hell Creek Formations, Garfield County, Montana, Unpublished Ph.D. thesis, Univ. Calif. Berkeley. 694 p.

Armstrong, R. L. 1968. Sevier orogenic belt in Nevada and Utah. Geol. Soc. Am. Bull. 79:429–458.

Armstrong-Ziegler, J. G. 1978. An aniliid snake and associated vertebrates from the Campanian of New Mexico. J. Paleontol. 52:480–483.

Baird, D. and J. Horner. 1977. A fresh look at the dinosaurs of New Jersey and Delaware. New Jersey Acad. Sci. Bull. 22(2):50.

Bakker, R. T. 1971. Ecology of the brontosaurs. Nature 229:172–174.

————. 1972. Anatomical and ecological evidence of endothermy in dinosaurs. Nature 238:81–85.

————. 1977. Tetrapod mass extinctions—a model of the regulation of speciation rates and immigration by cycles of topographic diversity, p. 439–468. *In:* A. Hallam (ed.), Patterns of Evolution. Elsevier, Amsterdam.

Baltz, E. H. 1965. Stratigraphy and history of Raton Basin and notes on San Luis Basin, Colorado–New Mexico. Am. Assoc. Petrol. Geol. Bull. 49:2041–2075.

————. 1967. Stratigraphy and regional tectonic implications of part of Upper Cretaceous and Tertiary rocks, east-central San Juan Basin, New Mexico: U.S. Geol. Surv. Prof. Paper 552:101 p.

Baltz, E. H., S. R. Ash, and R. Y. Anderson. 1966. History of nomenclature and stratigraphy of rocks adjacent to the Cretaceous–Tertiary boundary western San Juan Basin, New Mexico. U.S. Geol. Surv. Prof. Paper 524-D:23 p.

Barsbold, R. 1974. Saurornithoididae, a new family of small theropod dinosaurs from central Asia and North America. Paleontol. Polonica 30:5–22.

Bauer, C. M. 1916. Stratigraphy of a part of the Chaco River valley. U.S. Geol Surv. Prof. Paper 98-P:271–278.

Beland, P., and D. Russell. 1979. Ectothermy in dinosaurs: paleoecological evidence from Dinosaur Provincial Park, Alberta: Can. J. Earth Sci. 16:250–255.

Benton, M. J. 1979. Ectothermy and the success of dinosaurs. Evol. 33:983–997.

Bird, R. T. 1944. Did *Brontosaurus* ever walk on land? Nat. Hist. 53:60–67.

Bonaparte, J. F. 1978. El Mesozoico de America del Sur y sus tetrapodos. Opera Lilloana 26:1–596.

Borsuk-Bialynicka, M. 1977. A new camarasaurid sauropod *Opisthocoelicaudia skarzynskii* gen. n. sp. n. from the Upper Cretaceous of Mongolia. Paleontol. Polonica 37:5–64.

Brett-Surman, M. K. 1978. Phylogeny and paleobiogeography of hadrosaurian dinosaurs. Nature 277:560–562.

Brown, B. 1908. The Ankylosauridae, a new family of armored dinosaurs from the Upper Cretaceous. Am. Mus. Nat. Hist. Bull. 24:187–201.

————. 1910. The Cretaceous Ojo Alamo Beds of New Mexico, with description of the new dinosaur genus *Kritosaurus*. Am. Mus. Nat. Hist. Bull. 28:267–274.

————. 1914. Cretaceous-Eocene correlation in New Mexico, Wyoming, Montana, Alberta. Geol. Soc. Am. Bull. 25:355–380.

Brues, C. T. 1936. Evidences of insect activity preserved in fossil wood. J. Paleontol. 10:637–643.

Burgener, J. A. 1953. The stratigraphy and sedimentation of the Pictured Cliffs Sandstone and Fruitland Formation, Upper Cretaceous, of the San Juan Basin. Unpublished MS thesis, Univ. Ill., 59 p.

Butler, R. F., E. H. Lindsay, L. L. Jacobs, and N. M. Johnson. 1977. Magnetostratigraphy of the Cretaceous–Tertiary boundary in the San Juan Basin, New Mexico. Nature 267:318–323.

Carpenter, F. M. 1977. Geological history and evolution of the insects. Proc. 15th Int. Congr. Entomol.:63–70.

Case, T. J. 1978. Speculations on the growth rate and reproduction of some dinosaurs. Paleobiol. 4:320–328.

Clemens, W. A. 1973. The roles of fossil vertebrates in interpretation of Late Cretaceous stratigraphy of the San Juan Basin, New Mexico. Four Corners Geol. Soc. Mem.:154–167.

Clemens, W. A., and J. D. Archibald. 1980. Evolution of terrestrial faunas during the Cretaceous–Tertiary transition. Geol. Soc. France Mem. N.S. 59:67–74.

Clemens, W. A., J. A. Lillegraven, E. H. Lindsay, and G. G. Simpson. 1979. Where, when, and what—a survey of known Mesozoic mammal distribution, p. 7–58. *In:* J. A. Lillegraven, Z. Kielan-Jaworowska and W. A. Clemens (eds.), Mesozoic Mammals: the First Two-thirds of Mammalian History. Univ. Calif. Press, Berkeley.

Cobban, W. A. 1973. Significant ammonite finds in uppermost Mancos Shale and overlying formations between Barker Dome, New Mexico and Grand Junction, Colorado. Four Corners Geol. Soc. Mem.:148–153.

Cobban, W. A., E. R. Landis, and C. H. Dane. 1974. Age relations of upper part of Lewis Shale on east side of San Juan Basin, New Mexico. New Mex. Geol. Soc. 25th Field Conf. Guidebook :279–282.

Colbert, E. H. 1948. Evolution of the horned dinosaurs. Evol. 2:145–163.

Colbert, E. H., and J. D. Bump. 1947. A skull of *Torosaurus* from South Dakota and a revision of the genus. Acad. Nat. Sci. Phila. Proc. 99:93–106.

Coombs, W. P., Jr. 1975. Sauropod habits and habitats. Palaeogeog. Palaeoclimat. Palaeoecol. 17:1–33.

————. 1978. The families of the ornithischian dinosaur order Ankylosauria. Palaeontol. 21:143–170.

Cope, E. D. 1885. The relations of the Puerco and Laramie deposits. Am. Nat. 19:985–986.

Dane, C. H. 1936. The La Ventana–Chacra Mesa coal field. U.S. Geol. Surv. Bull. 860-C: 81–161.

————. 1946. Stratigraphic relations of Eocene, Paleocene, and latest Cretaceous formations of eastern side of San Juan Basin, New Mexico. U.S. Geol. Surv. Oil and Gas Inv. Prelim. Chart 24.

Dickinson, R. G., E. B. Leopold, and R. F. Marvin. 1968. Late Cretaceous uplift and volcanism on the north flank of the San Juan Mountains, Colorado. Colo. School of Mines Quart. 63:125–148.

Dodson, P. 1971. Sedimentology and taphonomy of the Oldman Formation (Campanian), Dinosaur Provincial Park, Alberta (Canada). Palaeogeog. Palaeoclimat. Palaeoecol. 10:21–74.

Efremov, I. A. 1940. Taphonomy: new branch of paleontology. Pan-Am. Geologist 74:81–93.

————. 1954. Nekotorie zamechaniya po voprosam istoricheskovo razvitiya dinozavrov (Some observations on the historical questions of the evolution of dinosaurs). Akademiya Nauk SSSR Trudy Paleontol. Inst. 48:125–151.

————. 1955. Zaxoronenie dinozavrov v Nemegtu (Yuzhnaya Gobi, MNR) (The burial of dinosaurs in Nemegt (Southern Gobi, Mongolian Peoples' Republic)). Vopr. Geol. Azii 2:789–809.

Erpenbeck, M. F. and R. M. Flores. 1979. Stratigraphy and depositional environments of the Upper Cretaceous Pictured Cliffs Sandstone and Fruitland Formation in the southwestern San Juan Basin, New Mexico. Geol. Soc. Am. Rocky Mtn. Section Abstracts 11:271.

Estes, R. 1964. Fossil vertebrates from the Late Cretaceous Lance Formation, eastern Wyoming. Univ. Calif. Publ. Geol. Sci. 49:1–180.

————. 1970. Origin of the recent North American lower vertebrate fauna: an inquiry into the fossil record. Forma et Functio 3:139–163.

————. 1976. Middle Paleocene lower vertebrates from the Tongue River Formation, southeastern Montana. J. Paleontol. 50:500–520.

Estes, R., and P. Berberian. 1970. Paleoecology of a Late Cretaceous vertebrate community from Montana. Breviora 343:35 p.

Estes, R., P. Berberian, and C. Meszoely. 1969. Lower vertebrates from the Late Cretaceous Hell Creek Formation, McCone County, Montana. Breviora 337:33 p.

Fagerstrom, J. A. 1964. Fossil communities in paleoecology: their recognition and significance. Geol. Soc. Am. Bull. 75:1197–1216.

Farlow, J. A. 1976a. Speculations about the diet and foraging behavior of large carnivorous dinosaurs. Am. Midl. Nat. 95:186–191.

————. 1976b. A consideration of the trophic dynamics of a Late Cretaceous large-dinosaur community (Oldman Formation). Ecology 57:841–857.

————. 1980. Predator/prey biomass ratios, community food webs, and the interpretation of dinosaur physiology, p. 55–83. *In:* E. C. Olson and R. D. K. Thomas (eds.), A Cold Look at the Warm-blooded Dinosaurs. Am. Assoc. Adv. Sci. Symposium.

Fassett, J. E. 1973. The saga of the Ojo Alamo Sandstone; or, the rock-stratigrapher and the paleontologist should be friends. Four Corners Geol. Soc. Mem. :123–130.

————. 1974. Cretaceous and Tertiary rocks of the eastern San Juan Basin, New Mexico and Colorado. New Mex. Geol. Soc. 25th Field Conf. Guidebook :225–230.

Fassett, J. E., and J. S. Hinds. 1971. Geology and fuel resources of the Fruitland Formation and Kirtland Shale of the San Juan Basin, New Mexico and Colorado. U.S. Geol. Surv. Prof. Paper 676:76 p.

Flores, R. M. 1979. Pictured Cliffs Sandstone–Upper Cretaceous distributary-channel, delta-front, and beach-bar deposits, southwestern San Juan Basin, New Mexico. New Mex. Geology 1:60.

Folk, R. L. 1974. Petrology of Sedimentary Rocks. Hemphill, Austin. 182 p.

Friedman, G. M., and J. E. Sanders. 1978. Principles of Sedimentology. John Wiley & Sons, New York. 792 p.

Gaffney, E. S. 1972. The systematics of the North American family Baenidae (Reptilia, Cryptodira). Am. Mus. Nat. Hist. Bull. 147:241–320.

Gazin, C. L. 1941. The mammalian faunas of the Paleocene of central Utah with notes on the geology. U.S. Natl. Mus. Proc. 91:1–53.

Gilmore, C. W. 1916. Vertebrate faunas of the Ojo Alamo, Kirtland and Fruitland Formations. U.S. Geol. Surv. Prof. Paper 98-Q:279–308.

———. 1919. Reptilian faunas of the Torrejon, Puerco, and underlying Upper Cretaceous formations of San Juan County, New Mexico. U.S. Geol. Surv. Prof. Paper 119: 71 p.

———. 1921. Discovery of sauropod dinosaur remains in the Upper Cretaceous of New Mexico. Science 54:274.

———. 1922. A new sauropod dinosaur from the Ojo Alamo Formation of New Mexico. Smithsonian Misc. Coll. 72(14):9 p.

———. 1935. On the Reptilia of the Kirtland Formation of New Mexico, with descriptions of new species of fossil turtles. U.S. Natl. Mus. Proc. 83:159–188.

———. 1942a. Osteology of *Polyglyphanodon*, an Upper Cretaceous lizard from Utah. U.S. Natl. Mus. Proc. 92:229–265.

———. 1942b. A new fossil reptile from the Upper Cretaceous of Utah. U.S. Natl. Mus. Proc. 93:109–114.

———. 1946. Reptilian fauna of the North Horn Formation of central Utah. U.S. Geol. Surv. Prof. Paper 210-C:29–53.

Goin, C. J., O. B. Goin, and G. R. Zug. 1978. Introduction to Herpetology (3rd edn.). Freeman, San Francisco. 378 p.

Gradzinski, R. 1970. Sedimentation of dinosaur-bearing Upper Cretaceous deposits of the Nemegt Basin, Gobi Desert. Palaeontol. Polonica 21:147–229.

Gradzinski, R., Z. Kielan-Jaworowska, and T. Maryanska. 1977. Upper Cretaceous Djadokhta, Barun Goyot and Nemegt Formations of Mongolia, including remarks on previous subdivisions. Acta Geol. Polonica 27:281–318.

Hay, O. P. 1908. The fossil turtles of North America. Carnegie Inst. Wash. Publ. 75:568 p.

———. 1910. Descriptions of eight new species of fossil turtles from west of the one-hundredth meridian. U.S. Natl. Mus. Proc. 38:307–326.

Horner, J. R. 1979. Upper Cretaceous dinosaurs from the Bearpaw Shale (marine) of south-central Montana with a checklist of Upper Cretaceous dinosaur remains from marine sediments in North America. J. Paleontol. 53:566–577.

Horner, J. R., and R. Makela. 1979. Nest of juveniles provides evidence of family structure among dinosaurs. Nature 282:295–298.

Kelley, V. C. 1950. Regional structure of the San Juan Basin. New Mex. Geol. Soc. 1st Field Conf. Guidebook :101–108.

Knowlton, F. H. 1916. Flora of the Fruitland and Kirtland Formations. U.S. Geol. Surv. Prof. Paper 98:327–333.

Kues, B. S., J. W. Froehlich, J. A. Schiebout, and S. G. Lucas. 1977. Paleontological survey, resource assessment, and mitigation plan for the Bisti–Star Lake area, northwestern New Mexico. Unpublished report to U.S. Bureau of Land Management, Albuquerque District Office. 1525 p.

Kues, B. S., T. Lehman, and J. K. Rigby, Jr. 1980. The teeth of *Alamosaurus sanjuanensis*, a Late Cretaceous sauropod. J. Paleontol. 54:864–869.

LaBrecque, J. L., D. V. Kent, and S. C. Cande. 1977. Revised magnetic polarity time scale for Late Cretaceous and Cenozoic time. Geology 5:330–335.

Langston, W., Jr. 1970. A fossil ray, possibly *Myledaphus* (Elasmobranchii: Batoidea) from the Late Cretaceous Oldman Formation of western Canada. Natl. Mus. Can. Publ. Paleo. 6:15 p.

———. 1974. Nonmammalian Comanchean tetrapods. Geoscience and Man 8:77–102.

———. 1975a. A Late Cretaceous vertebrate fauna from the St. Mary River Formation in western Canada. Athlon, Roy. Ont. Mus. Spec. Publ.:114–133.

———. 1975b. The ceratopsian dinosaurs and associated lower vertebrates from the St. Mary River Formation (Maestrichtian) at Scabby Butte, southern Alberta. Can. J. Earth Sci. 12:1576–1608.

Lapparant, A. F. de. 1957. The Cretaceous dinosaurs of Africa and India. J. Paleontol. Soc. India 2:109–112.

Lawson, D. A. 1972. Paleoecology of the Tornillo Formation, Big Bend National Park, Brewster Co., Texas. Unpublished MA thesis, Univ. of Texas, Austin. 182 p.

————. 1975. Pterosaur from the latest Cretaceous of west Texas: discovery of the largest flying creature. Science. 187:947–948.

————. 1976. *Tyrannosaurus* and *Torosaurus*, Maastrichtian dinosaurs from Trans-Pecos, Texas. J. Paleontol. 50:158–164.

Lindsay, E. H., L. L. Jacobs, and R. F. Butler. 1978. Biostratigraphy and magnetostratigraphy of Paleocene terrestrial deposits, San Juan Basin, New Mexico. Geology 6:425–429.

Lindsay, E. H., R. F. Butler, and N. M. Johnson. 1979. Reply to "Comment on 'Biostratigraphy and magnetostratigraphy of Paleocene terrestrial deposits, San Juan Basin, New Mexico' " by W. Alvarez and D. W. Vann. Geology 7:68–69.

————. 1981. Magnetic polarity zonation and biostratigraphy of Late Cretaceous and Paleocene continental deposits, San Juan Basin, New Mexico. Am. J. Sci. 281:390–435.

Lowe-McConnell, R. H. 1975. Fish Communities in Tropical Freshwaters: Their Distribution, Ecology, and Evolution. Longman, London. 337 p.

Lucas, S. G., and J. K. Rigby, Jr. 1979. Comment on "Biostratigraphy and magnetostratigraphy of Paleocene terrestrial deposits, San Juan Basin, New Mexico." Geology 7:323–325.

Mahabale, T. S. 1954. Ferns and palms as indicators of climate and palaeo-climate. Paleobotanist 3:33–37.

Mannhard, G. W. 1976. Stratigraphy, sedimentology, and paleoenvironments of the LaVentana Tongue (Cliff House Sandstone) and adjacent formations of the Mesaverde Group (Upper Cretaceous), southeastern San Juan Basin, New Mexico. Unpublished PhD thesis, Univ. of New Mex. 182 p.

Maryanska, T. 1977. Ankylosauridae (Dinosauria) from Mongolia. Palaeontol. Polonica 37:85–151.

Maryanska, T., and H. Osmolska. 1974. Pachycephalosauria, a new suborder of ornithischian dinosaurs. Palaeontol. Polonica 30:45–102.

Mateer, N. J. 1976. New topotypes of *Alamosaurus sanjuanensis* Gilmore (Reptilia: Sauropoda). Geol. Inst. Univ. Uppsala Bull., N.S. 6:93–95.

Matthew, W. D. 1937. Paleocene faunas of the San Juan Basin, New Mexico: Am. Phil. Soc. Trans., n.s. 30:510 p.

McGookey, D. P., J. D. Haun, L. A. Hale, H. G. Goodell, D. G. McCubbin, R. J. Weimer, and G. R. Wulf. 1972. Cretaceous System, p. 190–228. *In:* Geologic Atlas of the Rocky Mtn. Region. Am. Assoc. Petrol. Geol., Denver.

Morris. W. J. 1970. Hadrosaurian dinosaur bills—morphology and function. Los Angeles Cty. Mus. Contrib. Sci. 193:14 p.

Neill, W. T. 1971. The Last of the Ruling Reptiles Alligators, Crocodiles, and Their Kin. Columbia Univ. Press, New York. 468 p.

Odum, E. P. 1971. Fundamentals of Ecology (3rd edn.). Saunders, Philadelphia. 574 p.

Olson, E. C. 1952. The evolution of a Permian vertebrate chronofauna. Evol. 6:181–196.

————. 1966. Community evolution and the origin of mammals. Ecology 47:291–302.

————. 1971. Vertebrate Paleozoology. Wiley-Interscience, New York. 839 p.

Osborn, H. F. 1923. A new genus and species of Ceratopsia from New Mexico, *Pentaceratops sternbergii*. Am. Mus. Novitates 93:3 p.

Ostrom, J. H. 1961. A new species of hadrosaurian dinosaur from the Cretaceous of New Mexico. J. Paleontol. 35:575–577.

————. 1963. *Parasaurolophus cyrtocristatus*, a crested hadrosaurian dinosaur from New Mexico. Fieldiana: Geology 14:143–168.

———. 1964a. A reconsideration of the paleoecology of hadrosaurian dinosaurs. Am. J. Sci. 262:975–997.

———. 1964b. A functional analysis of jaw mechanics in the dinosaur *Triceratops*. Postilla 88:35 p.

———. 1970. Stratigraphy and paleontology of the Cloverly Formation (Lower Cretaceous) of the Bighorn Basin area, Wyoming and Montana. Peabody Mus. Nat. Hist. Bull. 35:234 p.

———. 1980. The evidence for endothermy in dinosaurs, p. 15–54. *In:* E. C. Olson, and R. D. K. Thomas (eds.), A Cold Look at the Warm-blooded Dinosaurs. Am. Assoc. Adv. Sci. Symposium.

O'Sullivan, R. B., and H. M. Beikman. 1963. Geology, structure, and uranium deposits of the Shiprock Quadrangle New Mexico and Arizona. U.S. Geol. Surv. Misc. Geol. Invest. Map I-345.

O'Sullivan, R. B., C. A. Repenning, E. C. Beaumont, and H. G. Page. 1972. Stratigraphy of the Cretaceous rocks and the Tertiary Ojo Alamo Sandstone, Navajo and Hopi Indian Reservations, Arizona, New Mexico and Utah. U.S. Geol. Surv. Prof. Paper 521-E:65 p.

Powell, J. S. 1969. Reptilian fossils and geology of uppermost Cretaceous deposits of the San Juan Basin, New Mexico. Geol. Soc. Am. Special Paper 121:546.

———. 1972. The Gallegos Sandstone (formerly Ojo Alamo Sandstone) of the San Juan Basin, New Mexico. Unpublished MS thesis, Univ. of Arizona. 131 p.

———. 1973. Paleontology and sedimentation models of the Kimbeto Member of the Ojo Alamo Sandstone. Four Corners Geol. Soc. Mem. :111–122.

Prasad, K. N. 1968. Some observations on the Cretaceous dinosaurs of India. Geol. Soc. India Mem. 2:248–255.

Reeside, J. B., Jr. 1922. Geological occurrence, p. 4–7. *In:* C. W. Gilmore, A new sauropod dinosaur from the Ojo Alamo Formation of New Mexico, Smithsonian Misc. Coll. 72(14).

———. 1924. Upper Cretaceous and Tertiary formations of the western part of the San Juan Basin Colorado and New Mexico. U.S. Geol. Surv. Prof. Paper 134:70 p.

Rigby, J. K., Jr., and S. G. Lucas. 1977. Fossil mammals from the Ojo Alamo Sandstone. New Mex. Geol. Soc. 28th Field Conf. Guidebook Supplement :55–56.

Rudwick, M. J. S. 1964. The inference of function from structure in fossils. British J. Phil. Sci. 15:27–40.

Russell, D. A. 1967. A census of dinosaur specimens collected in western Canada. Natl. Mus. Can. Nat. Hist. Paper 36:13 p.

———. 1970a. Tyrannosaurs from the Late Cretaceous of western Canada. Natl. Mus. Can. Publ. Paleo. 1:34 p.

———. 1970b. The dinosaurs of central Asia. Can. Geog. Journ. 81:208–215.

———. 1972. Ostrich dinosaurs from the Late Cretaceous of western Canada. Can. J. Earth Sci. 9:375–402.

———. 1975. Reptilian diversity and the Cretaceous–Tertiary transition in North America. Geol. Assoc. Can. Special Paper 13:119–136.

———. 1979. The enigma of the extinction of the dinosaurs. Ann. Rev. Earth Planet. Sci. 7:163–182.

Russell, D. A., and T. P. Chamney. 1967. Notes on the biostratigraphy of dinosaurian and microfossil faunas in the Edmonton Formation (Cretaceous), Alberta. Natl. Mus. Can. Nat. Hist. Paper 35:22 p.

Russell, L. S. 1934. Fossil turtles from Saskatchewan and Alberta. Trans. Roy. Soc. Can. 28(4):101–110.

———. 1964. Cretaceous non-marine faunas of northwestern North America. Royal Ont. Mus. Life Sci. Contrib. 61:24 p.

———. 1975. Mammalian faunal succession in the Cretaceous System of western North America. Geol. Assoc. Can. Special Paper 13:139–161.

Sahni, A. 1972. The vertebrate fauna of the Judith River Formation, Montana. Am. Mus. Nat. Hist. Bull. 147:321–412.

Shotwell, J. A. 1955. An approach to the paleoecology of mammals. Ecology 36:327–337.

———. 1958. Inter-community relationships in Hemphillian (mid-Pliocene) mammals. Ecology 39:271–282.

Silver, C. 1950. The occurrence of gas in the Cretaceous rocks of the San Juan Basin, New Mexico and Colorado. New Mex. Geol. Soc. 1st Field Conf. Guidebook:109–123.

Simpson, G. G. 1959. Fossil mammals from the type area of the Puerco and Nacimiento strata, Paleocene of New Mexico, Am. Mus. Novitates 1957:22 p.

Sinclair, W. J., and W. Granger. 1914. Paleocene deposits of the San Juan Basin, New Mexico. Am. Mus. Nat. Hist. Bull. 33:297–316.

Sloan, R. E. 1970. Cretaceous and Paleocene terrestrial communities of western North America. Proc. N. Am. Paleontol. Conv. I:427–453.

Spieker, E. M. 1946. Late Mesozoic and early Cenozoic history of central Utah. U.S. Geol. Surv. Prof. Paper 205-D:117–161.

———. 1960. The Cretaceous–Tertiary boundary in Utah. Proc. 21st Internat. Geol. Congr. 5(5):14–24.

Stanton, T. W. 1916. Nonmarine Cretaceous invertebrates of the San Juan Basin. U.S. Geol. Surv. Prof. Paper 98:309–326.

Steel, R. 1970. Saurischia, p. 1–87. *In:* Kuhn, O. (ed.), Handbuch der Palaoherpetologie 14. Gustav Fischer, Stuttgart.

Sternberg, C. H. 1932. Hunting Dinosaurs on the Red Deer River, Alberta, Canada. Published by the author, San Diego, 261 p.

Stokes, W. L. 1961. Dinosaur quarry near Cleveland, Utah. Proc. Utah Acad. Sci. 38:132–133.

Tracy, C. R. 1976. Tyrannosaurs: evidence for endothermy? Am. Nat. 110:1105–1106.

Tracy, J. I., Jr., and S. S. Oriel. 1959. Uppermost Cretaceous and lower Tertiary rocks of the Fossil Basin. Intermtn. Assoc. Petrol. Geol. 10th Ann. Field Conf. Guidebook:126–130.

Tschudy, R. H. 1973. The Gasbuggy core—a palynological appraisal. Four Corners Geol. Soc. Mem.:131–143.

Tweto, O. 1975. Laramide (Late Cretaceous–early Tertiary) orogeny in the southern Rocky Mountains. Geol. Soc. Am. Mem. 144:1–44.

Vakhrameev, V. A., I. A. Dobruskina, S. V. Meyen, and E. D. Zaklinskaja. 1978. Palaozoische und Mesozoische Floren Eurasiens and die Phytogeographie dieser Zeit. Gustav Fischer, Jena. 300 p.

Van Hinte, J. E. 1978. A Cretaceous time scale, p. 269–287. *In:* Cohee, G. V., M. F. Glaessner, and H. D. Hedberg (eds.), Contributions to the Geologic Time Scale. Am. Assoc. Petrol. Geol., Tulsa.

Van Valen, L., and R. E. Sloan. 1966. The extinction of the multituberculates. Syst. Zool. 15:261–278.

———. 1977. Ecology and the extinction of the dinosaurs. Evol. Theory 2:37–64.

Weimer, R. J. 1960. Upper Cretaceous stratigraphy, Rocky Mountain area. Am. Assoc. Petrol. Geol. Bull. 44:1–20.

Weimer, R. J., and C. B. Land. 1975. Maestrichtian deltaic and interdeltaic sedimentation in the Rocky Mountain region of the United States. Geol. Assoc. Can. Special Paper 13:631–666.

Wiman, C. 1930. Über Ceratopsia aus der Oberen Kreide in New Mexico. Nova Acta Reg. Soc. Sci. Upsaliensis, 4th ser. 7(2): 19 p.

———. 1931. *Parasaurolophus tubicen* n. sp. aus der Kreide in New Mexico. Nova Acta Reg. Soc. Sci. Upsaliensis, 4th ser. 7(5):11 p.
———. 1932. *Goniopholis kirtlandicus* n. sp. aus der Oberen Kreide in New Mexico. Bull. Geol. Inst. Univ. Uppsala 23:181–189.
———. 1933. Über Schildkroten aus der Oberen Kreide in New Mexico. Nova Acta Reg. Soc. Sci. Upsaliensis 4th ser. 9(5):34 p.
Wood, H. E., R. W. Chaney, J. Clark, E. H. Colbert, G. L. Jepsen, J. B. Reeside, Jr., and C. Stock. 1941. Nomenclature and correlation of the North American continental Tertiary. Geol. Soc. Am. Bull. 52:1–48.

APPENDIX 15.1

The major assumptions involved in the reconstruction of past communities, physical environments and ecosystems are:

ASSUMPTION 1—The assumption that it is possible to discriminate a slice of ecological time and thus recognize a network of interacting organisms (ergo community) is basic to many paleoecological endeavors. It may, however, be unwarranted, and what often is identified as a community may actually by a chronofauna (*sensu* Olson, 1952) of relatively short duration (a few thousands or tens of thousands of years).

ASSUMPTION 2—It is assumed here that communities existed in the past. Clearly extinct organisms directly or indirectly fed on each other and thus were metabolically linked, forming a community as defined in this paper.

ASSUMPTION 3—The problem of how faithfully a fossil assemblage mirrors the composition of a past community has been discussed at length in the literature of taphonomy, biostratinomy and paleoecology. Obviously no vertebrate fossil assemblage fully represents the composition of the vertebrate members of a past community (Efremov, 1940). However, when a relatively thin sequence of deposits is extremely fossiliferous, has been well sampled and contains significant numbers of autochthonous fossil remains it may be warranted to assume that the resultant fossil assemblage contains at least one specimen of a majority of the vertebrates that were present, and thus *qualitatively* represents the composition of the vertebrates in a past community.

ASSUMPTION 4, 5, . . . N—It is first assumed (ASSUMPTION 4) that it is possible to separate autochthonous from allochthonous fossil remains and thus discriminate members of the "proximal" community from members of "distal" communities. The procedure used here to achieve this employs reasoning (though not the quantitative methods) most fully elaborated by Shotwell (1955, 1958): members of the proximal community are generally those whose remains are both abundant and relatively well preserved (i.e., have a high degree of skeletal completeness and articulation). Thus, the number of specimens is not alone sufficient, and the quality of the preserved material has to be taken into account.

ASSUMPTIONS 5, . . . *N*—are subsequent suppositions involved in the reconstruction of the substrate and feeding preferences of each member of the proximal community (= autecology). Appendix 15.2 touches on some of these assumptions, the most basic of which is that preserved structures performed functions, in other words that adaptation existed (Rudwick, 1964).

ASSUMPTION N+1—It is assumed that sediments reflect past physical environments, the basic assumption of historical sedimentology. Many specific assumptions (not discussed further here), most of them actualistic, are necessary for the reconstruction of any environment from sediments, and these are analogous to those necessary for reconstructing the autecology of extinct organisms.

ASSUMPTION N+2—It is assumed that past ecosystems existed and that their nature can be discerned. This assumption follows from the assumption that communities existed (ASSUMPTION 2) and that physical environments can be discriminated from sediments (ASSUMPTION N+1).

ASSUMPTION N+3—The factor of time can confound an analysis aimed at discerning lateral variations in dinosaur community composition: if the communities examined were not coexistent, then the differences between them could be assumed to be largely the result of evolution, migration, and other historical factors. Hence time must, at least initially, be treated as an independent variable and play no role in the analysis even though historical factors were certainly of importance in determining dinosaur distribution (see later discussion in text).

APPENDIX 15.2

The inferred substrate preference and feeding habits of the organisms in the food webs (Text-figs. 15.5 & 15.9) are briefly discussed here. No vertebrates are included in the food webs that are not present in the Fruitland and Naashoibito assemblages, although some probably were present in the respective communities (e.g., lizards and mammals in the *Alamosaurus* community). Insects, "arthropods and worms," and mollusks plus gastropods (the latter two in the *Alamosaurus* community) are presumed to have been present and are included in the webs. Plants are treated as one unit, primarily a food source for the invertebrates and vertebrates.

FISHES:

Hybodonts—*Lonchidion selachos* is dentally similar to smaller species of the recent *Heterodontus* and thus possibly fed on mollusks and arthropods in freshwater (Estes, 1964).

Non-hybodont selachians—These are assumed to have been carnivorous like modern sharks and encompass forms of varying salinity tolerance and preference. Nobody has suggested, however, a good modern analogue for *Squatirhina* (Estes, 1964).

Myledaphus—Presumably a fresh to brackish water feeder on crustaceans and shellfish analogous to the modern ray *Hypolophus sephen* (Estes, 1964; Langston, 1970).

Pristids—Modern pristids are fresh to brackish water piscivores, and presumably Cretaceous ones were as well (Estes, 1964).

Sturgeons—In analogy to modern sturgeons, *Acipenser eruciferus* and *Paleopsephurus wilsoni* are inferred to have been euryhaline omnivores (Estes, 1964).

Amiids—*Amia fragosa* was probably a freshwater carnivore with a preference for mollusks and hard-shelled arthropods like modern *Amia calva* (Estes, 1964).

Gars—Cretaceous gars differ little from modern gars that are fresh to brackish water generalized carnivores (Estes, 1964).

Teleosts—Teleosts known from the Fruitland are few and represent probable nearshore marine (*Paralbula*) to possible freshwater (*Platacodon*) forms (Estes, 1964). The presence of scianids may indicate that a nearshore association of herbivorous, omnivorous and carnivorous teleosts similar to that found along some tropical coastlines today (e.g., Lowe-McConnell, 1975) was present in the *Parasaurolophus* community.

AMPHIBIANS:

Anurans—It is assumed that these were freshwater aquatic to semiaquatic insectivores like most modern frogs (Goin et al., 1978).

Urodeles—*Prodesmodon* and *Opisthotriton* were both probably aquatic to semiaquatic, with the former a durophage eating crustaceans and the latter primarily an insectivore like many modern salamanders (Estes, 1964).

REPTILES:

Baenids—These extinct turtles are judged to have been aquatic to semiaquatic freshwater forms with feeding habits ranging from molluscivory (especially *Thescelus*) to omnivory (L. S. Russell, 1934; Archibald, 1977).

Trionychids—Like modern soft-shelled turtles, Late Cretaceous trionychids are assumed to have been freshwater aquatic–semiaquatic carnivores (Goin et al., 1978).

Dermatemydids—This family is a wastebasket taxon, and thus simple analogy with modern *Dermatemys* is not justified. The "Dermatemydidae" includes one form with a somewhat domed carapace (*Adocus*) which may indicate terrestrial habits. Otherwise, "dermatemydids" are conceived of as a probable range of herbivorous, omnivorous and carnivorous semiaquatic to terrestrial turtles.

Lizards—Teiids and anguids are here lumped together and encompass forms that were probably voracious terrestrial carnivores (Estes, 1964).

Snakes—The burrowing snake *Coniophis*, like modern aniliids, probably fed on insects, amphibians and other small animals (Goin et al., 1978).

Crocodiles—Modern crocodiles have a wide range of salinity tolerances and eat a variety of fishes and tetrapods (Neill, 1971). The Cretaceous crocodiles presumably had a similar range of salinity tolerance and dietary preference and include *Brachychampsa*, a possibly molluscivorous form (Van Valen & Sloan, 1977).

Coelurosaurs—These include ornithomimids and coelurids and are thought to have been terrestrial carnivores (eating small animals and insects) and egg eaters (Steel, 1970; D. Russell, 1972).

Adult tyrannosaurs—The top carnivores of the *Parasaurolophus* and *Alamosaurus* communities, adult tyrannosaurs were terrestrial forms that fed on herbivorous dinosaurs, juvenile tyrannosaurs and may also have eaten carrion (Steel, 1970; Tracy, 1976).

Juvenile tyrannosaurs—These animals were probably generalized carnivores, the size of prey items taken being largely dependent on the growth stage of the tyrannosaur (Farlow, 1976a).

Adult ceratopsians—Ceratopsians have been conceived of as large terrestrial herbivores adapted to a diet of fibrous cycads and/or palms (Ostrom, 1964b). However, some probably were also semiaquatic (Dodson, 1971).

Adult hadrosaurs—More has probably been written about hadrosaur autecology than about any other group of dinosaurs. The consensus of various seemingly contradictory viewpoints (e.g., Ostrom, 1964a; Morris, 1970) is that various hadrosaurs had various substrate and feeding preferences ranging from semiaquatic herbivory and molluscivory(?) to terrestrial browsing on conifers, deciduous trees and shrubs.

Adult ankylosaurs—Presumably these were terrestrial herbivores though they may also have eaten some insects and carrion (Maryanska, 1977).

Adult sauropods—Coombs (1975) reviewed the equivocal evidence about sauropod habits and habitats and concluded that they were probably terrestrial herbivores (also see Bakker, 1971). Trackways (e.g., Bird, 1944) also indicate that at least some sauropods were semiaquatic.

Juvenile herbivorous dinosaurs—Juvenile hadrosaurs and ceratopsians in the *Parasaurolophus* community and juvenile hadrosaurs, ceratopsians, ankylosaurs and sauropods in the *Alamosaurus* community are lumped together. These forms probably ate a wide range of plant items supplemented in some cases (e.g., juvenile hadrosaurs) by some animal food.

MAMMALS:

Multituberculates—These are generally agreed to have been terrestrial and arboreal herbivores, possibly analogous to some modern rodents (Van Valen & Sloan, 1966).

Therians—Eutherians and didelphid marsupials were primarily insectivorous forms (though some omnivory cannot be excluded) that for the purposes of this analysis are not further separated into different adaptive types. Some or all may have been arboreal. Modern shrews and opossums are good analogues.